LARVAL GROWTH

CRUSTACEAN ISSUES

2

General editor:

FREDERICK R.SCHRAM

San Diego Natural History Museum, California

A.A.BALKEMA / ROTTERDAM / BOSTON / 1985

LARVAL GROWTH

Edited by
ADRIAN M.WENNER
University of California, Santa Barbara

CRC Press
Taylor & Francis Group
Boca Raton London New York

CRC Press is an imprint of the
Taylor & Francis Group, an **informa** business
A BALKEMA BOOK

1985

CRC Press
Taylor & Francis Group
6000 Broken Sound Parkway NW, Suite 300
Boca Raton, FL 33487-2742

First issued in hardback 2017

© 1985 by Taylor & Francis Group, LLC
CRC Press is an imprint of Taylor & Francis Group, an Informa business

No claim to original U.S. Government works

ISBN-13: 978-90-6191-294-1 (pbk)
ISBN-13: 978-1-138-44074-6 (hbk)
ISSN 0168-6356

Visit the Taylor & Francis Web site at
http://www.taylorandfrancis.com

and the CRC Press Web site at
http://www.crcpress.com

TABLE OF CONTENTS

FOREWORD

A pattern is established in the second and third volumes of *Crustacean Issues,* that of publishing expanded versions of the symposia of the annual meeting of The Crustacean Society. It must be pointed out, however, that these volumes are not mere redactions of the original oral presentations. As in Volume 1, authors that were in the symposiums were encouraged to prepare overviews of their topics. In addition, to ensure adequate coverage of the volumes' subjects, other authors were invited to prepare additional chapters. The resultant products offer a valuable contribution towards collating material to establish a benchmark upon which to launch a coordinated effort in future research on the subject of growth in crustaceans.

Although the 'symposium pattern' is established for *Crustacean Issues* to this point, and will persist in the immediately pending volumes, it is not necessarily a pattern that will continue. It is not our intention to publish all The Crustacean Society symposia, nor will we be restricted to publishing overviews based only on symposia. *Crustacean Issues* is intended to be a series that examines in some depth timely topics at issue in contemporary carcinology. How this series may develop, therefore, will depend on the course of research taken by carcinologists.

Frederick R.Schram

PREFACE

Fifteen years ago I considered gleaning scattered facts from the literature in order to write a small book on crustacean growth. When I broached that idea to a colleague, he replied, 'What do you plan to do, bind blank pages between two covers?' That reply deterred pursuit of such a project for these several years, even though it appeared that some valuable information had accumulated on the subject.

In fact, a great deal of progress has been made in the intervening 15 years. The environmental movement starting in the late 1960's led to much crustacean research in addition to that type of research germaine to environmental impact reports. The prominence of these animals in the aquatic habitat makes them ideal material for a host of both practical and basic studies. A knowledge of their growth and the factors which control growth is fundamental to an understanding of their role in nature.

When I was elected Chairperson of the Ecology Division of the Americal Society of Zoologists, Malcolm Gordon, my predecessor, indicated that outgoing chairs should sponsor a symposium in their speciality. In that sense, he organized and conducted the very successful symposium for the 1980 Seattle ASZ meetings, 'Theoretical Ecology: To What Extent Has It Added to Our Understanding of the Natural World?' With that as a precedent, I organized the symposium, 'Factors Influencing Crustacean Growth' for the 1982 ASZ meetings held in Louisville, Kentucky. In addition, a day-long series of contributed papers was devoted to the subject of crustacean growth at those meetings. The essence of most of those papers, supplemented by contributions from other investigators who could not attend the meetings, provided the basis for this volume.

Thanks go to the authors for providing the material and for yielding to suggested changes and abbreviations. Without them, there could have been no volumes. The American Society of Zoologists and the Crustacean Society graciously supported the symposium, an acknowledgement which includes both the Ecology Division and the Invertebrate Zoology Division of ASZ. Mary Wiley of the ASZ was particularly helpful on all arrangements, including the financial aspects with regard to funds provided by ASZ. The Department of Biological Sciences of the University of California, Santa Barbara assisted greatly with telephone costs, mailing expenses and duplication. Deborah Mustard of that department proved to be especially able and willing on a word processor, when time constraints did not permit manuscript exchange between the authors and me.

These volumes arose out of a symposium. That fact imposed constraints with respect to available time and with regard to the space available for publication. Despite the resultant restrictions, I feel privileged to have had the contact with these authors, to learn firsthand of recent developments in the subject, and to have been able to work with Frederick Schram. His patience, understanding, and cooperation have been most welcome. Finally, thanks go to A.T.Balkema, for his interest in marine biology and for his vision with respect to the need for these types of volumes.

Adrian M.Wenner
Santa Barbara, California

INTRODUCTION

Issues of larval growth are often encumbered by the confusion over the identity of larvae. Even after decades of study, progress is only now being made in identifying which larvae go with particular adults. In some groups (e.g. stomatopods) there is much work to be done before our knowledge of complete developmental sequences is adequate.

The emphasis on larval and postlarval growth in this volume deals with decapods. Many of the other crustaceans either do not have anamorphic development (e.g. peracarids), or their larvae may be part of a specialized parasitic life cycle (e.g. copepods and barnacles). Larval stages in the various decapod groups were assigned names before biologists knew to which group they belonged (see Felder, Martin & Goy). Gradually, developmental sequences were worked out within groups, and a wide variety of developmental patterns became evident, resulting in the confusing terminology which exists in this area (Gore). The topic of abbreviated development, by which decapods in various groups have become adapted to extreme environments (Rabalais & Gore) is herein supplemented by a review and analysis of the role of nutrition in growth and development through the larval stages (McConaugha). Crustacean larvae also apparently oscillate in their accumulation of biomass when exposed to particular pollutants. A pollutant, such as metal, can then be used to study the nature of that oscillation (Sanders, Laughlin & Costlow). Decapod larvae apparently are primarily carnivorous, irrespective of postlarval and adult diets.

Postlarval forms are a transition between the planktonic existence of most larvae and the diverse life of juvenile and adult decapods in a bewildering array of habitats. Much information now exists on morphological and anatomical changes which occur among decapods as they grow after metamorphosis from the larval form (Felder, Martin & Goy).

Much work remains to be done in the field of larval growth, and such holds great import for the eventual economic exploitation of crustaceans. This volume is devoted largely to decapods because that is the field in which most work has been done. Other volumes could be written on non-decapod forms, but such tomes must await more research to be done on development in these groups. In addition the potential for new discoveries concerning early ontogeny in crustaceans has barely been tapped.

Adrian M.Wenner

ROBERT H.GORE
Academy of Natural Sciences of Philadelphia, Pennsylvania, USA

MOLTING AND GROWTH IN DECAPOD LARVAE

No more acceptable service could be rendered to crustaceology than by showing . . . the gradual changes which take place in the different species . . . the subject being quite uninvestigated.

Hailstone & Westwood (1835)

ABSTRACT

Molting, staging and growth in decapod larvae are examined in the light of recent advances in larval development. The confusing plethora of terms occurring in the literature are redefined within two major categories: regular and irregular development. Within these headings are assigned the various ontological concepts of direct, abbreviated, advanced, accelerated, extended, precocious, retrogressive, and dimorphic or poecilogonous development. Aspects of growth within the egg, nauplius, prezoeal, and zoeal (or its equivalent) stages are treated simply using a computer-programmed synthesis derived from Brooks' law. Decapod development is best considered as a continuum extending from hatching as an imago through a series of few to many larval instars, often not assignable to discrete stages owing to vagaries in ecdysis as a consequence of a heterochronic developmental sequence imposed upon an isochronic molting series. Evolutionary implications and consequences are briefly considered in relation to this idea.

1 INTRODUCTION

Nem bilong Arthropoda i lusim ausait skin bilong em.

Melanesian Pidgin definition of ecdysis, Simons (1977)

In 1828 John Vaughan Thompson published a short series of Zoological Researches in which he wrote '. . . the Crustacea Decapoda then, indisputably undergo a metamorphosis, a fact which will form an epoch in the history of this generally neglected tribe, and tend to create an interest which may operate favorably in directing more of the attention of Naturalists towards them.' Thompson's observations, which showed that Bosc's (1802) *Zoea pelagica* was not an adult crustacean but its larva, were not completely original, having been preceded by those of the Dutch naturalist Slabber in 1778, and perhaps even earlier by the microscopist Leeuwenhoek (Gurney 1942, Williamson 1915). They were, however, far more accurate than either predecessor and provided a firm foundation and no little encouragement for subsequent studies on the larval development of the Decapoda (e.g. Bate 1878a, Couch 1843, Mueller 1863, and numerous others). Unfortunately, many of the earlier studies were rarely detailed or even complete, and were limited by the quality

1

of the optical equipment and the logistics of laboratory conditions of the day (Bate 1878a, W.Thompson 1836). At that time it was little suspected how many molts were required, or to what degree the general morphology of a larva would change prior to metamorphosis. In this regard, Thompson was also the first to note that various zoeae exhibited much morphological variation, and that subsequent molts caused them to change from 'younger stages' to older, and eventually through the megalopal stage to juvenile crabs (J.V.Thompson 1828, 1836, W.Thompson 1836). These historical early investigations have been sufficiently documented elsewhere (Costlow 1968, Gurney 1942, H.Williamson 1915) to preclude repetition here. It need only be noted that nearly all such studies were primarily directed towards description of the ecdyses, with little or no attempt at determination of any type of growth. To be sure, change in size was quite apparent, but actual measurements of the rate of such change, or the reasons for them, did not become commonplace until well into the twentieth century.

In this paper I review the general aspects of morphology, molting, and growth in the larvae of decapod crustaceans, and consider the various types of development exhibited in their ontogeny, as well as some of the factors responsible for its occurrence. But before proceeding to this it is helpful to briefly review concepts of larval development and the associated terminology.

2 LARVAL DEVELOPMENT IN THE DECAPODA

The larvae produced by decapods usually consist of a free-swimming or motile planktonic form that hatches from the egg either as a morphologically simplified stage termed a nauplius, as in the Penaeidea (e.g. Cook & Murphy 1971, Gurney & Lebour 1940), or in other decapods as a morphologically more complex stage termed a prezoea, zoea, naupliosoma, or phyllosoma (Aikawa 1929, Crosnier 1972, Fincham & Williamson 1978). After a period of growth, either by a process of posteriorly directed segmentization in the nauplius, or by a series of discrete developmental ecdyses in the zoeal (or in the Penaeidea, protozoeal) stages, a transformational point is reached and the larva molts to a postlarval stage.

The postlarval stage usually appears as a form substantially different in morphology from the preceding zoeal stages in most decapods. However, in the penaeidean and caridean shrimps this stage is often not abruptly distinguishable morphologically, but rather is gradually assumed through a series of continuing molts that pass imperceptibly into the final adult form (e.g. Shokita 1970a,b, 1977b). In the scyllaridean and palinuridean lobsters, and in the anomuran and brachyuran crabs, the postlarval ecdysis is a metamorphic molt producing a stage transitional between the larva and the juvenile (Costlow 1968, Knowlton 1974, Lyons 1980, Snodgrass 1956). The postlarval stage may appear as a miniature adult, as seen for example in atyid and bresiliid shrimp (D.Williamson 1970, Couret & Wong 1978), in astacid crayfish (Andrews 1907), in scyllaroidean lobsters (Robertson 1968a), or in some species of anomuran and brachyuran crabs (Lebour 1928, Hale 1931, Atkins 1954, Knight 1967, Gore 1970, Campodonico & Guzman 1981). In other Decapoda, the postlarva may bear little more than familial resemblance to the adult (e.g. Yang 1968, Webber & Wear 1981). In some nephropidean and eryonidean lobsters, and in many caridean shrimp, the larval and postlarval stages have partially merged, either in general morphology or in appendage form and function, so that differences between the stages often become a matter of degree rather than of kind[1] (see Selbie 1914, Bouvier 1917, Jorgensen 1925, Wear 1976, Fincham & Williamson 1978, Nichols & Lawton 1978).

Table 1. Families of Crustacea Decapoda in which the larvae or developmental type is unknown.
+ J.W.Martin (in litt.) has evidence that the psalidopodids may hatch as advanced zoeal-like stage, with
much yolk present, and well-developed pleopods and pereopods. *The Status of some members in these
families has been controversial and genera have been re-assigned by various authors. The familial rank
of the Tymolidae and Cyclodorippidae has also been questioned. Thus the larvae may or may not be
applicable to certain taxa within the families noted.

Family	Habitat and distribution
CARIDEA	
Procarididae	Anchialine pools; Ascension and Hawaiian Islands
Eugonatonotidae	Deep-sea benthic; Atlantic and Pacific Oceans
Gnathophyllidae	Shallow marine benthic; tropical seas
Physetocarididae	Deep-sea pelagic; western Atlantic
Psalidopodidae[+]	Deep-sea benthic; Atlantic, Indo-west Pacific Oceans
Stylodactylidae	Deep-sea benthic; Atlantic and Pacific Oceans
Thalassocarididae	Moderate marine benthic; Malay Archipelago
MACRURA	
Axianassidae	Shallow marine benthic; tropical west Atlantic
Callianideidae	Shallow marine benthic; tropical seas
Glypheidae	Moderate marine benthic; Phillipine Archipelago
Thaumastochelidae	Deep-sea benthic; West Indies and Japanese waters
ANOMURA	
Lomisidae	Shallow marine benthic; Australian waters
Pomatochelidae	Moderate marine benthic; tropical seas
Pylochelidae	Moderate marine benthic; tropical seas
BRACHYURA	
Belliidae*	Shallow marine benthic; temperate South America
Bythograeidae	Deep-sea benthic; hydrothermal vents, eastern Pacific
Cyclodorippidae*	Moderate to deep sea benthic; Atlantic and Pacific
Deckeniidae	Freshwater; tropical Africa
Gecarcinucidae	Freshwater; Asian and African tropical regions
Hexapodidae	Shallow marine benthic; tropical seas
Homolodromiidae	Moderate to deep-sea benthic; Atlantic and Pacific
Isolapotamidae	Freshwater; tropical east Asia
Mimilambridae	Shallow marine benthic; West Indies
Palicidae	Moderate marine benthic; tropical and temperate seas
Retroplumidae	Deep-sea benthic; Indian Ocean
Sinopotamidae	Freshwater; tropical eastern Asia
Sundathelphusidae	Freshwater; Malay Archipelago, Australia, New Guinea
Tymolidae*	Deep-sea benthic

At present, the larval stages for all but 28 of 105 families (following Bowman & Abele
1982), or about 75 % of the Decapoda, are known at least in part (D.Williamson 1967,
Sandifer 1974). Families in which the larvae remain unknown are listed in Table 1. In some
cases, for example, the several families in the freshwater brachyuran superfamily Potamoidea, development has been little studied but has been generalized as being direct in all
members based on studies of a few species.

The various larval and postlarval stages have received well over 70 different appelations
over the past 150 years (see, e.g. H.Williamson 1915, Gurney 1942, for listings). Williamson
(1969) suggested that most of these terms be dropped, a proposal generally accepted by
workers in the field, so that now only the terms nauplius, protozoea, prezoea, zoea, nauplio-

soma and phyllosoma (for the postembryonic larval stages), and megalopa, glaucothoe, puerulus, and in some cases mysis (for the postlarval stages) appear consistently in the literature (e.g. Raynor 1935, Robertson 1969a, Nyblade 1970, Roberts 1970, Gore 1971a,b, 1979, Scelzo & Boschi 1975, Gore & Van Dover 1980, Nayak 1981, et auct.). Williamson (1969, 1982) discussed each of these terms and they need not be considered further. The stages listed above can be divided into four main groups, based on their appearance in development: nauplius (section 4.3), prezoea, naupliosoma (section 4.4), protozoea, zoea, phyllosoma (section 4.5), and megalopa, glaucothöe, puerulus, mysis (section 4.6). These groups, as well as embryonic and egg development, will be addressed in a later section of this report. The postlarval phase will also be examined in detail by Felder et al. in Chapter 5.

The criteria for separating larval and postlarval phases depend on a combination of form and function. The naupliar phase (Fig.1A) is easily recognizable by its general morphology, but is also characterized by the development and use of the antennule and antenna in locomotion. In the protozoeal phase (Fig.1B), additional abdominal development occurs and locomotion shifts to the maxillipeds. In the prezoeal, prenaupliosoma, and naupliosoma phases (Figs.1C-E) locomotory ability is quite limited and requires abdominal flexure

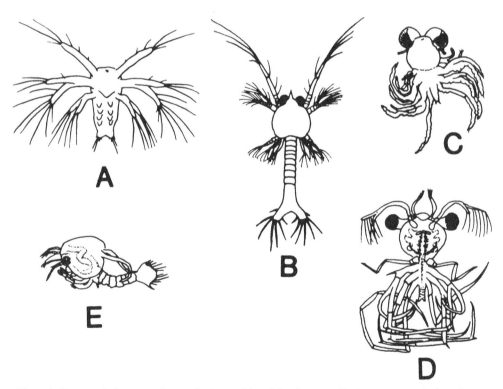

Figure 1. Some typical non-zoeal stages in decapod larval development. Each stage occurs before the zoeal stage or its equivalent is attained. (A) Nauplius (penaeidean shrimp) (after Cook & Murphy 1971). (B) Protozoea (penaeidean shrimp) (after Thomas et al. 1974a). (C) Pre-naupliosoma (after Desmukh 1968) and (D) Naupliosoma (scyllaroidean lobsters) (after Silberbauer 1971). (E) Pre-zoea (some caridean shrimp, anomuran and brachyuran crabs) (from Gore 1968). Not to scale.

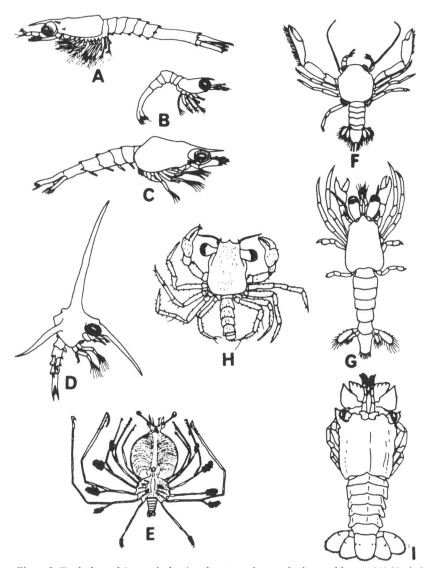

Figure 2. Typical zoeal (or equivalent) and postzoeal stages in decapod larvae. (A) Mysis (penaeidean shrimp) (after Cook & Murphy 1971). (B) Zoea (caridean shrimp) (from Gore et al. 1981a). (C) Zoea (anomuran crab) (after Provenzano 1978). (D) Zoea (brachyuran crab) (from Wilson & Gore 1980). (E) Phyllosoma (scyllaroidean lobster) (after Sims 1964). (F) Megalopa (anomuran crab) (from Gore 1968). (G) Glaucothoe (anomuran pagurid crab) (after Provenzano 1978). (H) Megalopa (anomuran crab) (after Gore et al. 1981b). (I) Pseudibacus (scyllaridean lobster; equivalent to puerulus in palinurid lobsters) (after Robertson 1968a). Not to scale.

because the maxillipedal natatory setae are either not yet, or only partially extruded. Locomotion in the zoeal and phyllosoma phase (Fig.2A-E) almost invariably occurs using cephalothoracic maxillipeds in which the natatory setae on the exopods are both well developed and numerous. Shokita (1973b) distinguished the first two zoeal stages of a Japanese freshwater shrimp *Macrobrachium shokitai* Fujino & Baba, on the basis of telsonal and abdominal flexure in swimming; only in zoea III were the pleopods, present but non-functional in stages I and II, sufficiently developed for swimming. This last stage was equated to the megalopa.

In the postlarval stages (megalopa, glaucothoe, puerulus, mysis, Figs.2F-I) the locomotory function is shifted to the abdominal pleopods. In the first juvenile stage and thereafter, locomotion is again thoracic, but this time employing the now well-developed pereopods (in reptant Decapoda, although some natant decapods also use pereopods). Most natant decapods continue to use pleopods; however, in swimming, the lobsters, some shrimps, and porcellanid crabs (Anomura) may use abdominal flexure as an escape mechanism when attacked.

It is important to note, as emphasized by Mayrat (1966) that the terminology distinguishing larval from postlarval phases is based primarily on functional morphology and bears little relationship to the actual maturational stage of the crustacean. Mayrat states, for example (and see 3.3.2.4) that giant phyllosoma larvae may often begin to develop primary or secondary sexual characters such as appendix masculinae, thus implying an incipient maturation even though the phyllosoma form is incontrovertibly 'larval' (i.e. pre-juvenile) in the generally accepted sense of the term. With this in mind we turn now to a consideration of ontogenetic concepts in decapod larvae.

3 DEVELOPMENTAL CONCEPTS AND EXAMPLES IN DECAPOD CRUSTACEAN LARVAE

The type of development most often seen in decapod larvae and postlarvae is termed anamorphosis (Knowlton 1974, Snodgrass 1956), but superimposed on this is a metamorphosis in which a radical change takes place both morphologically and functionally during the passage from larva to postlarva (Costlow 1968). This results in a preadult form that may have adult attributes in varying degrees. Some authors (e.g. Costlow 1968a, Costlow & Bookhout 1968) consider the molt from postlarva to juvenile (e.g. megalopa to first crab stage) to be the metamorphic ecdysis, while that from zoea to megalopa merely terminates the larval phase. Others (e.g. Knowlton 1974, Rice 1981b, Snodgrass 1956) agree with the concept that the metamorphic ecdysis occurs only at the last larval stage, producing a megalopa or its equivalent, which is itself transitional between larval and juvenile phases. According to Snodgrass (1961), 'A true metamorphosis involves the discarding of specialized larval characters. which allows the completion of adult development, and differs in degree according to the degree of abberation of the young from the adult structure'. The transition from zoea to megalopa in most Anomura and Brachyura (or phyllosoma to puerulus in spiny lobsters) fit this definition more precisely than do the Caridea, Penaeidea, Nephropidea, and Astacidea. It is also at the time of metamorphosis that the diecdysic larval cycle, comprised generally of a short intermolt period followed by a longer premolt period, changes to an anecdysic cycle, consisting predominantly of an intermolt phase with a shorter premolt period.

The diecdysis cycle is the type most often observed in rapidly molting crustaceans, including larvae, whereas the anecdysic cycle characterizes adults (Freeman & Costlow 1980). As in the adults, the molting cycle in larvae and postlarvae can usually be separated into early and late postmolt, intermolt, premolt, and ecdysis, although the distinctions between these phases may not always be clear.[2] In some groups (e.g. palinurid larvae), a setagenesis cycle also extends over the entire molt cycle, beginning at premolt, so that it may be possible (given sufficient observations) to determine the phase of the molting cycle from setal morphology (see Dexter 1981, and chapters in this volume).

Not all decapods have a complete larval development. In some the developmental type is epimorphic, as seen in species in normally long-developing families such as the Atyidae, Alpheidae, or Hippolytidae (Caridea) that have almost, or completely, dispensed with the pelagic zoeal phase (Dobkin 1965a, Makarov 1968a, Benzie 1982). In other groups, such as the freshwater crabs or crayfishes, the young hatch more or less directly into the parental image (Hailstone & Westwood 1835, McCann 1937, Bott 1969, Hobbs 1974, Pace et al. 1976). In still others, for example some pinnotherid and ocypodid crabs, a pelagic stage exists but is of extremely short duration, lasting in some cases no more than 36-48 hours (Goodbody 1960, Rabalais & Cameron 1983) before attaining post-larval stage.

Coupled with these concepts is the recently proposed idea of a critical period, an interval of time from hatching and immediately thereafter, during which time some hormonal or physiological activity attains a level essential to continued development or functioning, including the ability to molt (Costlow 1963a, Paul & Paul 1980). In fact, this period is so important that Costlow (1963a:221-222) stated 'If, by natural or artificial means, the stimulus for a specific process dependent on the Critical Period is delayed, subsequent activation of the process is not possible, or must await the next period of change in the level of activity.' The critical period thus has great importance in staging variations seen in some larvae (3.1.1.3).

3.1 *Isochronicity, heterochronicity, retrogressive development, and poecilogony*

Three important factors related to time, and the appearance of numerical and morphological stages, are influential in larval development. Throughout ontogeny development may be isochronal, i.e., having a relatively constant (even if short) intermolt period during the larval phase, with the appearance of characters more or less at the same time in a larval sequence (Knowlton 1974). If the intermolt period is of irregular length, ontogeny is said to be heterochronal, i.e., characters appear at different times than they normally would during a sequence (Rothlisberg 1979). Both types have repeatedly been observed (e.g. Boyd & Johnson 1963, Costlow & Fagetti 1967, Knight 1967, Baba & Moriyama 1972), although not necessarily specifically mentioned as such.

A third phenomenon that may occur is retrogressive development (sensu Kurata 1969). In this situation regression of larval characters is seen to a varying degree as development proceeds toward the postlarval phase. The resultant megalopa, for example, tends towards having fewer, or a minimal number, of zoeal features but may still carry over into the postlarval stage one or more larval attributes. There is considerable evidence that heterochronicity is in large measure influential in producing or enhancing retrogressive development; specific examples will be provided in section 3.1.2.

In many decapod species, conspecific individuals living under different environmental conditions can produce larvae having vastly different modes of development, showing either differences in morphology, duration of development, molting frequency, growth, or a combination of these, unlike that seen in normal developmental sequences (Brooks & Herrick 1892, see also H.Williamson 1915, Gurney 1942, Knowlton 1970, 1973, for more discussion). This process, termed poecilogony, has been observed in several families and seems to be an adaptive response to certain combinations of environmental conditions perceived to be adverse. Specific examples will be given in 3.1.3.

Table 2. First appearance of selected morphological characters in decapod larval development. Note: These are general characters in an average situation. Great variation appears both intragenerically and intraspecifically in some larvae. In abbreviated development, many of the character sequences become compressed backward toward zoea I; in extended development, character appearance may be delayed or mature more slowly in later stages (i.e. greater than instar 6); see text for further discussion.

Character	Stage I	Stage II	Stage III	Stage IV	Stage V	Stage VI and beyond
Eyes	Sessile	Stalked				
Antennule				Endopod bud develops gradually, aesthetascs appear		
Antenna				Endopod bud appears, enlarges throughout development		
Mandible			Palp primordia	Palp primordia	Palp bud or developed	
Maxillipeds (exopod setae)	4 setae	6 setae	8 setae	10 setae	12 setae	14 or more
Pereopods			Primordia	Buds enlarge	Segmentation appears	Well-developed
Abdomen	5 somites	5 somites	6 somites			
Pleopods			Primordia	Buds	Segmentation appears	Setation may be present
Tailfan			Uropod buds appear	Uropods enlarge, may appear functional		
Processes	7 + 7 processes (most movable, some fixed)	8 + 8 processes	Number may increase and/or some other movable processes become fixed or modified			

3.1.1 *Isochronic and heterochronic development and growth*

The concept of heterochronicity has long been of interest to biologists, especially those concerned with growth. The subject has been investigated by de Beer (1951) in relation to larvae, and more generally by Gould (1979). Briefly, an organism exhibits somatic growth as it progresses toward reproductive maturity. If somatic growth becomes accelerated so that adult size is attained before reproductive maturity is realized, then an adult-larva is produced with many (or all) of the adult morphological attributes except gonadal maturity. If this is carried far enough, developmental aspects will be continued in the descendant form, producing a recapitulation of ontogenetic stages (Table 2). This results in a type of Haeckelian recapitulation in which the terminal stages of an organism are gradually condensed (Gould 1979).

On the other hand, if reproductive maturity occurs before the requisite somatic growth has been attained, then a larval-adult can occur. This process, termed paedomorphosis, produces an organism having larval attributes that may be super-imposed on gonadal maturity in the soma. According to Gould (and his definitions differ somewhat from those of de Beer) two important mechanisms function in paedomorphosis: progenesis, in which reproductive maturation is accelerated considerably before the requisite adult stage; and neoteny, in which adult somatic growth is retarded but reproductive maturity occurs more or less normally. The two concepts are quite similar and relative to whether the perspective is forward from the larvae or backward from the adult. As Gould points out (and de Beer earlier had noted) several evolutionarily interesting combinations of adult and juvenile morphological and maturational stages can be attained. Some examples will be given later.

It is now well established that both isochronic and heterochronic development may occur within the larval stages of numerous decapods. In many instances larval growth in early stages may be nearly constant up to some certain point (e.g. phyllosoma V, Saisho 1966; zoea III in Caridea, Makarov 1968a), and then irregularly long intermolts begin to occur. Later stages thus change from being isochronal to heterochronal. In some cases the duration of instars may shorten before lengthening again in a given individual. Boschi & Scelzo (1970) found the first two zoeal stages of *Corystoides chilensis* Milne Edwards & Lucas lasted eight and seven days, respectively; the next two stages were isochronic and lasted five days each, and the last stage lengthened to nine days. Baba & Moriyama (1972) observed a similar situation in *Helice tridens tridens* De Haan and the subspecies *H.t.wuana* Rathbun. In the latter, the first two zoeal stages differed in length (eight, seven days) from the next two stages (four days each). Gore & Scotto (1982), studying *Cyclograpsus integer* H.Milne Edwards, recorded the average duration of zoea I at 25°C as six days, zoea II and III at four days, and zoea V and megalopa at five days. *C.integer* sometimes had six zoeal stages at this temperature and stage VI lasted four days. At lower temperature (20°C) zoea I-VI (no zoea V molted to megalopa) had modal values of 10, 8, 8, 8, 8, 12 days. In other species development appears to be more or less shortened in the intermediate zoeal stages, but no consistency is observed. For example, Costlow & Fagetti (1967) showed that in *Cyclograpsus cinereus* Dana zoea I and II lasted about the same length of time (4-5 days), zoea III lasted 2-4 days, and zoea IV from 1-6 and zoea V from 5-8 days. But the overall trend was still a decrease in duration, followed by an increase.

3.1.1.1 *Asynchrony in the molting cycle in larvae.* To understand the effects of heterochronicity and isochronicity it is necessary to consider some recent hypotheses on molting and growth in decapod larvae. Costlow (1962) was among the first to suggest that the hormones that control molting in decapod larvae differ from those controlling growth, basing his conclusions on results obtained with de-eyestalked megalopae of *Callinectes sapidus* Rathbun. Comparing a possible molting hormone in crustacean larvae with that known in insects, Costlow (1963a,b) also noted that an increase in juvenile hormone either prolongs larval development or prevents metamorphosis, so that it seemed that by decreasing this hormone development might be accelerated. Boyd & Johnson (1963), commenting on the production of substages and their survivability in the galatheid crab *Pleuroncodes,* cautioned that the speed of larval molting could be accelerated only to a certain degree before the larvae became 'physiologically or hormonally exhausted', and died. Earlier, Gurney & Lebour (1941) jointly provided a good discussion with several examples of how variation in molting, even within the egg, might produce a more advanced first stage larva than normal, especially in some oceanic caridean shrimps in the genus *Systellaspis.* This conclusion agreed with observations made earlier by Fraser (1936) who concluded that no fixed stages (instars) really occurred in euphausiid shrimps, and all specimens in a school grew with variable rapidity. However, because molting occurred at more or less the same time, the resultant larvae were of different sizes and development (see also Heegaard 1963).

Studies by Le Roux (1963) provided additional information. Le Roux defined three periods in larval development of the caridean shrimp *Hippolyte inermis* Leach: 1) passage through four zoeal stages and the acquisition of fundamental larval characters; 2) passage through a variable number of additional stages (from 1-4), during which time the larvae presented little (but nevertheless progressive) modification (but growth during this period allowed attainment of the size 'indispensable' for the appearance of the last stage); and 3) a last zoeal stage wherein morphological characters stabilized and metamorphosis could proceed.

In 1965, Costlow suggested that molting had priority over regeneration of appendages in larvae and that a regular pattern of molting is superimposed on the development of morphological features (morphogenesis), that may itself have priority over the normal pattern of development. Costlow postulated a molting endocrine system for controlling ecdysis, and a related morphological endocrine system which controlled zoeal morphology. The effects of any overriding by one or the other of these systems would produce several variations in molting. Provenzano (1967b) supported this concept and stated that the variability of larval instars resulting from this type of independence was probably of strong adaptive significance. It is easily seen, for example, that the potential to suspend metamorphosis is the first step toward neoteny (sensu de Beer 1951), and this in turn may lead to speciation (see discussion in Gurney & Lebour 1941).

But evidence was often conflicting. Roberts (1971) demonstrated that in *Pagurus longicarpus* Say metamorphosis was seemingly an all or nothing event, at least in the glaucothoe, and this postlarval stage must molt whether gastropod shells suitable in size and shape for the new hermit crab phase were present or not. Knowlton (1965) also believed that growth and molting were at least semi-independent processes, and Rice (1968) showed that growth is not an essential prerequisite for molting because many species with abbreviated development molted with little growth being observed. Schatzlein & Costlow (1978) found a similar situation in *Emerita talpoida* (Say), where growth, but not molting, stopped in this mole crab at the sixth zoeal stage, but the species doubled in size at each molt that occurred thereafter.

In an excellent synopsis, Knowlton (1974) hypothesized that the zoeal stage has three priorities: 1) stay alive; 2) molt; 3) undergo growth and eventually metamorphosis. He pointed out that variability in instars numbers and morphology was a strong indication of the semi-independent functioning of developmental processes and their differential control by various environmental parameters.

We should observe caution in coming to any conclusions regarding the reasons for these processes. The 'need' for 'a larger coat' is not the prime reason that molting takes place, even though the statement seems (teleologically) attractive. Growth and larger size results from, but is not a prerequisite for, molting. It is all to easy to impose human order or mathematical precision where nature has placed none (see 3.1.4). Nevertheless, as Fincham has emphasized (1979), the molt does remain the definitive event in decapod larval history. Because most larvae molt in nature in the water column, and so are essentially weightless, the resulting increase in size is probably as much dependent on how fast the newly deposited cuticle hardens as on the amount of water taken up by cuticular expansion. Morphogenesis will be almost permanently arrested once the exoskeleton hardens, and the instar then becomes 'fixed' in our eyes.

3.1.1.2 *Obligatory and non-obligatory stages.* Makarov (1968a,b) discussed abbreviated and direct development, especially in relation to the caridean shrimp families Crangonidae and Hippolytidae, and stated that, in general, all Caridea have obligatory, or non-obligatory stages. The first three stages seem to be obligatory, but later stages can vary; no individual need pass through all of them. But the longer the larval developmental cycle, the more non-obligatory stages will be involved. Referring to direct, or short-development species, Makarov suggested that these groups show such development in stages which would be obligatory in long-development forms. The non-obligatory stages occur primarily at the end of development, and Makarov considered these the first steps in the evolution of abbreviated development. The second step in the process is seen in Caridea that have obligatory stages which move backward toward the point of hatching, so that eventually these stages may be passed in the egg. As a consequence, the later non-obligatory stages have now become obligatory (because they have been condensed toward eclosion) and the previously obligatory stages are thereby eliminated. The end result in this process is a delay of hatching and a greater development in the egg. In Makarov's scheme, all Decapoda should tend toward direct development as they progress evolutionarily.

Williamson (1976) and Rice (1980b) both agreed that the 'ancestral zoea' probably passed through a series of stages (six or more). Rice disagreed with Williamson, however, in that he did not consider shortened development as being primitive, because too many advanced features appeared in any of the present day larvae to be consistent with a plesiomorphous condition. Rice acknowledged the possibility of condensation of characters and stage suppression, and indicated tentative acceptance for the possibility of a subsequent subdivision of the ancestral number of stages ('perhaps . . . two') resulting in a secondarily increased number in some families. He pointed out 13 characters which led him to believe that (in brachyuran crabs) evolution traversed several 'quite independent lines, rather than a single one leading from the most primitive families to the most advanced' (Rice 1980b: 301). It thus remains an unsettled question as to whether Makarov's scheme or Williamson's (with Rice's modification) will prove the more correct.

3.1.1.3 *Hormones, molting, and evolution.* Costlow (1963a, 1968) reviewed larval development and metamorphosis in decapod larvae, providing a valuable, albeit brief, update of ontogeny and ecdysis. He pointed out several difficulties in arriving at a satisfactory thesis on larval molting, noting for example that the sinus gland and X-organ complex is often not recognizable in zoeae until late in their development (cf. Hubschman 1963) and may not become functional until the post-larval phase. The X-organ/sinus gland complex is considered the site of the molt-inhibiting hormone that regulates molting by slowing or delaying the frequency. On the other hand, the Y-organ is present in at least some zoeae in the later stages, and it is this complex that produces the molting hormones and maintains molting frequency. Costlow was careful to note, however, that it is not necessarily associated with morphological development.

McConaugha (1980) and McConaugha & Costlow (1981) also investigated the cyclical activity of the Y-organ and the production of β-ecdysone in the molting process in larvae. They showed that hormonal titers increased rapidly during early premolt, reaching a peak at late premolt, and then dropped precipitously just prior to ecdysis. The changes in β-ecdysone titer seemed to follow a specific pattern allowing appropriate coordination and integration of the premolting events in larval life. It follows, therefore, that interruption of this coordination (e.g. by 'stress' of some sort) may either cast the larva prematurely into premolt, or perhaps delay the buildup of hormonal titer, thus prolonging the stage inordinately.

Anger & Dawirs (1981) proposed two supplementary concepts in an attempt to explain

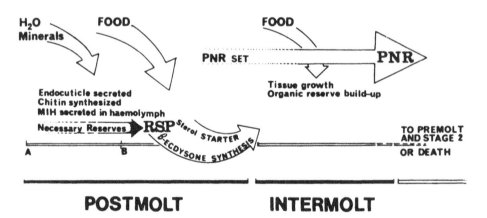

POSTMOLT **INTERMOLT**

Figure 3. The initiation of molting in a typical decapod larva. The postmolt period (the point directly after hatching in this case), occupies approximately 10 % of the total stage, and during this time water and minerals are assimilated, chitin synthesized, the endocuticle secreted, and molt-inhibiting hormone (MIH) secreted into the haemolymph. With intake of sufficient food the cycle proceeds toward intermolt. As the necessary energy reserves are accumulated the Reserve Saturation Point (RSP) is attained, at which time an hypothesized sterol 'starter' and β-ecdysone synthesis begins. The future Point of No Return (PNR) may also be set at this time. Continued food intake allows tissue growth to take place and organic reserve buildup. Toward the end of the intermolt period the Point of No Return is reached, at which time the zoea must molt, or remain in stage and eventually die. If organic reserves are sufficient, premolt is initiated and ecdysis to the next stage occurs. Without food (or insufficient food) all reconstruction ceases and protein is eventually used for energy, leading ultimately to death (based on Anger & Dawirs 1981).

the initiation of the larval molting cycle. These were the Reserve Saturation Point (RSP) and the Point of No Return (PNR). The RSP is that point in larval development when sufficient nutritional reserves allow molting to the next stage, regardless of the availability of further food supplies. The PNR reflects directly from this, because it is at this time that the zoea must molt if it is to continue to develop (Fig.3). Otherwise, the larva may merely live on without further ecdyses, or perhaps enter into a terminating series of molts with little or no further development, and eventually die (see Broad 1957b, Anger et al. 1981). Anger & Dawirs postulated several critical periods during the larval intermolt cycle (Fig.3, 4) and suggested that food is the limiting factor in the initiation of further molting substages. Kon (1979) also recognized these general effects but did not go into detail.

It is now known that in order to produce a variety of ecdysial effects in a larval stage, manipulation of the Y-organ and its secretions (either by environmental effects or through human intervention such as eyestalk ablation) is necessary. This apparently happens frequently in nature, resulting in anomalies in molt frequency and morphogenesis which occur at different times among different species. As Costlow held, and a perusal of the vast literature on larval development shows, there is little consistency in molting frequencies in

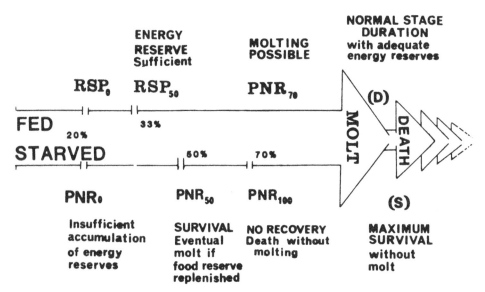

Figure 4. Critical points during a decapod larval molting cycle in fed and starved larvae. RSP_0 indicates the point occurring approximately 20 % into the cycle at which insufficient energy reserves have been accumulated in either fed or starved zoeae. The point at which approximately 50 % of the fed larvae can molt (RSP_{50}) occurs about one-third the way through and successful molting may be accomplished by about 70 % of the larvae at approximately ¾ into the cycle (PNR_{70}). The Point of No Return where 50 % of the larvae may survive and molt if provided sufficient food (PNR_{50}) occurs about half way through the cycle. Death without further molting will occur if insufficient food reserves have been accumulated in starved larvae, or if the larvae are not fed for approximately 70 % of the molting cycle, at (PNR_{100}). Sufficiently fed larvae will molt at (D), the point indicating the normal duration of the molting cycle. Maximum survival (S) without molting may continue for a variable length of time after the normal ecdysial point, and depends on the decapod species (based on Anger & Dawirs 1981).

many species. But through a combination of X- and Y-organ influences, larval development and duration of stages can be modified. This modification undoubtedly plays an important role in the evolution of many decapods, because it is at this stage in the animal's history that saltational mutation not only may occur, but seems to have the best possibility of being fixed. This concept is so important that we shall examine it in some detail.

In 1979 Matsuda provided a theory that attempted to explain morphogenesis in the Arthropoda, by postulating that varying titers of juvenile and molting hormones would trigger metamorphosis. Valentine (1981) in a summarizing paper pointed out that in insects, for example, the molting hormone ecdysone would induce a larval molt in the presence of large amounts of juvenile hormone (JH), a pupal molt with less JH present, and an imaginal molt in the absence of JH. Moreover, JH in addition to being an inhibitor, also causes precocious egg development, which results in developmental neoteny. As originally proposed by Matsuda, any mutation that reduces or stimulates the production of JH at unusual levels, or during unusual times, can result in abnormal metamorphosis. Valentine's statements in this regard are worth quoting in full: 'Unusual environmental stimuli can alter the level of hormone secretion and produce abnormalities similar to those resulting from mutations. The morphological consequences of the developmental abnormalities can . . . (range) . . . from changes in limb size and number, and wing reduction and loss in insects, to maturation of larvae with body plans distinct from those of adults'. Valentine also stated, 'If a population is subjected to a novel environment that induces unusual hormone production, many similar abnormal phenotypes will result. This is a nongenetic origin for a novel morphology. These novel individuals would presumably have no special trouble in reproducing, assuming they happen to be well adapted. The (aberrant) morphology could thus be propagated. Later genetic changes could then fix the new morphology in phylogeny. These genetic changes might be favored by micro-evolution because they would stabilize a phenotype with large adaptive value.'

The above mechanism can explain the processes of heterochronic and isochronic development in crustacean larvae (see also de Beer 1951). The influence of these factors on other developmental modes and the subsequent terminology that has arisen to describe these modes will be examined in later sections. To appreciate the full extent and significance of work on these mechanisms the reader is directed to studies by Broad (1957b), Costlow (1962, 1963b, 1968), Kurata (1962, 1969), Knowlton (1965, 1973, 1974), Makarov (1968a,b), Rothlisberg (1979), Fincham (1979), McConaugha (1980), Rice (1980b) and Dalley (1980). Although not all of these authors necessarily addressed the concepts of isochronicity and heterochronicity specifically, their work is important in the context of understanding the whole of the concepts involved.

3.1.2 *Retrogressive development*

This type of development, although noted by many earlier authors and first defined by Kurata (1969), is a combination of regular developmental growth and a retardation of certain developmental phases. Normally in development the various larval characters become more and more suppressed, and adult-like characters become more and more enhanced as the zoea progresses toward metamorphosis. However, because of the effect of heterochronic growth, some of the larval characters may be prolonged beyond the point where they would normally disappear and hence be carried over into the postlarval stage. The development then becomes paedomorphic (at least in part) with the maintenance of larval features perhaps even beyond the postlarval stage and into the juvenile (imago) stages. De Beer (1951)

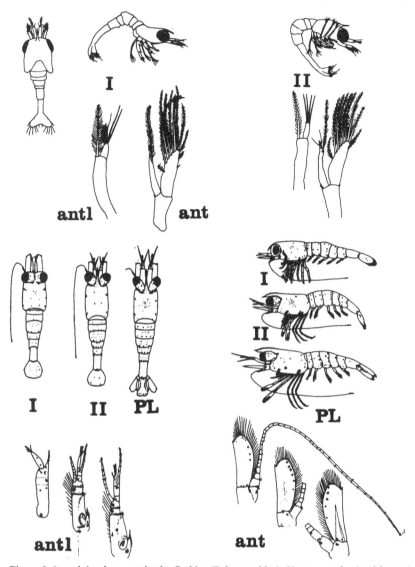

Figure 5. Larval development in the Caridea (Palaemonidae). Upper, regular (multi-stage) development showing antennular (antl) and antennal (ant) development in the first two free-swimming pelagic zoeal stages of *Periclimenes pandionis*. Lower, abbreviated (advanced) development in the two 'zoeal' stages of *Macrobrachium asperulum*. Compare antennular and antennal development. The two 'zoeal' stages are actually benthonic and except for retention of a larval-type telson and antennule in the first stage (retrogressive development), are essentially imagos undergoing nearly direct-type development (upper: from Gore et al. 1981a; lower: after Shokita 1977b). Not to scale.

categorized a similar type of development under the headings of neoteny and paedogensis (see Figure 5).

In addition to noting retrogressive development in his earlier studies, Kurata (1969) compiled a large amount of data on majid crab larvae, and noted that in *Hyastenos diacanthus* (de Haan) the megalopa exhibited only partial development of postlarval characters, with the rostrum, for example, being longer (i.e. more zoea-like) in less mature megalopae. Kurata also considered the presence of a dorsal spine on the carapace as a zoeal remnant. This same type of spine has also been occasionally seen in megalopae of *Libinia emarginata* Leach (Johns, pers. comm.), although normally the postlarva is smooth (Johns & Lang 1977), as well as in some cancrid and ocypodid crab postlarvae (Ingle 1981); (see Chapter 3 for *Uca*). Atkins (1954) found one abnormal megalopal stage in *Pinnotheres pisum* (Pennant) that had a zoea II telson on a megalopal abdomen; the species has only two zoeal stages but required six weeks to attain metamorphosis. Konishi (1981b) noted a similar development in *Pinnaxodes mutuensis* (Sakai), another pinnotherid crab in which the fourth zoeal stage exhibited megalopal features in telson furcae and inner spines, but the megalopae bore long zoea-like rostral and lateral spines on the carapace. All of these forms died without further molt. Warner (1968) illustrated a megalopal stage with a zoeal telson in the mangrove tree crab *Aratus pisonii* H.Milne Edwards, and Williamson (1965) described a megalopa of *Paramola petterdi* (Grant) that carried extremely elongate lateral and dorsal spines, so that the stage appeared as a megalopal form in a zoeal carapace.

Some caridean shrimp may also exhibit retrogressive development. In the freshwater atyid *Caridina denticulata ishigakiensis* Fujino & Shokita development is almost direct, and the single zoea appears very nearly like the postlarva in general morphology but lacks uropods on the telson. The remaining 'zoeal' characters are advanced and drop out in later postlarval stages (Shokita 1976). Kurata (1964a) described the larva of *Pandalopsis coccinata* Urita that hatch in an advanced stage, bearing antennae, partially mobile eyes, six abdominal somites and a clearly separated telson, and mandibles, maxillae, and pereopods of adult form. Gurney (1937b) described the postlarva of a species of *Chlorotocella* (Pandalidae) that had many larval features, including a mandibular palp, arthobranchs, epipods, and underdeveloped pleurobranchs. In the alpheids, Knowlton (1970, 1973) found one brood of *Alpheus heterochaelis* Say that simultaneously exhibited third zoeal and postlarval characters in stage III. The premature postlarval characters included functional walking legs, but with setose exopods. The uropodal endopod remained rudimentary, and 'typical' for stage III. Other specimens exhibited an 'incomplete metamorphosis' from zoea III to postlarva, possessing a fully developed tailfan and setose pleopods (postlarval), but retaining exopods on all but the last pair of pereopods (zoeal). None of these specimens molted to another stage either.

There are often other gray areas in which larval stages could conceivably be considered as extremely advanced and perhaps precociously molted, so that their characterization as postlarvae becomes a question of semantics. Dobkin (1965a) recorded a 'pseudolarval' stage of short duration, having biramous pleopods with appendices internae but no setae, segmented pereopods with rudimentary exopodites, and non-hooded eyes, in the snapping shrimp *Synalpheus brooksi* Coutiere. These pseudolarval stages hatched from the same egg mass as the more advanced postlarval stages, but did not molt to that stage until several hours later. Powell (1979) noted that in *Desmocaris trispinosa* (Aurivillius), a freshwater African palaemonid shrimp, the chief differences between stage I and II were the presence of enclosed, instead of free, uropods, respectively; the stages were otherwise imagos of the adult, and larval development was considered completely suppressed. Continued development brought refinement of adult features already present, and ultimately sexual maturity at about stage X.

In *Metapenaeus burkenroadi* Kubo (Penaeidae) the first two postlarval stages retain a definitely pelagic existence and exhibit larval features while in this mode of life, even though other juvenile characters are also present (Kurata & Pusadee 1974). The authors considered these instars as the time when '. . . degeneration of larval structures and the construction of the adult structures (is) occurring at the same time'. Dakin (1938) also noted retrogressive development in *Penaeus plebejus* Hesse, with the postmysis stages having seven rostral teeth (a larval character) but an adult type of telson, in contrast to others which had eight rostral teeth (more advanced) but with a non-adult type of telson. The latter remained in the plankton, while the former were considered more able to assume a benthic existence.

In the Anomura, Reese & Kinzie (1968) showed that the terminally added zoea V in *Birgus latro* (L.) exhibited a mixture of larval and glaucothoeal characters that included zoeal-shaped telson and zoeal armature on the fifth abdominal somite, but a segmented antennal flagella, well-developed thoracic appendages, and setose pleopods of the glauco-thoe. Provenzano, in a series of studies on pagurid larvae, noted several variants of retro-gression in the glaucothoe stages of *Calcinus* (1962a), *Coenobita* (1962b), *Trizopagurus* (1967b) and *Petrochirus* (1968a). These consisted of extra stages beyond the normally recorded number, which generally appeared less developed than equivalent pre-metamorphic zoeae, and usually possessed more zoeal-like features. Provenzano (1962a) considered these stages a '. . . result of non-uniform rates of internal growth coupled with a more or less regular molting cycle', and was often able to trace the less advanced stages as far back as the third instar after hatching. Apparently few of these larval characters were carried over into the postlarval stage, indicating that the metamorphic molt comprised a 'catch-up' period so that development proceeded normally thereafter. However, few studies have fol-lowed life histories beyond the postlarval phase, so that data on possible retardation in juvenile or adult phases remain almost unknown.

3.1.3 *Poecilogony and dimorphism*
Much of the data have been obtained from caridean and thalassinidean shrimps, including freshwater species of the former. Webb (1919), for example, obtained two different larval types from the second and later zoeal stages in *Upogebia deltura* Leach and *U. stellata* Montague from the British coast. In a complex series of molts the third zoeal stage could metamorphose either into a postlarva, or go on to a fourth zoeal stage which then molted to a postlarva of a different form than the first. Shenoy (1967) noted a similar situation in *U. kempi* Sankolli from India which molted through two postlarval stages, the second being bimorphic in that the maxillipeds and pereopods either possessed or lacked a distinct fringe of hairs. Non-fringed second stage postlarvae could molt into fringed third postlarvae.

Gurney (1942) provided a detailed discussion on poecilogony and dimorphic develop-ment (in freshwater atyid shrimp) but remained skeptical of its occurrence, although agree-ing that variation in instar number and morphology might occur. The earliest studies by Brooks & Herrick (1892) were also viewed with skepticism by Knowlton (1970, 1973) who nonetheless held that some alpheid shrimp may exhibit two different developmental modes; one an abbreviated or nearly direct development, the other with longer ontogeny encompassing several stages. Other data come from *Palaemon debilis* Dana, which has two different types of postlarvae, one with supraorbital spines, a rounded telson margin and reduced exopods on the pereopods; and the other without supraorbital spines, a telson bearing a median spine, and degenerate exopods on all pereopods (Shokita 1977a). In *Hip-polyte inermis* (Leach) development consists of either five zoeal stages and a postlarvae, or up to 8, 9 or 10 zoeae followed by a postlarva (Le Roux 1963). The latter type of development may be a form of additive molting resulting in supernumerary larval stages (see 3.3.2.5).

Goy & Provenzano (1978) described a bimorphic seventh zoea in the laomediid shrimp *Naushonia crangonoides* Kingsley, that possessed a single medial spine on the telson instead of the usual paired setae, and consistently varying antennular and appendage setal formulae, thus recalling variation similar to that seen in *Palaemon debilis* noted above.

Little dimorphism has been observed in brachyuran crabs. Williamson (1965) described a type of variation in the larval carapace of *Latreillia australiensis* Henderson, in which a denticulate posterolateral fold and carapace margin was seen in all larvae of some hatchings, no larvae in others, and in a mixture of larvae in still other rearings. In this example and in many others, the immediate explanation that comes to mind is variation induced by fluctuating environmental parameters. But as yet there have been no studies which have deliberately tried to provide dimorphic development, although the abundance of known cases would seem to offer ample and fertile field for such explorations.

3.2 *Terminology of developmental types*

We turn now to the various developmental modes that incorporate or result from the previously discussed concepts of heterochrony, isochrony, and retrogressive development. In the Decapoda, larval development proceeds along a continuum from essentially the lack of a free-living larval stage, through a larval phase of brief duration, to that of greater and greater temporal extent. Within this continuum three major categories are distinguished: regular development, abbreviated development and extended development. None of these is mutually exclusive of the other within any particular decapod taxon, and each is relative to one another, so that what appears as regular development in one taxon could conceivably be considered abbreviated in another, or extended in a third. In the sense that one or more types may occur as a matter of course in the ontogeny of a species, none can be considered abnormal.[3]

The point in a group where development changes from abbreviated to extended is without real meaning, because the question that remains unanswered is 'relative to what?' Knowlton (1973) categorizes development in the snapping shrimp *Alpheus heterochaelis* (Say) as extended if having four or more instars, abbreviated with three or fewer, and direct if without larval instars. In this sense, development is either regularly extended, regularly abbreviated, or regularly direct.

On the other hand, development may consist of a variable number of stages either within an individual, a species, or as a comparatively dissimilar number of stages within a higher taxon, for example between species in a genus, or genera within a family. In such cases development has often been broadly classified as irregular, but the term so broadly applied is without much importance. It is, instead, more important to view irregular development as a correlative categorization, contrasted to regular development, because shortening or lengthening of ontogeny may occur at any time, as opposed to the more or less fixed ontogeny seen in regular development. The important distinction seen is that between a regular development of consistent and repetitive stages, even if differing within a genus, and irregular development producing an inconsistent larval series, often as a consequence of environmental and/or genetic factors (sensu Sandifer & Smith 1979). In such cases the diecdysic larval cycle is altered from an isochronic to a heterochronic modality, resulting in accelerated, delayed or sporadic development. These developmental modes are summarized in Table 3 and each will be considered in more detail below.

Table 3. Comparison of molting modes, size, structural relationships, and resulting developmental types in decapod larvae. *Exhibition of one or more types of developmental modes depending on species or environment.

Molting mode	Size of each instar	Structural features	Molting frequency	Resulting morphogenetic mechanisms	Development
REGULAR	Different	Different	Maintained	Progress with each stage	Variable*
Abbreviated	Different	Different	Interrupted or maintained	Often saltational within a series, or from larva to adult	Advanced or direct
IRREGULAR	Mode-dependent	Variable	Maintained/interrupted	Mode-dependent, with or or without progress	Variable*
Accelerated	Different	Different	Interrupted or maintained	Saltational	Individually condensed
Skipped staging	Different	Different	Maintained	Saltational; stages with advanced morphology, especially penultimate and ultimate instars	Accelerated
Combinatorial molt	Different	Different in parts	Maintained or interrupted	Maintained or accelerated in part; within series stages may be advanced; ultimate stage typical or with saltational features varying within instar	Condensed; retrogressive; accelerated; sometimes direct
Precocious	Different	Different	Maintained	Saltational; anatomically adult, functionally so?	Retrogressive, or larval-adults
DELAYED	Same or different	Mode-dependent	Interrupted	Usually inhibited	Delayed
Mark-time	Same	Usually the same	Maintained	Inhibited in series; a single typical ultimate stage	Delayed, often quite lengthy
Intercalated staging	Same or different	Usually different	Interrupted	Inhibited or accelerated within series; usually a typical (sometimes advanced) penultimate stage	Variable*; usually delayed (substages)
Terminally additive staging	Same or different	Same or different	Maintained	Inhibited only at termination of series; a series of typical penultimate or ultimate stages occur	Delayed
Sporadic	Usually different	Usually different	Interrupted	Saltational or retarded, often non-sequential	Variable*
EXTENDED	Different	Different	Maintained	Progress slowly with occasional minor saltation in intermediate or later stages; morphogenesis may be suppressed or retarded	Lengthy but fairly regular

We do not know whether at some point in their evolutionary history all decapods passed through an identical, fixed, number of stages, and if they did what that number was. Moreover, we have no evidence (and probably never will) whether any decapod taxon today exhibits the most 'primitive' number of stages, although many authors consider the Penaeidae as a prime candidate, owing to the retention of naupliar and protozoeal stages in their

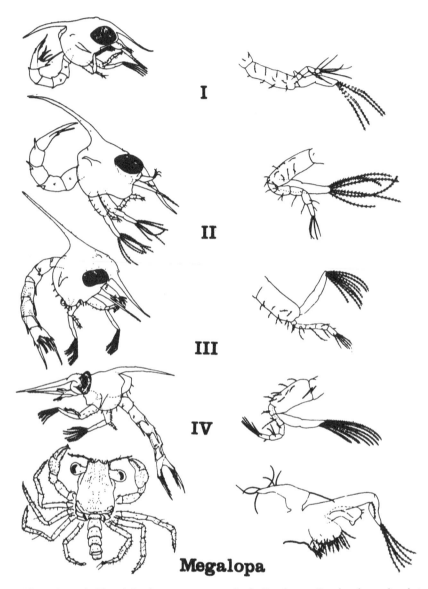

Megalopa

Figure 6. Typical larval development sequence in the Brachyura. Zoeal and megalopal stages (left), maxilliped 1 (right). Note regular increase in exopodal natatory setae from stage to stage, first appearance of sixth abdominal somite and pleopod primordia in stage III of the zoeal series, and the structural change in the maxilliped in the megalopal stage (from Gore et al. 1981b).

development. With this in mind, perhaps the best way to visualize the relationship between regularly abbreviated, extended, and direct developmental modes is to return to the idea of a continuum and follow the various types of staging, providing appropriate examples and nomenclature. The pivotal point is the concept of regular development which will be considered next.

3.2.1 *Regular development*

Regular development is defined as 'the numerically predominant type of ontogenetic growth and developmental staging that consistently recurs in a particular decapod group' (cf. Broad 1957a). Regular development in some freshwater brachyurans (e.g. Pseudothelphusidae) is direct, with imagos of the adult produced at hatching and no free swimming phase requiring a metamorphic molt (see authors cited in Moreira 1912). In other brachyurans (Fig.6), for example the marine Xanthidae, development regularly consists of four zoeal and a megalopal stage (Andryszak & Gore 1981). In some Anomura (e.g. Porcellanidae) two zoeae and a megalopa occur, whereas in others (Paguridae) four zoeal stages are the norm. In the Macrura Natantia (e.g. Penaeidae) some genera have five naupliar and three each of proto-zoeal and mysis stages (Cook & Murphy 1971, Dakin 1938, Thomas et al. 1974a,b), but others may possess three, four, or up to eight regularly occurring naupliar stages, but a variable number of mysis stages (Dobkin 1961, Gurney 1927, Heegaard 1953). Palinurid and scyllarid lobsters regularly undergo an extended type of development before attaining the final phyllosoma stage (Johnson & Knight 1966, Robertson 1968a, Phillips et al. 1979), and it could just as well be stated that the larval development of other decapods is abbreviated relative to that seen in the scyllaroideans. (A detailed compilation of developmental modes in 75 families of decapods is available from the author on request.)

3.2.2 *Abbreviated development*

Abbreviated development is somewhat of a vague catchbasket term, because it has been employed in both regular and irregular developmental modes. It may be defined as 'a larval sequence of shorter duration than that normally occurring in a preponderance of related species in a taxon, resulting in fewer numbers of morphologically discrete instars, a reduced duration for these, or both'.

In abbreviated development the larvae may exhibit a wide range of morphological variation, leaving the egg as morphologically 'early' zoeae, often following a prezoeal stage, or hatching in a morphologically advanced state, equivalent to a penultimate or ultimate zoeal stage, or even a postlarval stage. Whatever the case, abbreviated development assumes that the longer developmental sequences, i.e. those recorded more often in a particular taxon, more accurately reflect the general ontogeny within that taxon, than does the shorter (i.e. abbreviated) developmental sequences. Thus, the occurrence of only two zoeal stages in the semi-terrestrial grapsid crab *Geosesarma peraccae* (Nobili), or three zoeal stages in *Sesarma reticulatum* (Say) are considered abbreviated development within a generic complex that most often undergoes four or five zoeal stages before postlarva (see Yatsuzuka 1957, Costlow & Bookhout 1960, 1962, Soh 1969, Baba & Miyata 1971).

Under the subheading of abbreviated development may be placed two infraheadings: direct development and advanced development. These subheadings characterize the different forms produced at eclosion; viz., an imago (direct) or a zoeal stage morphologically equivalent to a penultimate or ultimate zoea, but which may undergo additional ecdyses prior to metamorphosis (advanced).

3.2.2.1 *Direct development.* Direct development may occur regularly, as noted earlier in some freshwater Brachyura and astacidean crayfish, or it may form a type of abbreviated development, differing from the norm usually seen in a taxon. The important distinction is that no free-swimming larval stages are produced, and the young hatch more or less in the form of the adult, possessing distinct juvenile morphology. Examples occur in many marine and freshwater caridean shrimps, including the Alpheidae (Brooks & Herrick 1892, Dobkin 1965a), Palaemonidae (Gurney 1938b, Powell 1979), Atyidae (Mizue & Iwamoto 1961, Shokita 1973a, 1976, Benzie 1982), some stenopodidean shrimps (Kemp 1910a,b, cf. D.Williamson 1976), in thalassinidean shrimps (Gurney 1937b), at least one paguridean crab (Dechancé 1963), possibly in a galatheid crab (Fage & Monod 1936), in several dromiacean crabs (Montgomery 1922, Hale 1925), xanthoid crabs (Wear 1967), at least one majid crab (Rathbun 1914), and in all known freshwater crabs in which development has been ascertained (Smalley, pers. comm.). Examples from these and other groups will be examined in detail by Rabalais & Gore (this volume) and need not be considered further here.

In direct development the postlarval or pre-juvenile instar that is produced has approximately the same relationship to the adult as the pre-zoeal stage has with the zoea (see 4.4), with the important exception that a discrete ecdysis from the megalopa or juvenile-like stage takes place. Morphological change of some type usually occurs, and the cast exuvium is not simply a cuticular sheath as seen in the prezoea. In some cases the imago is nearly complete and only adds or refines some adult limbs or mouthparts (Benzie 1982) or completes telson morphology (Dobkin 1968, Shokita 1973b). In other cases, however, the assignment of stages becomes quite difficult because each form more or less resembles the

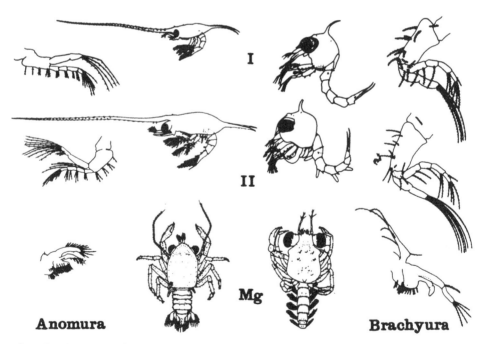

Figure 7. Advanced development in larval stages of Anomura (Porcellanidae) (left) and Brachyura (Majidae) (right). Compare the exopodal setae on maxilliped 1 in either group. Note also the appearance and structure of the pleopods in stage II in either group, but the relatively great dissimilarity in maxilliped 1 in the megalopal stages, although the overall appearance of the instars are superficially similar (from Gore 1979, and Wilson et al. 1979). Not to scale.

adult but carries one or two larval characters that drop out as molting toward sexual maturity continues (see 3.1.2). This type of paedomorphic development seems to be confined primarily to caridean shrimp species. Further information will be provided in Rabalais & Gore (this volume).

3.2.2.2 *Advanced development.* In this type of abbreviated development, the young hatch as zoeae, but in a state more developed that that seen in their congeneric relatives, so that larval development is often foreshortened in comparison, both durationally and ecdysially, and a reduced number of morphologically advanced instars obtains before the metamorphic molt to the postzoeal stage. The entire developmental sequence is regular, without supernumerary or skipped stages in the series, such as occurs in accelerated development. Examples of anomuran and brachyuran crabs having advanced development are provided in Figure 7.

The reduced number of zoeal stages in these species often have varying degrees of maturational development, exhibiting characters that are normally seen only in later stages of their congeners. A good example is seen in the lithodid crab *Lithodes maia* (L.) described by MacDonald et al. (1957). The first stage is quite advanced, having mobile eyes, mandibular palps and rudimentary pereopods, but the second stage 'showed nearly all the characters associated with the last (fourth) stage in the Pagurinae' according to these authors.[4] The early appearance of more mature developmental characters such as these, coupled with the shortening or elimination of later stages, results in a very rapid larval development and subsequent attainment of the juvenile phase. As with direct development, the young may remain in the vicinity of the parent owing to the short time spent in the plankton.

In many cases, the morphological characters normally appearing in later larval stages are shifted backward or compressed into the first and second stages, while developmental duration remains more or less unchanged. Thus, pleopod buds, mandibular palps, well-developed antennules, segmented pereopods, six abdominal somites, and even movable eyes, all of which usually appear in later, or even ultimate and penultimate zoeal stages of longer developing species, make their appearance in first or early stages. However, when compression of morphological features unites with elimination of instars and shortening of duration in the remaining stages, a highly advanced larva is the result. Numerous examples of this type of development are considered in Chapter 3.

3.2.3 *Extended development*

This category is established for those species which regularly pass through a large number of morphologically discrete instars, so that a higher number of stages, coupled often with increased duration within these stages, results in a larval development lasting from several months to nearly a year. Extended development is seen primarily in the pelagic caridean shrimp families Oplophoridae and Pandalidae, in some genera of the Palaemonidae (especially in a few estuarine and freshwater species), some Alpheidae, in several species of freshwater atyid shrimps, in the Penaeoidea and Sergestoidea to a large extent, and as a matter of course in the scyllaridean and palinuridean lobsters.

In species undergoing extended (and delayed q.v.) development, the molting sequence seems to take precedence over morphogenesis; or, alternatively, the morphogenetic cycle becomes delayed or even suppressed, so that the larvae pass from one stage to the next often with little change in form, only slight or no increase in size, or both. In extended development morphogenesis proceeds slowly but on a more or less regular basis with size increase. This results, in some groups such as the oceanic Caridea, in the gradual assumption of the adult form during a lengthy development. Other groups, such as the penaeid shrimps or spiny and spanish lobsters, pass through a relatively long series of one form before changing to a second form, and finally undergo a discrete metamorphosis into a postlarval form still different from that preceding it.

An important consideration in distinguishing between species having extended, as opposed to delayed, development is that in the former a change in environmental or nutritional conditions rarely, if ever, shortens development by causing elimination of stages, whereas in the latter the regular developmental sequence may be 'extended' by addition of supernumerary stages, or 'shortened' by elimination of these same stages, depending on how beneficial the conditions are. In short, the contrast is between regular and irregular development.

3.3 *Irregular development*

Irregular development is applied to those species which occasionally deviate in number or duration of larval stages from that seen in the preponderance of related species. Within this framework can be assigned the subcategories of accelerated, delayed and sporadic development. In accelerated development, hatching produces a morphologically early zoeal stage, often proceeding from a prezoeal stage, and development is condensed by the elimination of one or more intermediate or terminal stages in ontogeny. Delayed development is only similar in that a prezoea or morphologically early zoeal stage is produced at hatching, but subsequent insertion of intercalated or supernumerary stages delays attainment of the postlarva phase. In sporadic development the number of zoeal stages may vary both intra-specifically or interspecifically, and without apparent consistency, so that no generally fixed number of instars can be predicted.

3.3.1 *Accelerated development*
This type of development is among the most interesting and complex in decapod larval ontogeny. In accelerated development, the larval series in individuals within a species, or species in a genus, becomes shorter than in their regularly molting counterparts, owing to direct elimination, probably as a consequence of environmental adaptation, of one or more otherwise regularly occurring stages within the ontogenetic series. As might be expected, the difference between advanced and accelerated development is often rather vague, and the two types may intergrade in some species or genera. In fact, accelerated types of development occurring over evolutionary time probably account for the presence of advanced larval stages, thus clearly exhibiting the effects of heterochronic development on isochronic growth, especially if the eliminated stages become permanently lost in ontogeny.

Within accelerated development, several variations have been recorded. These include skipped staging, combinatorial staging (a form of retrogressive development), and precocious development (a form of heterochronic growth involving compression of instar morphology). The distinction again is emphasized that this type of staging is relative, especially if the addition of intercalated stages (a form of extended development, see section 3.3.3) is known to occur. The important difference is what transpires in the preponderance of cases. Thus, if the majority of larvae in a species undergo N stages, but some few pass through N−1 in completing their development, then accelerated development is occurring. However, as will be emphasized later, nearly all developmental data have been obtained

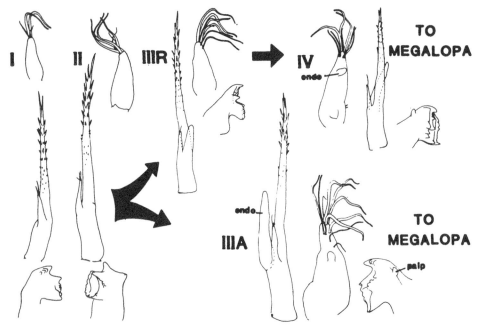

Figure 8. Development of antennule (top), antenna (middle), and mandible (bottom) in an accelerated developmental series in the Brachyura. Note general similarity of appendages in stages I and II, and increasing complexity beginning with regular (R) and advanced (A) stage III. Endopodal buds (endo) undergo differential growth and mandibular palps (palp) appear earlier or later, depending on whether stage III proceeds to stage IV (regular development) or to megalopa (accelerated development) (from Gore et al. 1981b). Not to scale.

under laboratory conditions, so that it becomes quite difficult to say whether stages are being subtracted (N vs N−1), or actually added (N−1 becoming N) in the plankton. A typical effect on morphology is seen in Figure 8.

Broad (1957a) discussed instars in *Palaemonetes pugio* Holthuis and showed that the rate, rather than the sequence of development, could account for the 'out of sequence' appearance of some appendages over others. Broad cited Fraser's earlier study (1936) on euphausiids in support, noting that Fraser hypothesized that the time of molting could shift forwards or backwards, even though development remained, to all extents, continuous. Broad argued that the presence of larvae of varying size and form but with identical molting history provided support for the independence of molting frequency from development as discussed earlier (but see Hartnoll & Dalley 1981). Al Kholy & Mahmoud (1967) believed that the first free larval stage in *Synalpheus biunguiculatus* (Stimpson) was actually the second, because the first was passed in the egg, and the spinal formula, uropods and fifth pereopods of the numerical first zoea were morphologically equivalent to zoea II of other alpheids. These authors noted that Gurney (1938b) had suspected as much when he alluded to the possible speeding of metamorphosis in his description of the same species. Knowlton (1973) noted a similar phenomenon in *Alpheus heterochaelis,* with some broods hatching as zoea II instead of zoea I. Scelzo & Lichtschein de Bastida (1979) and Thiriot (1973) provided discussions for species in the Brachyura, as did Makarov (1968a,b) for some Caridea, and Rice (1980b) for the Decapoda in general.

In many instances, however, development is more or less consistent only to a given instar, suggesting that if variation is to occur it cannot take place before a particular maturational level is attained. Thus, Costlow (1965) noted that the first four zoeal stages in *Callinectes sapidus* Rathbun were morphologically consistent, but that wide ranging variation in instar number and morphological complexity could happen thereafter. Saisho (1966) found that growth in phyllosomas of *Panulirus* were nearly constant until stage V; thereafter ecdyses were generally of longer duration and stages VI-XIII were considered to be abnormal in some degree. Knight (1967) also found great individual variation in both relative growth and appendage setation in *Emerita rathbunae* Schmitt, and that the number of stages possessed by any individual could be determined after stage IV only by following its molting history. Goldstein & Bookhout (1972) and Roberts (1969) have also seen markedly saltatory increases in growth between stages II-III of the hermit crab *Pagurus prideauxi* Leach, and IV-V of the portunid crab *Bathynectes superbus* (Costa), respectively. Roberts suggested that this might prove to be a normal situation in portunid larvae but data from other studies are not wholly supportive. Of equal importance, however, is that this type of saltatory increase has not always been taken into account when larval stages are being reconstructed from planktonic specimens, so that intermediate stages (or substages) are inferred owing to the apparently great size discrepancies between what are actually sequential stages (see e.g. Sandifer 1973b). But intermediate stages may indeed be present in some species and may also be skipped, a situation that will be examined next.

3.3.1.1 *Skipped staging.* Skipped staging, in which one or more instars may be bypassed by a species as it proceeds toward metamorphosis, is probably the most common type of accelerated development. It appears more often than not in those forms which normally undergo greatly extended larval development, such as seen in the Caridea or Scyllaridea (e.g. Little 1968, Sandifer 1972, 1973b), but it is not uncommon in other decapod groups. Here again, because regular development in some groups may be quite sporadic (e.g. as in some Caridea) both acceleration and extension of development become relative concepts, as noted in other ontogenetic modes. Often, definitions of staging become a matter of convenience (Johnson 1960, Robertson 1969a, Phillips et al. 1979) and the prevailing number of stages will determine whether stages are being skipped or intercalated. For example, Kurata (1964a) studying *Pandalus* development, wrote that there was 'a tendency . . . within (the) genus to reduce the number of zoeal stages, largely by metamorphosing at the earlier state of development'. He noted that the number of zoeal intermolts was highly variable, both intra- and interspecifically. Johnson (1968a,b) succinctly summarized the situation in scyllarid lobsters by stating that the phyllosoma instars may be grouped to form a given number of 'stages', but that because of variation it would be possible to increase or decrease this number by allowing more leeway in the definition of these same instars.

Although staging in other groups is usually less numerous than seen in the Caridea and Scyllaridea, it is still rather variable. In the Anomura the numbers of skipped stages may be as few as one or up to four or more. Examples are seen in the Paguridea and Coenobitoidea (Provenzano 1962a,b, Gore & Scotto 1983). Lang & Young (1977) showed that *Clibanarus vittatus* (Bosc) produced either a regular, or an advanced stage IV zoea; the former molted to stage V; the latter was larger, morphologically more complex than stage IVa, and molted directly to glaucothoe. Reese & Kinzie (1968) determined that the coconut crab *Birgus latro* required either three or four zoeal stages before attaining glaucothoe, but did not delineate advanced or regular stage types. However, Gore (1979) found advanced stages in the larvae of *Galathea rostrata* (A.Milne Edwards), with stage V being skipped if stage IV was suitably advanced.

For a long time, it was thought that the Brachyura would only rarely exhibit skipped staging. However, Costlow (1965), in an exhaustive study on the larval variability of the blue crab, stated that the species underwent a variety of accelerated developments, exhibiting skipped stages (e.g. zoea IV molting to a stage morphologically equivalent to zoea VI, and then to a zoea VIII), as well as several other types of rapid development. In the Xanthidae, several authors (Andryzak & Gore 1981, Gore et al. 1981b) drew attention to skipped staging in various genera, but data are otherwise sparse for many other brachyuran families.

Often there is little consistency in whether a stage will be deleted or not, and environmental factors seem to play an important role in this respect. Le Roux (1970) demonstrated that *Palaemonetes varians* (Leach) has a variable number of molts, with stage IV suppressed at culture temperatures greater than 30°C, so that the postlarva appears after stage III. At lower temperatures (ca.17-20°C) ontogeny was extended and five zoeal stages were seen before postlarva. Interestingly, the postlarval morphology from stage III or stage IV larvae was similar, suggesting that ontogentic changes were already underway in zoea I or II. Knowlton (1972) briefly summarized this in *Alpheus,* too.

3.3.1.2 *Combinatorial staging.* In this type of staging the zoeal or megalopa stage exhibits a combination of characters from the preceding or succeeding instar, so that the resulting stage appears generally more, or less, advanced than the normally numeric stage. Combinatorial staging is often seen in conjunction with skipped stages (e.g. Knowlton 1970) and the individual instar shares a combination of features from the stage passed over (q.v.). Another effect of combinatorial staging is its relationship to retrogressive development. In the latter case, the zoeal characters are carried forward into the postlarval stage (Knowlton 1973) but the postlarval characters tend to outweigh the zoeal features. Perhaps the best example of combinatorial staging was provided by Costlow (1965) in his study on *Callinectes sapidus.* He noted that up to zoea IV the instars were morphologically consistent, but thereafter several variations could occur. The combinatorial characters often consisted of the antenna and maxilliped of a later stage and the abdomen of an earlier stage in a particular instar. Costlow noted that the 'more advanced characteristics . . . invariably (were seen) in the anterior portion of the zoea; i.e. in the stage VI-VII zoea the anterior portion resembled stage VII . . . while abdominal segments and development of pleopod buds were . . . stage VI'. This type of advanced anterior development might be retained throughout the zoeal development, or the 'laggard' morphology might catch up in later stages (cf. Hartnoll & Dalley 1981).

According to Costlow, in a combinatorial molting situation, the molting frequency within a series is maintained, but the morphological mechanisms are accelerated or retarded either anteriorly or posteriorly resulting in an instar exhibiting heterochrony in its anatomical features. As noted earlier, if the morpho-endrocrine system becomes out of phase with the molting endocrine system, several novelties in instar development can occur (refer, for example, to Tables 2 and 3 cited earlier).

The presence of a combinatorial stage within a larval series does not seem to be detrimental to larval survival. Costlow (1965) noted that at least 30% of the larvae reared in the laboratory that showed some form of combinatorial features (e.g. zoea V molting to zoea V/VI, VI/VII, etc.) attained megalopal stage. Other authors (Rice & Provenzano 1970) stated that the fifth and sixth stage may be combined in planktonic specimens of the crab

Homola barbata (Fabricius), so the phenomenon is not necessarily an artifact induced by laboratory culture.

3.3.1.3 *Precocious development.* If such morphology plasticity is maintained within a species via its larvae, then a possible consequence of either skipped staging, or combinatorial molting, is the considerable shortening of development time spent in the plankton, especially if the temporal duration for other instars in the series remained more or less as before. Precocious attainment of stages lays the groundwork for evolutionarily important ramifications, including neoteny or paedomorphosis at one end, and abbreviated development (at least) at the other. Both ultimately result in the acceleration of postzoeal maturity. The evidence now available from the literature on larval development strongly suggests that these erstwhile 'aberrant' situations are not unusual at all; indeed, they may in fact be the normal situation in nature, instead. These concepts will be considered at length by Rabalais & Gore (this volume).

3.3.2 *Delayed development*
As noted previously under the general heading of extended development (3.2.3), there is often a fine line between what appears as extended regular development and irregular delayed development. The distinctions may become blurred partially because of the staging exhibited by the larva, and partially because of the semantics involved. For example, a panulirid lobster normally undergoing 12 or more phyllosoma stages is said to have an extended regular development. If one or more stages or substages are inserted within the sequence so that attainment of the postlarval stage is pushed farther ahead, its development can then be considered delayed. As a general rule, if a long, regularly occurring molting sequence within a species produces morphologically discrete instars at each ecdysis, each showing some advancement over the previous instar, then development can be considered extended. On the other hand, if a molting sequence, regardless of its usual duration, produces one or more instars showing little or no morphological progress toward a more advanced stage, then development is delayed.

Three subcategories are included in delayed development, viz. mark-time molting, intercalated staging, and terminally additive staging. All lengthen the pelagic phase of a species. As pointed out above, some overlap in definitions for each of these categories cannot be avoided, and a particular species might experience all three types of staging in its larval history.

3.3.2.1 *Intercalated stages.* The simplest type of delayed development is through the insertion of intercalated stages, i.e. extra stages between those normally occurring in a preponderance of developmental sequences within a species. Intercalated stages may appear as distinct instars, morphologically discrete from the preceding stage, but differing in some manner from what would normally be the following stage. Variation may occur in size, in appendage morphology, setal formulae, anatomical development, or in all of these to some extent. In laboratory culture intercalated stages may be overlooked if a complete history for each individual larva is not maintained. They may, however, be discovered in retrospect once comparisons are made between developmental features among specific instars. At this time the differences become quite noticeable and the stage can be seen to be more advanced, or otherwise disagreeing in one or several features, from other larval stages in the series.

Intercalated stages may molt into a normal following stage, or they may bypass or skip the subsequent stage and molt to a form morphologically equivalent to the next stage thereafter. It was instars of several species which exhibited just such morphology that first drew the attention of naturalists toward them. Lebour (1928, 1930a) suggested, based on a reconstruction of planktonic larval series, that some stages might be inserted. Yet as early as 1879, Faxon had postulated the occurrence of 'abnormal stages' in larval development. Since then, intercalated stages have been recorded in numerous groups including Penaeidea (Dakin 1938), Nephropidea (Templemann 1936), Thalassinidea (Sandifer 1973b,c), Anomura (Le Roux 1966b), and Brachyura (Yatsuzuka 1962). Knowlton (1974) provided a list of suprageneric taxa in which supernumerary stages were recorded and discussed some reasons for their occurrence. We shall return to these reasons subsequently.

In defining intercalated stages the distinction is emphasized that these stages occur within the normal developmental series, i.e., after zoea I and before the last zoeal stage. Another type of intercalated staging occurs when an extra instar or instars are added to what would normally be the end of a typical larval progression; these are distinguished as terminally additive stages (see section 3.3.2.5) warranting a separate consideration.

It will also be seen here, as elsewhere, that overlap in definition is unavoidable. Thus, an intercalated staging sequence may consist of a single added instar, a series of instars only slightly divergent morphologically, and in size from each other ('substages' of some authors, see section 3.3.2.2), a longer series of instars showing little or no change in size or morphology (marktime staging, q.v.), or one or more instars of noticeable morphological variation and size appearing at the end of a larval sequence (terminal additive molting). To further complicate the definitions, some anomalous intermediate stages may be added within a series of terminally additive molts, so that substages at the end of a larval sequence are defined (e.g. Carli 1978).

It is now generally agreed that intercalated stages are a response by a larva to adverse conditions involving food deprivation, temperature, salinity, and pollution, in the main (e.g. Kurata 1968, Goldstein 1971, Lumare & Gozzo 1972, Rochanaburanan & Williamson 1976, Goy & Provenzano 1978). These factors disrupt the hormonal cycle that controls molting and growth. Although these factors have been studied individually in their effects on larvae in the laboratory, it seems more probable that the action of one may lower the threshold of tolerance in a larval stage to another. As Mirkes et al. (1978) noted, 'The question of synergism of stresses in the natural . . . environment cannot be ignored'. The most noticeable reaction to such stress (other than death) is seen in the disruption of an ordinarily consistent molting cycle.

3.3.2.2 *The question of substages.* Growth in crustacean larvae was believed to be discontinuous (Shen 1935, Rice 1968), with size or maturational change occurring only at molting, as opposed to continuous growth exhibited, for example, in vertebrates (see Bonner 1965 for complete discussion; also Green 1963, Lockwood 1967, Warner 1967). However, Lebour (1930b, 1940, 1943, 1950) noted polymorphic stages in planktonic zoeae of several species of anomuran and brachyuran crabs, and postulated substages based on the growth of certain appendages during the intermolt period. Marukawa (1933) supposedly observed three substages in the glaucothoe of *Paralithodes brevipes* (H.Milne Edwards & Lucas) according to Sato (1958). Bourdillon-Casanova (1960) added support in her study on Mediterranean decapod larvae, and Boschi et al. (1967) noted substage-type growth in larvae of the porcellanid crab *Pachycheles haigae* Rodriguez da Costa. The latter authors stated that the passage from one substage to the next was very quick, occurring in one day, and was indicated primarily by a size increase and advanced development of the maxillipeds and pleopods. Wear (1964a) too, had difficulty in separating an ecdysial substage in the New Zealand porcellanid *Petrolisthes novaezelandiae* Filhol. Other authors (e.g. Kurata 1964b, Knight 1966, Gore 1968, Roberts 1968, Shepherd 1969, Gonor & Gonor 1973a, Huni 1979a) declared that no actual ecdyses were occurring, but that growth was more or less continuous within the lightly chitinized cuticles of some of the appendages. Le Roux (1966a) decided that in the interstage growth of maxillipeds, pereopods and pleopods in another porcellanid, *Pisidia longicornis* (Pennant) '. . . passage d'une forme a l'autre s'effectue toujours par simple croissance'.

The idea of epicuticlear growth was not new, having been advanced by Lebour herself (1943), who thought it possible that the various interstages were merely faster growing individuals, or perhaps a result of variable or even unsuitable planktonic conditions (see also Bordereau 1982 for another example). It should also be noted that Wear (1964a,b) based his substage reconstruction primarily on planktonic materials of *P.novaezelandiae* and *P.elongatus* (A.Milne Edwards), as did Bourdillon-Casanova (1956) and Le Roux (1961, 1966a).

Substages with actual ecdysis do occur in laboratory culture. Boyd & Johnson (1963) and Fagetti & Campodonico (1971) found from four up to nine such instars, each differentiated by a molt, and distinguished by an increase in size and setation, uropodal development, and antennal scale growth in the galatheid genus *Pleuroncodes*. Johnson & Lewis (1942) found a morphologically intermediate stage ('lower stage IV') between stage III and V in the larval development of the hippidean mole crab *Emerita talpoida* Say. At present, intermolt growth is known for numerous species, primarily in the anomuran crabs, especially in the larvae of the Galatheidae, Porcellanidae, Hippidae, Albuneidae, Paguridae, Coenobitidae, Diogenidae, and Lithodidae (see Gonor & Gonor 1973a). According to Goy & Provenzano (1978) over 15 families of Decapoda exhibit some form of variation in their larval development, mostly occurring in the Thalassinidea and in the Anomura.

Curiously, there has been little record of substage or intermolt growth in the Brachyura. Hart (1935) studied the larvae of *Pinnotheres taylori* Rathbun which has an abbreviated type of development, and found that the thoracic appendages became fully developed during the second (final) stage before the next molt to megalopa. Goldstein (1971) described interstages between zoea II and III, III-IV, and IV-V in *Macropipus marmoreus* (Leach), all bearing extra setae on the maxillipeds and thus anticipating (in part) the setation that would appear in the following stage.

3.3.2.3 *Mark-time molting.* This type of molting has long been known in laboratory cultures of various decapod larvae. A zoeal stage may enter a sequence of molts in which very little change in morphology takes place, although each succeeding instar may show some slight increase in size. It is as if the animal is 'marking time' but not proceeding toward morphological maturity. The larva may continue in this mode for an indeterminate period of time (more than 30 instars in some caridean shrimp, Provenzano, pers. comm.) and eventually die, or it may suddenly resume an apparently normal molting mode, exhibit normal morphogenesis and complete its development.

Scyllarid and panulirid phyllosomas are well known for this type of ecdysis. But the phenomenon has been seen in many other groups, including oceanic caridean shrimps of the genus *Rhynchocinetes* (Gurney & Lebour 1941), galatheid crabs (Boyd & Johnson 1963), estuarine palaemonid shrimp (Dobkin 1971b), some freshwater atyid shrimp (Hunte

1979a,b), among others. Ngoc-Ho (1976) listed at least nine species of Palaemonidae from fresh, estuarine and marine waters which exhibited marktime molting, in the genera *Palae-monetes* and *Macrobrachium.*

Kurata (1964a) noted that *Pandalopsis coccinata* Urita may pass through stages VI or VII twice, and pointed out that 'This irregularity is frequently observed in other species of decapods . . . (and) . . . is unavoidable for the larvae of decapods with ecdysis'. Le Roux (1963) hypothesized that the 'stades transitoires' that he recorded in *Hippolyte inermis* (Leach) between the regular stage IV and the last zoea before the postlarva were a consequence of variable feeding. He noted that these transitory stages were practically impossible to distinguish morphologically.

There is another variation within the marktime motling mode which was noted by Diaz & Costlow (1972). They found that in the larvae of *Ocypode quadrata* (Bosc) a prolonged duration between zoea IV and V occurred which they could not explain. It might be analogous to the marktime mode seen in the Caridea and Scyllaridae, except that few Brachyura seem to undergo repetitive molting with little or no change in morphology or size.

There is no question that the marktime molting sequence will lengthen the planktonic stage of a species. Why this type of molting has been evolved and retained in some groups, but not in others, is an enigma. Also unexplained is why some members of a genus (e.g. *Macrobrachium*) are able to molt in this fashion, and others are not. If, according to Knowlton (1974), low food levels are responsible in that growth and morphogenesis cease altogether, while molting continues in a more or less normal frame, then it would appear that the larvae of many of the species just discussed exist in a nearly perpetual state of food deprivation. This supports Thorson's (1950) contention that the larvae of most pelagic crustaceans living under natural conditions 'would starve'. Thus, a marktime molting mode is simply a maintenance mode, sensu Knowlton, with any food energy first utilized toward keeping the larva alive. Once maintenance requirements are met then growth, and eventually morphogenesis, can proceed.

3.3.2.4 *Giant larvae.* References to giant larval forms are among the most interesting and intriguing appearing in the decapod larval literature. These forms, usually zoeal or phyllosoma, in some cases glaucothoe, or even caridean mysis, are oversized stages that are periodically captured in plankton tows. Their very great size, usually from 25-150 % larger than their counterparts, draws immediate attention of naturalists to them, and several authors have treated these larvae in detail.

Guerín (1830) was among the first to describe giant phyllosomas (3.5-7 cm long) of scyllarid and palinurid lobsters, and several later authors (Richters 1873, Johnson 1951, Sims 1964, 1965b, Sims & Brown 1968, Robertson 1968b) all noted the remarkable size, some of which may attain over 80 mm in length in the Indo-Pacific Ocean (Robertson 1968b). Sims & Brown (1968) noted that no gills were present in the 69 mm form they described, and postulated that if gills developed in the manner usual with phyllosomas, i.e. about two-thirds of the way through larval development, then 'one can only speculate on the size it may attain before metamorphosis'. In an earlier paper, Sims (1965b) stated that in the later stages of *Parribacus* the only differences to be observed are in size, and that stages XI and XII were probably equivalent to what he called a giant phyllosoma in his 1964 paper.

Other lobster-like genera also have giant stages. Members of the polychelid genera *Eryo-nicus* and *Eryoneicus* were long thought to be adults of an otherwise little known deep-living lobster, because their relatively advanced development (Faxon 1895) and large size (40-62.5 mm in length) exceeded that known for allegedly mature adults of some other polychelid genera (e.g. *Polycheles,* see Selbie 1914). Bate (1882) suspected that *Eryoneicus* would subsequently prove to be a larval form, but the presence of developing male gonopods in Faxon's and Selbie's specimens was considered strong evidence for adulthood, even though no ovigerous females in the same size range were ever obtained. Bouvier (1905a) followed this same line of reasoning in describing a 35 mm specimen of *Eryonicus spinoculatus* as an adult. It is now known, however, that *Eryonicus* and *Eryoneicus* are larvae of adult genera in the Eryonidae (Sund 1925) and the fact that male gonopod development occurs in some specimens is taken as another example of possibly neotenous development. Recent studies on some adults has shown that males are morphologically mature at a rather smaller size than females, so that the somewhat precocious development of male organs in the larvae is continued in the later, larger benthic stages. The eggs borne by the female are extremely numerous (numbering in the thousands) and of very small size (less than 0.5 mm), and they probably produce zoeae (cf. Bernard 1953). The interested reader is directed to the latter publication for a complete summary of this fascinating group.

Coutiere (1907) provided a long and detailed discussion of similarly sized larvae of some caridean shrimps that, although exhibiting adult-form pereopods and pleopods, still retained well-developed exopods of the larval stages. These larval-adults resembled their true adult counterparts in nearly every feature but sexual maturity,[5] and Coutiere postulated that they were a result of two different types of development. In the one, a normal adult will be produced; in the other a larval adult. The latter might be obtained from either a normal adult, or as a consequence of delayed development from a subadult. Either type might continue to grow indefinitely until conditions became right for metamorphosis, at which time (conceivably) a reduction in size might obtain similar to that seen in some insects when they metamorphose to the adult imago. This situation is not uncommon in many decapod larvae, where the zoeal stages continue to grow until metamorphosis at which time the post-larva that is produced is noticeably smaller than the predecessors (see e.g. Gurney 1935, Kwan & Uno 1969; Ali Kuni 1967 described a similar phenomenon in the larva and post-larva of Stomatopoda). According to Coutiere, some of the adults would be 'acromegalics', attaining a maximum compatible size in regard to organs of nutrition, but otherwise retaining an infantile aspect, and finally dying (or being eaten?) without reproducing. Gurney & Lebour (1941) and D.Williamson (1970) recorded a similar situation in *Rhynochocinetes,* with some specimens molting to postlarvae at about 8.4 mm length, but others still appearing as larvae at this size, and still others over 16 mm long (collected from the plankton) that showed few morphological differences from counterparts half their size. Williamson suggested that the species may be able to delay metamorphosis for several molts until conditions are favourable.

Other giant larvae have been noted in the Stenopodidae. Lebour (1941) recorded some giant larvae metamorphosing to postlarval stage at 21 mm length, but larvae 31 mm long of some other species were observed in plankton collections. Williamson (1976) listed fully developed stenopodid larvae ranging from 17-31 mm, as well as a specimen of 19 mm that lacked pleopods, thereby implying that additional stages were to follow. Williamson considered these forms clear examples of teleplanic larvae as defined by Scheltema (1972).

Although some anomuran and brachyuran crabs produce large larvae these may be gigantic only in relation to other zoeae or megalopae, and not forms of retarded development or delayed metamorphosis.[6] For example, Wilson (1980) and Wilson & Gore (1980) described late stage zoeae of *Euchirograpsus* and *Plagusia* (Grapsidae) that were over 4 mm in carapace length. Al Kholy (1963) gave the size of the last zoea of *Tetralia glaberrima* (Herbst) as 7.0 mm, and the large zoea and megalopae of the ocypodid genus *Ocypode* are well known, with the latter attaining carapace lengths greater than 9 mm (Crane 1940). On the other hand, the carapace lengths of *Orithyia sinica* (Linne) are not overly large, but the greatly elongated dorsal and rostral spines extend the overall size up to 14 mm (Hong 1976).

There is some evidence, however, for giant larvae in the Anomura. The genus *Glaucothöe* was originally established for a purportedly adult form but later proved to be the giant post-larval stage of a pagurid crab. Bouvier (1891, 1905b) commented at length on these interesting forms, and citing the lack of several appendages suggested that these large instars (some reaching 20 mm in length) were simply regular postlarvae of hermit crabs that continued to molt because they found no suitable substratum on which to settle. Gurney (1942) considered the problem briefly and pointed out that the glaucothöe form bore all the characters seen in other pagurid postlarvae; he thought the giant specimens were probably attributable to as yet unknown deep-sea pagurid crabs.

I have seen an extremely large megalopal form of the shallow water porcellanid crab *Polyonyx* (greater than 3 mm carapace length) that seems to belong to *P.gibbesi* Haig. The megalopal stage of this species normally does not exceed about 1.4 mm, and the first crab about 1.6 mm (Gore 1968), so the specimen collected from the Gulf of Mexico exceeds these dimensions considerably.

It is hard to see how these larval and postlarval stages exhibit much adaptive significance. Although Sims & Brown (1968), for example, considered giant phyllosoma larvae as not uncommon, and references in the literature to other giant larvae suggest the same, it seems intuitively obvious that such large, planktonic, feebly swimming stages would form ideal prey for many of the surface-feeding and midwater larval and adult fishes. Such forms may never, or only rarely, attain metamorphosis. It would be interesting to determine the prevalence of giant larvae in fish stomachs, but their lightly chitinized nature probably renders them quickly digestable.

3.3.2.5 *Terminally additive staging.* This type of staging shows similarities to the mark-time molting mode in that supernumerary stages are produced. It is also similar to the intercalated molting mode, but differs from both by adding any supernumerary stages in a sequence following the regular molting series, and not within it. In this respect terminally additive molting is a type of marktime molting that occurs only at the end of a developmental sequence, with the production of relatively similar stages, differing chiefly in size and only to a minor extent in morphology. In intercalated staging (e.g. Yang 1971) the inserted instar is often advanced in features over its numerically equivalent counterparts. A noticeable feature that often occurs in terminally additive molting is an extension of development for an indefinite period.

Terminally additive stages have been seen in a number of decapods. Costlow & Bookhout (1959) recorded a terminal zoea VIII that occasionally appeared in *Callinectes sapidus,* although seven zoeal stages was the normal instar number before postlarval molt. They noted that most, but not all, of the eighth stages died. Wilson (1980) found a sixth stage terminal molt in the grapsid crab *Euchirograpsus americanus,* which presumably would molt to megalopa. The sixth stage was seen in a series consisting of zoeae I-IV, an intercalated (semi-advanced) zoea V, then zoea VI; the latter resembled a typical zoea V in morphology and differed from the semi-advanced zoea V in the series. This sequence provides a good example of the mark-time effect at the end of a development. Goy & Provenzano (1978) recorded a terminally additive molting sequence in *Naushonia crangonoides* which usually required six, but could take seven stages to attain postlarva. Their study was of additional interest because they found the terminally added seventh stage in plankton samples too, so it was not attributable to a laboratory artifact. These authors also recorded one specimen in the laboratory that molted three additional times, but never attained postlarva.

The Palaemonidae are well known for having terminal as well as mark-time molting sequences, but neither mode is mandatory within the family. Ngoc-Ho (1976) pointed out that no marktime molts were seen in the terminal stages of a *Macrobrachium* he reared, but the extension of terminal stages commonly occurred in *Palaemonetes*, a genus also exhibiting mark-time molting. In the Hippolytidae, *Eualus pusiolus* occasionally adds a seventh zoeal stage, and *E.occultus* may add three or more, if the usual mode of development is six zoeal stages (Pike & Williamson 1961).

Many of the pandalid shrimp also have terminally additive staging, but Kurata (1967a) noted these are often difficult to discern owing to the great variation in staging that normally occurs within the genus. For example, *Pandalus pacificus* averages nine stages in development but may have from 18-19 intermolts recorded. Rothlisberg (1980) stated that within the genus *Pandalus* larval development is quite variable, ranging from 2 to 13 stages, so that later stages (e.g. 7-13) may possibly be terminally added. But data are ambiguous in this respect. Rothlisberg noted, however, that two types of larval development seem to happen in *Pandalus:* fast, with four to seven zoeal stages, cheliped buds and pleopods in zoeae II-IV, pereopods 1-5 present and 1-3 functional; and slow, with nine to 13 stages, no chelipeds or pleopods in zoea III to V, and pereopods appearing as buds for the most part. The slow type of larval development may be a terminally additive molting situation now fixed in the genome of the species, which the fast developmental mode may have at one time incorporated some type of a terminally additive situation which has now been dropped.

The initiation of terminal ecdyses in a series may take place quite early in larval development. Provenzano (1962a,b) described terminally additive stages in two hermit crabs, *Calcinus tibicen* and *Coenobita clypeatus*, and found that the ultimate zoeal stage in a longer developing series was not necessarily more developed than the same stage in a shorter (non-terminally additive) series, but the preceding stages were less developed than their regularly molting counterparts. Provenzano was able to distinguish differences in morphology as far back as zoea III in those series which added one or more molts to a normal (i.e. six staged) series.

There is also evidence that initiation of terminally additive stages is a result of various forms of stress. Bookhout & Costlow (1974) and Christiansen et al. (1977) showed that the extra stage VI (in *Menippe mercenaria*) or V (*Rhithropanopeus harrissi*) appeared when the larvae were exposed to the insecticide Mirex or to stressful salinities in the presence of methoprene, a juvenile mimicing hormonal analog. Stressful temperatures, or temperature-salinity combinations alone may also induce a terminally additive molt in hermit crabs (Provenzano 1968a), scyllarid lobsters (Robertson 1968a) and xanthid crabs (Scotto 1979). Robertson pointed out that most terminally additive phyllosomas died, whereas those larvae undergoing a mark-time mode (i.e. within, instead of at the end of, a sequence) lived and attained postlarva. Yet Ong & Costlow (1970) in studying *Menippe mercenaria*, considered the terminally added molt a rare situation and could not correlate it to any salinity-temperature combinations.

3.3.3 *Sporadic development*

This type of development is best seen in the caridean shrimps, especially the Pandalidae, where the number of stages varies within individuals of a species, and between species in a genus (Kurata 1964a). An analogous situation also occurs in some brachyurans (e.g. Portunidae) with different species in the same genus occasionally exhibiting different numbers of larval stages, so that their familial developmental pattern can only be classified as variable (Hong 1974, Greenwood & Fielder 1979, Iwata & Konishi 1981). The variations in development may be produced by a combination of intercalated, mark-time, or terminally additive stages appearing either singly or in concert within the developmental series of a particular species. Sporadic development is seen in the laboratory when physical parameters fluctuate, or food supply varies in quantity or quality. Whether sporadic development consistently occurs in nature is not yet known but it would not be surprising, given the size variations seen among many planktonic stages within the same species and the plasticity observed in larvae in the laboratory. Although often uncritically characterized as being abnormal, this type of development is perhaps better considered merely another response by the larvae to varying conditions in their environment. These concepts are nonetheless quite important and will be considered in more detail in this first section of the volume.

4 MOLTING AND GROWTH IN DECAPOD LARVAE

Paraphrasing Aikawa (1929), Kurata in his classic (1962) paper on crustacean growth stated that the life cycle of a crustacean comprises three phases: the larval, the juvenile, and the adult phase. He noted that growth in arthropods in general is superficially not a continuous process, although protoplasm formation and cellular division and addition probably is (see also Bocquet-Vedrine 1982). Passano (1960) pointed out that exuviation, the most notable feature in crustacean growth, is not really an interruption of the growth cycle, but instead a conspicuous manifestation of a continuous cycle of physiological events that make up the overall growth sequence. Moreover, growth is not always constant or even consistent in crustaceans, and there are many references in the literature noting that smaller specimens often increase more in size at each molt than do their larger, older counterparts.

Three types of information are employed in growth studies; viz., size frequencies, molting intervals (separated by tagging specimens), and the increase in size per molt in conjunction with molting frequencies. The problem of obtaining growth data is compounded, however, in larvae because of their small size and delicate nature. In general, the trend has been to utilize size increases based on preceding and succeeding body sizes, in a given larval series. Each molt increment is considered to be static, although Hudjinaga (1942) provided some evidence that the nauplius, at least, may not grow via a series of steps, and other authors have noted substage and other variational growth in many larval forms (see section 3). In any growth study, however, there must be an initial point, and in decapod larval development this can be selected either at the egg, or the pre- or first zoeal stage, i.e., the newly hatched larva. Instar measurements may then be accumulated to show size increases which may follow an arithmetic or geometric trend. In some cases, however, size may remain the same or even decrease, as Kurata (1962) and other authors have shown (Lloyd & Young 1947, Travis 1954). These variations appear as a consequence of staging and will be discussed at length in the following section. For our purposes, we shall confine our discussion to aspects of molting and growth within the several larval phases that occur in decapod ontogeny in this section.

Several different axioms have been proposed to account for crustacean growth. One of the simplest is called Brooks' Law which states that at each molt a crustacean will increase in linear dimension by a fixed percentage multiple of its previous size. By dividing the growth measurement at stage N+1 by that observed at stage N, a percentage value, termed the growth factor, can be calculated. A modification of this method was proposed by Hiatt (1947) which involved a log-log transformation to give:

$$\text{Log } n + 1 = a + b \log n,$$

where n + 1 = the postecdysial dimension, n = the pre-ecdysial dimension, a = x axis intercept, and b = a constant termed the growth coefficient;

the latter denotes the rate at which the amount of increase at successive moltings will vary, with b greater than, or less than 1 indicating positive or negative allometry, respectively, and if equal to 1 indicating isometry. Several other proposed methods are beyond the scope of the present paper. For explanations and examples of Dyar's, Przibram's, Hanaoka's, Ito's and other so-called 'rules', see Kurata (1962) and Rice (1968). Many of these concepts will be treated in more detail in other chapters in this volume, as well.

Gurney (1942) believed that the growth factor in decapod larvae would, as a generality, be about 1.26, although some variation around this value was to be expected. This presumption was based in part on Brooks' calculations on larvae collected from the plankton, in which a value of about 1.25 was obtained. Rice (1968) investigated these claims in some depth and showed that marked variation was seen in several groups, and growth factors could range from less than 1 to 2.7, but that Brooks' 'Law' was more or less obeyed with individual species that underwent an exponential type of increase. Much additional data have become available since Rice's pioneering study, and these factors will be examined again for some of the phases where sufficient data are present.

Table 4. Mean egg area and calculated growth factors in selected families of decapod Crustacea. Egg area calculated considering egg width and length measurements from literature as equivalents of major and minor axes of an ellipse and calculating area accordingly. For purposes of volume egg height was arbitrarily set at 1, because no meristic data were available.

Family	Larval stages	Egg area (mm^2)		Mean growth factor	Percent increase
		At deposition	At eclosion		
Atyidae	0-10+	0.25	0.36	1.36	36%
Palaemonidae	0-10+	0.95	1.23	1.30	30%
Alpheidae	0-9+	0.52	1.36	2.86	186%
Paguridae	1-8	0.23	0.40	1.85	86%
Galatheidae	4-5	0.19	0.26	1.41	41%
Porcellanidae	2-5	0.29	0.37	1.44	44%
Albuneidae	3-6	0.30	0.45	1.50	51%
Majidae	0-2	0.26	0.35	1.33	34%
Cancridae	5	0.09	0.14	1.43	43%
Portunidae	3-8+	0.08	0.11	1.49	49%
Xanthidae	0-6	0.17	0.25	1.49	50%
Grapsidae	2-6?	0.10	0.13	1.33	33%
Pinnotheridae	1-5	0.13	0.16	1.34	34%

4.1 *Growth in the egg*

Growth in decapod larvae, in the strictest sense, begins within the egg. A search of the literature revealed numerous measurements for general egg sizes, but relatively few that distinguished between size at deposition compared to that just prior to eclosion. The available data, including calculated growth rates, are listed in Table 4 for selected families. It can be seen that deposited eggs in those species that carry them on the abdominal pleopods can increase in size 30% to nearly 200% from the time of deposition to the time of hatching (e.g. Gurney & Lebour 1941, Guaita 1960, Elofsson 1961, Greenwood 1965, D.Williamson 1965, Davis 1965, 1966, Dobkin 1968, Glaister 1976). This increase is attributed to several factors, including embryonic expansion, and to osmotic swelling via water uptake. The latter situation can result in a volumetric increase of 1.5-3.0 times the original size (Davis 1965, 1966, Herring 1974a,b). Such increases may be homologous with size increases seen later during molting sequences, in which newly deposited cuticle also expands via water uptake, producing a larger instar. An embryonic membrane has been observed in some of these species, in addition to the egg membrane, and the former surrounds the developing egg-nauplius (e.g. Fielder et al. 1975, Wear 1967).

There seems to be little relationship between egg size (based on area, extrapolated from length and width measurements) and the number of larval stages present in a family. For example, the mean area at eclosion in the Porcellanidae and Atyidae is the same, but the former almost never has more than two zoeal stages, while the latter shows great variation in instar numbers. Families with relatively numerous larval instars (Cancridae, 5; Xanthidae, usually 4-6) show discrepancies in mean egg size at eclosion (0.14 vs 0.25 respectively). Mean growth factors also show poor correspondence in general. The alpheid shrimps, with a mean growth factor (MGF) of 2.86, indicative of a 186% increase from egg size at hatching to eclosion within species in the family, differ from the Palaemonidae having a MGF of 1.30 and only a 30% overall increase from depositional to eclosional area. Yet both have (in general) numerous larval stages. In the latter two families direct or nearly direct development also occurs, and these eggs (where measurements are available) are almost invariably quite large but do not show as high a rate of growth as do, say, eggs from longer developing forms (Table 4; but see Rabalais & Gore, this volume, for additional data).

What is interesting, however, is the rather restricted ranges seen in the percentage increase over such widely divergent and taxonomically diverse families. Although the overall range is great (30-186%) the general percentage increase seems to be about 30-50%, suggesting that embryonic growth and osmotically mediated water uptake can only proceed so far before size becomes restricted. Of additional interest is the fairly close agreement in depositional areas, which implies that regardless of egg size (i.e. linear dimensions) egg area (and perhaps egg volume) will range around a rather restricted set of values from about 0.10 to about 0.30 mm^2. Unfortunately, the paucity of data in the literature allow little more than speculation as to the reasons (if any) for these values, and much more study is needed before sound conclusions can be drawn.

4.2 *Embryonic growth*

Although the data from the majority of studies indicate that embryonic development within any given species is more or less fixed, and follows a set maturational pattern from first formation to zoeal stage (cf. Mayer 1877), nevertheless, there is some evidence that differential embryonic development may occur. Temperature is one parameter responsible for this and can exert its first effects on embryonic development, which then carries over into larval ontogeny. Christiansen (1973), for example, recorded two females of the majid crab *Hyas coarctatus* Leach collected on the same day in February, both carrying developing eggs. In one female the larval eyespots and heartbeat were visible, but they hatched three months later; in the other the eggs were undeveloped at time of collection but hatched 14 days later. A similar situation was recorded by Wear (1974) in several species of brachyuran crabs (*Cancer, Corystes, Maia*). All had an incubation period of ten months or more, so that eggs laid in May or July remained more or less undeveloped until October or November, matured slowly over the winter, and hatched in March or April of the following year. The same phenomenon is seen in the midwater caridean shrimp *Ephyrina* that apparently hatches eggs after an incubation period of eight to ten months (Herring 1974b), or in the bathyal caridean *Glyphocrangon* which hatches after 11 months (Dobkin 1965b, Provenzano 1967a). Shallow water shrimps are not excepted either, and Knowlton (1972) stated that temperature acclimation may occur in eggs of *Alpheus heterochaelis* (Say) that could conceivably induce hatching aberrations or alter developmental time, lengthening the latter at 20°C, or shortening it at 30°C. On the other hand, Ingle & Rice (1971) noted that a female *Corystes cassivelaunus* (Pennant) that was acclimated in the laboratory at 15°C produced eggs which underwent rapid embryonic development, resulting in the hatching of zoeae more advanced over those produced at the prevailing, but lower, natural temperatures of about 10°C at the end of February. Apparently an optimal temperature effect was observed. However, the subsequent effects such high and low temperatures have on larval development have been little documented.

Salinity, or salinity and temperature in combination, can also affect larval size and morphology at hatching. Late stage eggs of *Macrophthalmus hirtipes* (Jacquinot) were significantly larger at 18‰ than at 36‰ salinity, and the hatched zoeae had longer dorsal spines at the latter salinity (Jones & Simons 1982). The larger larvae were also more viable than their smaller and less saline-exposed counterparts, again suggesting that an optimal salinity condition was operating on the developing embryos in the eggs.

The question that next comes to mind is whether differential embryonic development affects larval survival and maturational rates. It would seem so in some species, based on Pandian & Katre's (1972) observation that larvae of *Macrobrachium idae* Heller that hatched on the first night were larger, heavier, and better able to tolerate starvation conditions than larvae hatched on the following nights. The latter zoeae had decreased swimming speeds and exhibited a shift in metabolism from protein to fat, suggesting overtaxation of larval energy reserves. It has also been noted (Gore, unpubl.) that subsequent broods produced by the same female in some anomuran and brachyuran crabs often exhibit a lower viability in both numbers of surviving larvae and duration of their development (see also Yang 1967, unpubl.).

4.3 *Growth in the nauplius*

Fritz Mueller (1863b) first noted the nauplius as a stage in *Penaeus*. However, Bate (1878b), in a rather rambling expository paper, criticized Mueller's ipso facto concept of wanting to determine the link between the nauplius and the zoea as being less important than determining that between the nauplius and the parent. Bate (citing Du Cane 1839) fell into the same trap as Hailstone & Westwood (1835), viz., disbelieving the fact of a subsequently proved developmental stage because another species showed exception to its general applicability.

The nauplius is the earliest motile stage within decapod crustacean development. Foxon (1934, 1936) regarded the nauplius as a precociously developing head, but with the appendages modified for swimming and feeding, so that the latter were undeveloped in regard to their final form. Mayrat (1966) considered the naupliar stage more or less as a free embryo, an interesting concept considering the relatively great growth and degree of morphological change that the stage undergoes. Free naupliar stages are restricted primarily to the penaeoidean shrimps in the Decapoda, although some scyllaridean lobsters produce a stage reminiscent of it, termed a naupliosoma or prenauplisoma (Gilchrist 1916). These stages, however, are not true naupliar stages but a type of pre-zoea, and so their relationships will be left to the next section.

The naupliar stage is restricted to within the egg in the more advanced anomuran and brachyuran crabs, and caridean shrimps, and most scyllaroidean lobsters and thalassinoidean shrimps. Most of these forms hatch as a zoea or its equivalent (q.v.) although some develop substantially further than this. Dissection of the egg, for example, reveals a series of discrete phases that often can be equated to phases of naupliar development, without naupliar freedom, of course. The developing zoea forms ventral side uppermost on a large proteinaceous yolk. In many cases the zoea hatching in these forms retains yolky material within the dorsal region of the carapace, suggesting that the zoeal exuvium forms around the yolk remnant of the eggs. However, a nauplius cuticle has also been observed (e.g. Wear 1967).

Although a series of discrete naupliar stages characterizes the penaeid-like shrimps, growth seems to be gradual, with posterior abdominal somites, thoracic appendages, and prototelson appearing over a series of at least three, usually four to five, and up to eight distinct nau-

Table 5. Some naupliar, protozoeal, and mysis stage growth factors in selected penaeideans. Number of instars in each stage in parentheses. See text for author references.

Species	Nauplius Total	Mean	Protozoea Total	Mean	Mysis Total	Mean
Hymenopenaeus muelleri	2.25 (6)	1.18	2.70 (3)	1.64	1.41 (3)	1.19
Metapenaeus affinis	1.39 (5)	1.09	2.16 (3)	1.47	No data available (3?)	
Metapenaeus burkenroadi	1.55 (6)	1.09	2.08 (3)	1.44	1.32 (3)	1.15
Parapenaeopsis acclivirostris	1.50 (5)	1.11	2.10 (3)	1.45	No data available (3?)	
Penaeus aztecus	1.42 (5)	1.09	2.69 (3)	1.64	1.30 (3)	1.14
Penaeus duorarum	1.50 (5)	1.10	2.42 (3)	1.56	1.30 (3)	1.14
Penaeus esculentis	1.73 (4)	1.20	2.06 (3)	1.44	1.41 (3)	1.19
Penaeus setiferus	1.59 (5)	1.22	2.47 (3)	1.57	No data available (3?)	
Penaeus stylirostris	1.45 (5)	1.10	2.70 (3)	1.67	1.28 (3)	1.13

pliar stages (e.g. Pearson 1939, Dobkin 1961, Shokita 1970b, Cook & Murphy 1971). Naupliar growth rates vary with the species involved, but are invariably slower and with an overall smaller increase in size than in the protozoeal stages (Table 5). This is explained by the generally smaller increments measured between the various naupliar phases, compared to those in the protozoeal stages. Even where total naupliar growth is rather large (e.g. *Hymenopenaeus muelleri,* 2.25), the mean growth (1.18) remains substantially smaller than total or mean growth in the protozoeal stages (2.70, 1.64).

It can also be seen from Table 5 that naupliar growth within the series shows no consistent trends, and between stage growth may increase in a general manner (e.g. *Metapenaeopsis affinis*), fluctuate (*Penaeus esculentus*), or show a decrease-increase trend (*Parapenaeopsis acclivirostris*).

An interesting saltational increase in size is seen, however, in the molt from last naupliar to first protozoeal stage, resulting in nearly a doubling of size for most species. This can be attributed to the maturational development of the abdomen and thorax, which become more or less completely defined, and the appearance of a carapace. As noted by Cook & Murphy (1971) a radical change occurs over this ecdysis, with concomitant appearance of fixed compound eyes, segmentation of the cephalo-thoracic appendages, truncation and dentition in the mandible (plus loss of exites and endites), and a general appearance of increased functionality in the entire larva.

These developments are exceeded only by the more radical change that will occur in the molt from protozoea to mysis. In these instars (which may be variable in form, duration, and number, although three is considered common), the animal gradually progresses toward the adult imago. In some cases, the passage from mysis stage to postlarval stage is so vague as to be undefineable. Retrogressive development may occur, with the erstwhile postlarva still carrying some mysis stage features (e.g. pereopodal exopods, incompletely formed telson and uropods, immature pleopodal development). As noted by Coutiere (1907) and Gurney (1942) aspects of sexual maturation (at least anatomically) may also appear in instars that otherwise seem functionally immature. As noted in the section on giant larvae, these forms may continue to molt for an extended period in the plankton. Felder et al. (this volume) will consider the post-larval stages and their heterochronic development further.

4.4 *Growth in the prezoea*

The prezoeal stage (Aikawa 1929), a brief and relatively non-locomotory stage, was observed and figured as early as 1792 by Cavolini (see H.Williamson 1915) and was considered an integral part of larval development by early authors (e.g. Thompson 1828, Dujardin 1843). When it occurs, it appears just prior to the zoeal stage and is usually thought of as 'pre-pelagic', but the stage is in no way 'pre-zoealike' in a developmental sense, because the morphological changes that occur during its progression are primarily non-ecdysial. Consideration of this stage has been complicated by confusion with the term 'protozoea' (H.Williamson 1911, 1915, Connolly 1923), a term now generally restricted to a distinct developmental swimming phase in the Penaeoidea (see Gurney 1926). There is, however, no equivalent prezoeal stage in the Penaeoidea, but a prezoea has been recorded in caridean shrimp, in a related morphological form in the Scyllaroidea, and in numerous anomuran and brachyuran crabs and their relatives (Table 6).

For a long time debate centered on whether the prezoea was an artifact of laboratory

Table 6. Decapod families in which a prezoeal stage, or its equivalent, has been recorded.

Family	Selected author(s)	Notes
STENOPODIDEA		
Stenopodidae	Bate 1888	Extracted from egg
CARIDEA		
Nematocarcinidae	Kemp 1910a	Extracted from egg
Atyidae	Joly 1843	cf. Gurney 1942
Campylonotidae	Pike & Williamson 1966	Prezoeal cuticle present
Palaemonidae	Greenwood et al. 1976	
Alpheidae	Gore (unpubl.)	Non-viable laboratory hatching
Hippolytidae	Gurney 1942	
ASTACIDEA		
Nephropidae	Wear 1976	Cuticle unarmed
THALASSINIDEA		
Callianassidae	Devine 1966	
PALINURA		
Palinuridae	Desmukh 1968	Naupliosoma
Scyllaridae	Crosnier 1972	Pre- and naupliosoma
ANOMURA		
Diogenidae	Provenzano 1978	Prezoea viable in lab
Lithodidae	Sars 1890	Stage figured and discussed
Paguridae	Goldstein & Bookhout 1972	Viable developmental stage
Chirostylidae	Pike & Wear 1969	Prezoea develops to zoea I
Galatheidae	Lebour 1930b	Embryonic cuticle figured
Porcellanidae	Gore 1968	Illustrated and discussed
Hippidae	Johnson & Lewis 1942	May survive several days
Dromiidae	Wear 1970b	Completely described
BRACHYURA		
Dynomenidae	Rice 1981a	Dissected from egg
Cymonomidae	Wear & Batham 1975	Advanced development
Leucosiidae	Rice 1980a	Poor survival
Majidae	Yang 1967	Prolonged survival
Hymenosomatidae	Boschi et al. 1969	
Atelecyclidae	Lebour 1928	Brief discussion
Cancridae	Buchanan & Millimann 1969	A normal but rapid stage
Corystidae	Ingle & Rice 1971	Embryonic cuticle in egg
Pirimelidae	Bourdillon-Casanova 1960	Prezoeal affinities noted
Geryonidae	Perkins 1973	Stage lasts 'less than 1 hour'
Portunidae	Sandoz & Rogers 1944	Implicated as aberrant
Goneplacidae (?)	Lebour 1928	Implied to occur
Xanthidae	Chamberlain 1961	Very short; salinity-induced?
Gecarcinidae	Moreira 1912	Figured as zoeae
Grapsidae	Wilson 1980	Viable stage in laboratory cultures
Mictyridae	Cameron 1966	Bypassed in active females
Pinnotheridae	Hart 1935	Cuticle shed on hatching
Ocypodidae	Diaz & Costlow 1972	Non-viable when present

culture (e.g. Sandoz & Rogers 1944, Costlow 1965, Goldstein 1971), or was merely an aberrant, prematurely hatched stage that might occasionally be found in nature (Gurney 1938a, 1941; see also Roesijadi 1976, Ally 1975 for comparative examples). In other cases (e.g. Haynes 1976, Ingle & Clark 1980, Konishi 1981a,b) no prezoea was recorded during hatching.

Data support all of those contentions. As a generality, prezoeae often occurred when culture conditions were unfavorable (Poole 1966), and under more favorable conditions zoeal stages would hatch. For example, the number of prezoeae observed correspond in part to decreasing salinity in laboratory hatches of *Callinectes sapidus* (Sandoz & Rogers 1944). Mir (1961) found that eggs stripped from three species of *Cancer* and kept under refrigeration two to three days before hatching produced mostly prezoeae. In a similar situation, Rice (1980b) suggested that *Ebalia nux* hatched as a prezoea, possibly because the female was kept at about $4°C$ below the ambient temperature ($9°C$) from which she was collected. Other conditions seemingly favorable to the production of prezoeae were crowding of eggs on the female abdomen (Davis 1965), infestation by a fungus or bacteria (Sandoz & Hopkins 1944), lowered salinity (Sandoz & Rogers 1944), or perhaps antibiotics and growth supporters added to culture water (Williams 1968).

Prezoeal eclosion is often extremely rapid, so that in many cases the stage was not seen because the cuticle itself was often sloughed off directly at the moment of hatching, or at such a short time thereafter that unless the observer actually followed eclosion closely the thin and diaphanous cuticle could be easily overlooked (see discussion in Chamberlain 1961; also Ingle & Rice 1971, Greenwood et al. 1976, Provenzano 1978). Miyake (1935) stated that the larvae of *Pinnotheres latissimus* Burger (= *P.parvulus* Stimpson? fide Muraoka & Konishi 1977) hatched as prezoea but remained on the abdomen of the female where the molt to first zoea took place. Perhaps for these reasons many authors (e.g. Webb 1921, Raja Bai Naidu 1951, Rajabai 1959, Pike & Williamson 1966, Noble 1974, Kakati & Sankolli 1975a,b) failed to note or distinguish a prezoeal stage, or stated that it occured only intermittently and was seemingly not a consistent stage in decapod larval development (Sandoz & Hopkins 1944, Sims 1965a, Sankolli & Shenoy 1968, Srinivasagam & Natarajan 1976).

Because prezoeal stages were rarely collected in the plankton (but see Gurney 1926), the view was that they were not a true planktonic stage (cf. also Roberts 1970). They were considered probably not important insofar as determining the larval development of the various species (e.g. Gore 1968); but, as the accumulated evidence shows (Table 6), the prezoeal stage occurs as a regular stage in too many species to be dismissed merely as a laboratory artifact, an aberrantly hatched stage, or an unimportant phase in larval development (see also Buchanan & Millemann 1969, Guiata 1960).

4.4.1 *The prezoeal cuticle*

The prezoeal cuticle is not equatable to a true exuvial covering, but is instead a simple sac or sheath that surrounds the zoeal stage just prior to hatching (Faxon 1882, Connolly 1923). The exact period in embryonic development when this cuticle begins to be laid down is unclear, but there is no doubt that the prezoeal stage terminates the embryonic phase (Hart 1935) and can be dissected out from unhatched, ripe eggs (see Wear 1967).

Even though this cuticle is not a true exuvium, it nevertheless differs substantially from the zoeal stage that it encloses (Buchanan & Milliman 1969). Usually empty or partially filled setal sheaths are discernible on the antennules, antennae, and along the telson margins,

suggesting the occurrence at one time of setae or processes now lacking, or at least noticeably modified from those appearing in the following zoeal stage (Gurney 1924, Hart 1935,
Gurney & Lebour 1941). The prezoeal stage may not exhibit all of the armature seen in
the zoeal stage. Rostral or dorsal spines may be lacking (Buchanan & Millimann 1969), or
antennular and antennal cuticular armature may not be present (Wear 1968, 1976). The
latter may be consistently absent in some families (Hymenosomatidae, Goneplacidae, Pinnotheridae, Ocypodidae, Grapsidae and Gecarcinidae in the Brachyura), but may be present
in many others (Xanthidae, Wear 1970a,b).

One consequence of the various forms of armature seen in the prezoeal cuticle has been
to encourage speculation on phylogenetic relationships, because the features exhibited by
the prezoeal cuticle must be remnants of those seen when the prezoeal stage itself was a
free-swimming stage, before it became condensed into the embryonic phase (Conn 1884).
Gurney (1942) provided arguments supporting the prezoeal stage as the equivalent of a
nauplius stage in those decapods that do not exhibit naupliar development. Wear (1967,
1975) considered the prezoea equal to the metanauplius and not a nauplius in zoea-hatching
forms, noting that the metanauplius was once a free-swimming larval stage that is now confined to the embryonic phase in those same forms.

As noted earlier, the prezoeal stage was also equated morphologically to the protozoeal
stage of the penaeoidean shrimps, but the relationships of the two forms are more as ontogenetic remnants and are not in any way conterminous. Although both prezoeal and protozoeal stages are endpoints of naupliar development, and both (in a sense) continue the
pelagic development of a species, they differ nonetheless in two major points. The naupliar
phase giving rise to caridean, anomuran, and brachyuran prezoeae remains in the egg (in
many cases nondifferentiable as to staging). The prezoeal stage in these groups is not a discrete stage differing greatly from the zoea, so that the term 'prezoea' should only be considered in the temporal sense as a phase prior to the zoea. On the other hand, in the Penaeoidea the naupliar-protozoeal phases are entirely pelagic and derived from one another by
a series of discrete molts. In other words, there is no true molt from the prezoea to the
zoea, whereas a true ecdysis does occur in the nauplius-protozoea ontogeny.

4.4.2 *Prezoeal eclosion*

Prezoeal hatching takes place from the prezoeal cuticle, and not from the egg shell as was
commonly suspected (see Davis 1964). In the change from the prezoea to first zoea, however, no true ecdysis occurs; instead the prezoeal cuticle is shed and, at the same time or
shortly thereafter, the setae and spines of the various zoeal appendages and telson are extended (Hyman 1925, Lebour 1928, 1930b, MacDonald et al. 1957, Davis 1964; and others).
Any zoeal carapace spines also become extended, probably through hydraulic pressure or
an osmotic differential across the zoeal cuticle (Wear 1965b,c, 1966, Gore 1968, for illustrations). Failure to emerge completely from the egg at this time results in the death of the
animal (Davis 1964, 1965, Diaz & Costlow 1972, Powell 1979). In some cases there seems
to be a selective advantage in the retarded erection of the various carapace spines. Gore
(1968) pointed out that unless the larvae of some commensal porcellanid crabs hatched as
prezoeae, their enormously lengthened rostral and posterior carapace spines would either
become entangled with one another, or wedged within the tube of the polychaete worm
host, preventing their escape into the plankton. This adaptation may be as much fortuitous
as evolutionary, because nearly all other porcellanid crabs (most of which are not commensal with polychaete worms) also possess a clearly marked prezoeal stage of limited duration

(see Gonor & Gonor 1973b, Yaqoob 1974, 1977, 1979, and authors cited therein).

The prezoeal stage can be considered numerically the first stage in larval development of those decapods that possess it, even though the stage itself is not a true planktonic form. It is probably best not considered the true first morphological instar, however, because the resultant zoeal form (which is functional both natatorily and nutritionally, whereas the prezoea is not), possesses the requisite pelagically oriented morphology once the carapace spines are erected and the feeding and natatory setae are extruded on the various appendages. The prezoea can neither swim well nor feed adequately until this occurs.

As always, there are exceptions. Within the palinurid and scyllarid lobsters prepelagic forms termed prenaupliosoma and naupliosoma (Gilchrist 1913, 1916) have been observed (Desmukh 1968, Robertson 1969a, Mohammed et al. 1971, Silberbauer 1971). These forms seem to have a limited capability for swimming, the prenaupliosoma by arhythmic body contractions similar to that seen in prezoeae, and the naupliosoma by the antenna, as in the nauplius stage. Both prenaupliosoma and naupliosoma molt very soon thereafter into the truly pelagic phyllosoma stage (a zoeal form) characteristic of the scyllaroidean lobsters. Von Bonde (1936) described a prenaupliosoma stage in *Jasus lalandii* (H.Milne Edwards) that molted to a naupliosoma eight hours after hatching, and then to a phyllosoma, so that this last stage becomes the third larval instar in the ontogeny of *Jasus*, instead of the first. Crosnier (1972) held the naupliosoma to be a regular, and the first, stage in the development of *Scyllarides herklotsi* (Herklots). On the other hand, Robertson (1969a) did not accept the prenaupliosoma as a viable stage, although he considered the naupliosoma as valid, but of short duration in scyllarid lobsters.

As far as is known no growth occurs within the prezoeal or naupliosomal stages, although few data are available. This is not unexpected given the relatively brief duration of these stages, which last from 15 seconds (Chamberlain 1961) to several hours at most (Lebour 1928, Crosnier 1972). It is known that egg expansion occurs probably from the moment of deposition onward, and as pointed out earlier, the final egg size (which includes the as yet unhatched prezoea) is usually substantially larger than when first deposited (see Table 4). Growth within the embryonic phase, also noted earlier, probably cannot be separated into a 'pre'-prezoeal and a prezoeal phase, and it is probably of little importance to do so. What distinguishes prezoeal (and indeed all embryonic) growth from that seen in the zoeal and postzoeal phases is that the former is, to all extents and purposes, continuous, but the latter is discontinuous. For it is at hatching of the prezoeal stage that anamorphic growth begins and proceeds via the subsequent larval stages toward the final larval phase, and eventually into metamorphic growth (see Felder et al., this volume).

4.5 *Growth in the protozoeal, zoeal, and phyllosoma stages*

To examine growth within the zoeal and zoeal-like stages in the Decapoda a series of growth factors was calculated using data on carapace length (cl), or total length (carapace length + abdominal length; tl) compiled from values stated in the literature. Over 1 400 entries were assembled and run through a specially designed computer program to determine within-stage growth (zoea n + 1/zoea n), total growth (zoea n/zoea 1 or nearest equivalent), and mean growth (average size increase within stage). Because of the wide variety of means used to record larval size, some pooling of data was inevitable. Thus, a measurement of dorsal spine to rostral spine tip was considered a carapace length measurement similar to one measuring from anterior of eye to posterior lateral margin. Similarly, a measurement includ-

Table 7. Comparison of mean total growth, and mean incremental growth factors in larvae from selected families of decapod crustaceans. Note: Calculations based on total available measurements, whether development is completed or not; total growth based on earliest and latest available stage described; all data from literature descriptions, none taken from illustrations even if available.

Family	Growth factors Mean total	Mean incremental		Number of species in calculations
CARIDEA				
Alpheidae	1.99	1.13		6
Atyidae	1.96	1.09		8
Bresiliidae	4.80	1.22		1
Crangonidae	1.41	1.13		3
Hippolytidae	1.66	1.12		11
Oplophoridae	1.56	1.10		3
Palaemonidae	2.19	1.17		26
Pandalidae	2.31	1.17		7
Pasiphaeidae	1.68	1.14		3
Processidae	3.11	1.15		2
Rhynchocinetidae	3.65	1.15	$\bar{X} = 1.14$	1
MACRURA				
Nephropidae	1.85	1.27		13
Palinuridae	12.58	1.33	$\bar{X} = 1.30$	11
Scyllaridae	9.26	1.32		2
Axiidae	1.06	1.06		1
Callianassidae	1.28	1.08	$\bar{X} = 1.16$	2
Laomediidae	4.08	1.26		1
Upogebiidae	1.18	1.08		1
ANOMURA				
Albuneidae	2.41	1.33		5
Coenobitidae	1.72	1.16		2
Diogenidae	1.94	1.21		13
Dromiidae	1.48	1.26		4
Galatheidae	1.99	1.22		4
Hippidae	3.32	1.33		4
Lithodidae	1.16	1.06		2
Paguridae	1.66	1.18		24
Porcellanidae	1.52	1.52	$\bar{X} = 1.25$	32
BRACHYURA				
Atelecyclidae	2.82	1.30		1
Calappidae	1.95	1.40		1
Cancridae	2.36	1.26		4
Corystidae	2.80	1.29		2
Geryonidae	2.29	1.32		1
Goneplacidae	4.50	1.65		1
Grapsidae	2.59	1.28		18
Hapalocarcinidae	3.50	1.38		1
Homolidae	6.81	1.38		1
Leucosiidae	1.70	1.19		5
Majidae	1.29	1.29		39
Ocypodidae	2.68	1.29		7
Parthenopidae	2.73	1.22		1
Pinnotheridae	1.52	1.21		13
Pirimelidae	1.62	1.17		1
Portunidae	2.58	1.29		17
Raninidae	3.66	1.22		2
Xanthidae	1.99	1.26	$\bar{X} = 1.30$	24

ing rostral spine tip to telson was placed equivalent to other total length measurements incorporating anterior margins of carapace to telson tip. The amount of error introduced by the non-similarity of such measurements was negligible, and allowed more data to be used than would otherwise have been possible. As such, this growth program is simply an extension of the measurements that are incorporated as Brook's Law, and provide the simplest and most direct means of determining zoeal growth.

Growth within the zoeal or zoeal-equivalent stages in Decapoda is quite variable. There appears to be little relationship among the various families in the average total, or mean, growth factors (Table 7) and within the family little consistency is seen. This is not unexpected, given the great range of variation in larval sizes, and the fluctuating duration and modes of development employed by a particular group. However, there does seem to be a related size value within the framework of higher taxa. The Caridea, for example, exhibit an overall mean growth factor of about 1.14, the Anomura 1.25, the 'lobsters' about 1.30, the thalassinids about 1.07, and the Brachyura about 1.30. And although a comparison of total growth to mean growth factors may show no noticeable consistency of values in a family, within a species these values are often similar when compared against data obtained from cl vs tl. Thus, Rice's (1968) compilation of growth factors that pooled these values requires no apology because it seems that growth in a zoeal stage usually proceeds along a durational continuum incorporating constraints on size increase, for the most part, so that any particular stage will have a maximum value which it cannot exceed. These data indicate that carapacial and abdominal growth (i.e., total length values) are nearly, or completely isometric in many instances. For example, in *Callinectes sapidus* with a prezoeal and five zoeal stages (data from Churchill 1942), total and mean growth factors using either cl or tl are 3.41, 1.36, and 3.39, 1.36, respectively. When eight zoeal stages are recorded, the total growth and mean growth factors based on tl decrease to 3.06 and 1.17, respectively, and reflect both the larger instar size, and greater instar numbers. This situation is also seen in the Nephropidae, Coenobitidae, Diogenidae, Dromiidae, Lithodidae, Paguridae, and many brachyuran families. On the other hand, where discrepancies appear in total cl growth vs total tl growth, these are often the result of increased abdominal length owing to the addition of the sixth somite during the third or later zoeal instar. In some cases (e.g. some leucosiid zoeaé) the total growth factors differ (cl = 1.78, tl = 1.73) but the relative mean growth factors are nearly identical (1.21-1.20). In such cases measuring error may have crept in.

Another type of discrepancy is seen in growth factors of laboratory reared as opposed to planktonic larvae, in which the planktonic specimens invariably exhibit larger values in both total growth and mean growth. *Lepidopa myops* (cf. Knight 1970) values were 2.02 and 1.26 in the laboratory, and 2.50 and 1.36 in planktonic specimens. Robertson (1968a) noted a similar size difference in the laboratory reared larvae of *Scyllarus americanus* (5.78, 1.34) against planktonic specimens of the same species (7.19, 1.39).

However, too much significance should not be attached to observed similarities or differences in mean growth factors, because the average relative size increase of instars is not necessarily indicative of similarity in absolute sizes. Within a familial compilation among species having a similar number of zoeal stages, the total growth factor might differ, often by a factor of 2 or more (e.g. in the Grapsidae, *Euchirograpsus americanus,* 4.60, cf. Wilson 1980; vs. *Sesarmops intermedius* 2.15, cf. Fukuda & Baba 1976). This, of course, reflects the different sizes of the zoea at hatching as opposed to the final size at metamorphosis (e.g. *Euchriograpsus,* 0.75 to 3.45 mm; *Sesarmops,* 0.73 to 1.57 mm). Both species pass through at least five zoeal stages, but the relative growth rate is very much greater in *Euchirograpsus* than in *Sesarmops*. On the other hand, similarity in total growth factors between species only shows that relative overall growth is similar but gives no indication of within-series growth rates. Thus, *Sesarma dehaani* (cf. Baba & Miyata 1971) and *Parasesarma plicatulum* (cf. Fukada & Baba 1976) have respective total and mean growth factors of 1.86, 1.23 and 1.85, 1.23, but instar growth factors in the former were more uniform at each increment (1.24-1.22-1.23) than in the latter (1.26-1.34-1.09). Clearly, growth from stage 2 to 3 was greater in *S.dehaani*, and in stage 3 to 4 in *P.plicatulum*. Data such as these merely reflect the great variation in zoeal size increases within the species, with no consistent trend observable. In some cases (viz., in the Grapsidae) relative size increases at a decreasing rate as development progresses (e.g., *Chiromantes bidens*), in others a size increase is followed by a relative decrease (*Holometopus haematocheir*) (cf. Fukuda & Baba 1976), while in still others a slow size increase, then a decrease, followed by a saltatory increase in the last zoeal instar is seen (*Cyclograpsus integer*, cf. Gore & Scotto 1982).

The lobsters also form a particularly interesting assemblage with some total growth rates being quite large (21.6, *Jasus*; 24.3, *Panulirus*; 37.0, *Justitia*), while the mean growth rates (in comparison) are substantially reduced (1.36, 1.30, 1.49, respectively). Thus a large increase in size occurs from hatching to final larval stages, but the incremental growth within each instar increases about 1.3 to a maximum of 2.0 times (tl) per molt.

Finally, we may again consider the penaeidean shrimps. As noted earlier naupliar growth is slow, with very little size increase over the 3-8 stages. Available data show that with the molt to protozoea an almost doubling in size occurs, and over the mysis stages ranges from about 1.4 to over 2 times, and protozoeal + mysis growth factors range from 2.6 to 6.08. Major growth thus does not occur in penaeideans until the protozoeal stage is attained, and incremental growth is also highest at this time. Mean growth in the mysis stage is less than in the protozoeal stages, ranging from about 1.13 to 1.41 as compared to 1.60 to 1.67 in the latter.

Space does not allow further consideration of these factors. However, additional data are being compiled and the subject of larval growth factors will be treated at length in a subsequent paper presently in preparation. The data supplied above are in general agreement with those first considered by Rice (1968), although the addition of new values, especially in the Caridea and Anomura, has changed the general overall values, as might well be expected.

4.6 *Postlarvae*

4.6.1 *Growth in the postlarval stages*
The final stage in decapod development prior to the juvenile or prepubertal mode is a transitional metamorphic instar that often bears a combination of morphological characters from the previous zoeal stages and from those precursing the eventual adult. It is during this transitional stage that the planktonic larva changes both its form and habit in preparation for assuming a benthic existence in the bottom dwelling species. Even in those pelagic species of caridean shrimp, some morphological changes are noted, most of these being associated with pleopod, gill, and pereopod development, and the regression or loss of maxillipedal exopod locomotion. The postzoeal stage is at the same time both highly specialized and highly modified, exhibiting characters which are not yet completely functional in the adult, while almost (but not completely) losing characters previously of service in the zoeae. Lebour (1928) thought the brachyuran megalopa '. . . decidedly a very specialized

larvae', and she could not find anything primitive in it. Rice (1981b) believed that the metamorphic molt between the megalopa and the juvenile stage was almost as dramatic a change as that between the zoea and the megalopa. He also held that the megalopa was morphologically intermediate and not well adapted for life as a plankter or a member of the benthos. This is not always strictly the case, as seen in the porcellanid crabs, which produce a megalopa that often very much resembles the adult and seems eminently adapted for at least a limited pelagic existence.[6]

The transitional stage here considered will be called postlarval, with the understanding (cf. Dakin 1938) that the meaning of this distinction is strictly morphological. Without going into detail (because the subject will be investigated in Felder et al, this volume) that stage is metamorphic, usually is accomplished at one molt in the Anomura and Brachyura, may require several progressional molts in the Caridean and Penaeidea, and does not always exhibit great morphological differentiation during its passage.

An important criterion in distinguishing larval (i.e., zoeal) stages from postlarval stages is the functional change in locomotion, with the maxillipedal or cephalothoracic mode switching to the pleopodal or abdominal mode. However, this distinction is not always clear. Rothlisberg (1980) stated that the use of the term megalopa was not appropriate for the caridean shrimp *Pandalus jordani* Rathbun, because the later zoeal stage often used pleopods to some degree during swimming, thus blurring the functional morphological definition. Rothlisberg preferred the appearance of a mandibular palp as a more positive indicator of a postzoeal mode of development.[7] Wear (1965a) emphasized that the megalopa of the New Zealand porcellanid crab *Petrocheles spinosus* Miers lacked pleopods entirely and thus may be unable to swim in the plankton; he noted that no megalopa of this species had yet been collected. Wear also pointed out that the zoeae of the genus *Petrocheles* are the only known decapods in which pleopodal development is entirely suppressed in both the larval and postlarval life. However, a metamorphic ecdysis occurs in both *Petrocheles* and *Pandalus,* with a distinct change occurring in morphology from obviously postlarval characteristics.

Some caridean shrimps also evidence this indistinction. Haynes (1976) studied the larval development of the pandalid shrimp *Pandalus hypsinotus* Brandt, and wrote 'Depending on one's definition of 'megalopa', it may be valid to consider stage VII . . . as the megalopa; or one may consider stages IV through VII are all megalopal, or the term 'megalopa' is not strictly applicable to *P.hypsinotus.*'

4.6.2 *The number of postlarval stages*
The number of stages is variable and is a function of the particular decapod group in which it occurs. The change, as previously noted, may be quite gradual or quite rapid. In most decapods, in fact, the postlarval stage is confined to a single instar, although exceptions still occur. Dakin (1938) showed that the progression from mysis (postlarval) to postmysis (juvenile) stage in *Penaeus plebejus* Hesse took place probably over a single molt. The available literature suggests that penaeids usually require three or more such stages (mysis) before a juvenile postmysis is discernible, so that postlarval development (and growth) is not applicable to any one stage. Similarly, in the scyllaridean and palinuridean lobsters the pseudibacus or puerulus stage (respectively) is quite distinct from the ultimate phyllosoma (a zoeal stage), but the morphological changes leading toward the juvenile form are often gradual and measured. The same situation obtains in many carideans. Shokita (1977a) for example, found that two types of telson morphology existed in the postlarval stages of *Palaemon debilis* Dana, one in which the posterior margin was rounded, the other in which

it was drawn into a median point. Shokita also noted that in those species which undergo direct or nearly direct development (q.v.) the postlarval stage is very similar to the juvenile.

Anomuran and brachyuran crabs usually exhibit a distinct, single molt, but some evidence in the literature suggests that this does not occur exclusively. At least two megalopal stages have been recorded by a number of authors, using a variety of observational criteria. Cano (1891) recorded two megalopae in the grapsid crab *Pachygrapsus marmoratus* (Fabricius) and in a species identified as *Xantho*. Aikawa (1937) stated that two megalopae were present in another grapsid, *Plagusia dentipes* (de Haan), because he observed three different modes of body length in his planktonic material, and in *Charybdis bimaculata* (Miers).[8] Lebour (1944) found a megalopa of *Percnon gibbesi* H.Milne Edwards (another grapsid) bearing typical uropods, suggesting that this was a megalopa 'in the second stage'. Wear (1967) cited the xanthid crab *Pilumnus vestitus* Haswell as having two megalopae because the first stage postlarva had no setae on the pleopods, but the second stage did. Hale (1931) who originally described these stages considered the second stage a juvenile. In another study, Hyman (1925) referred to Birge's (1883) paper which alleged that *Neopanope sayi* Smith passes through at least four molts before the megalopa assumed the first crab stage, with the animal '. . . now los(ing) the power of swimming and crawl(ing) about near the tide line'. Hyman thought this unlikely, and evidence suggests that Birge was observing juvenile forms. However, Crane (1940) also suggested that two megalopal stages were occurring in some species of ghost crab (*Ocypode*) because of differences observed in carapace sizes.

Students of larval development have assessed these statements with varying degrees of skepticism. Lebour (1944), for example, noted that no molts were obtained in Cano's studies so that double megalopal stages remained uncertain. On the other hand, Hyman (1925) was most emphatic in denying Birge's observations, while at the same time (Hyman 1924, 1925) holding that the appearance of a rostral spine in one postlarva, and the lack of this spine in the following 'postlarva' of *Pachygrapsus marmoratus* was evidence of two megalopal stages. Similar arguments were advanced in his studies on *Xantho*, but as can be seen from his illustrations, he was confusing a juvenile stage with a megalopa.

Other authors postulated a skipped megalopal stage because the resultant morphotype from the ultimate zoeal molt bore such a close resemblance to the adult of the species. Faxon (1879) suggested that the porcellanid crab *Polyonyx gibbesi* Haig, and the pinnotherid crab *Pinnixa sayana* Stimpson bypassed the postlarval stage and molted directly to the young crab. Smith (1880) concurred, stating that in the case of *P. sayana* the postzoeal form was not a megalopal stage equivalent to that seen in *Pinnixa chaetopterana* Stimpson, which certainly implies that he could distinguish the differences between a megalopa and a first crab stage. Gore (1968), however, refuted the case in *Polyonyx gibbesi*, and one suspects that *P. sayana* will be shown to have a megalopal stage as do others in the Pinnotheridae.

The circumstances surrounding extra megalopal stages nonetheless remain an open question. Costlow & Bookhout (1960) compared data from zoeal staging and believed that dietary deficiencies might produce an extra megalopal stage. Their argument was strengthened in light of later experiments by Costlow (1963a,b) in which early zoeae with both eyestalks ablated produced two sequential megalopae in their development. However, Le Roux (1980) conducted similar ablation experiments on the porcellanid crab *Pisidia longicornis* (L.) which has only two zoeal stages, and found that as long as sufficient food was present no significant alteration occurred in the second (i.e. last) zoeal stage, and the single megalopal stage that occurred differed only in being an accelerated instar. Le Roux's work supports the hypothesis that the sinus gland can become functional (at least in *P. longicornis*) by the last zoeal instar.

Although these examples are admittedly based on abnormal results caused by surgical modification,[9] they nonetheless suggest that some type of environmental or physiological stimulus acting in the same manner on the zoeal sinus gland could produce the same result in nature. Many of the intriguing morphological variations exhibited by other megalopae that were considered earlier under the heading of retrogressive development (see section 3.1) may be a consequence of some sort of similar environmentally induced trauma. Whether any selective advantage accrues to the individual or species having a supernumerary postlarval stage remains unclear, and all the arguments advanced both for and against extra zoeal stages apply equally well here. This brief discussion will suffice to introduce the reader to the possibilities inherent in such staging, and these possibilities will be explored at length in Felder et al. (this volume).

5 SOME PHILOSOPHICAL CONSIDERATIONS AND UNANSWERED QUESTIONS

The foregoing review of the knowledge on growth and molting in decapod larvae has examined much data, provided answers to some questions, but at the same time raised many more issues. For example, does the number of larval stages determine the rate and duration of postlarval molting? Or survivability? Or size of any subsequent stage? What effect does the larval cycle have on the attainment of the pubertal molt? How does increasingly warmer (or colder) waters affect larval development in those forms transported from one climatic region to another? Is there a relationship between duration of any individual within a stage and its future viability? Does phenetic adaptation occur in ovigerous females that influences the intermolt duration of their larvae? Do larvae hatched at different times of the year have different amounts of yolk reserves, thereby affecting their potential developmental ability? Do zoea require carnivory to start or continue development? Does herbivorous feeding really delay development in a manner similar to starvation, and if so, what adaptive mechanism is operating in those species which can apparently metamorphose without carnivory?

It will eventually have to be decided whether the uniform, unvarying controlled conditions in the laboratory, or the fluctuating, multi-parametered effects that occur and comprise the natural environment are less optimum; i.e., which are more influential in producing the more 'normal' larvae. Put another way, the extra stages often seen in laboratory studies may just be standard in the oceanic environment, while the allegedly optimal conditions in the laboratory may, in fact, result in shorter larval developments and consistently eliminate 'extra' instars.

But we have to ask, does it really matter whether growth is regular (and therefore orderly to our eyes), or irregular (and therefore more important in an evolutionary sense)? It is apparent from the foregoing discussions and to anyone who has studied the now voluminous literature on larvae that decapod ontogeny is not the simple and uncomplicated phenomenon it was once thought to be. As our investigations become more sophisticated and our knowledge increases, we are finding that the larvae not only can respond to a bewildering array of conditions imposed in our laboratories, but that they are probably able to respond in some way to all of the permutations and combinations of conditions to which they are subjected, in vitro and in vivo. And they are doing so with the vigor derived from nearly 300 million years of successful evolution. The very fact that larvae can tolerate, and even metamorphose, in conditions of modern industrial pollution with which they have had little if any evolutionary experience is remarkable. Moreover, given the observed plasticity in decapod larvae, there may be no such thing as a 'normal' larva, per se; viz., all larvae (excluding, of course, those deformed by pollutants or other factors) may be normal for the particular environment in which they occur. Indeed, this very plasticity may well account for the overall success of the decapod Crustacea as a group. To paraphrase Bate (1878a) one wonders if Bosc was not a little prescient when he took in mid-Atlantic the small animal which he christened *Zoe*, the generic epithet derived from the Greek word for 'life'.

6 ACKNOWLEDGEMENTS

This paper has benefitted from, and would not be possible without the help, criticism, advice and patience of numerous people. At the Academy of Natural Sciences of Philadelphia, I extend my appreciation and thanks to the staff and research assistants in the Department of Malacology, particularly Drs G.M.Davis, A.E.Bogan, E.Hoagland and R.Robertson. Mr E.E.Gallaher, T.Lane and Ms C.Liu patiently tolerated my attempts to come to an understanding with the computer; Dr S.L.Poss helped de-bug a recalcitrant program. I also take this opportunity to thank once again my two able research assistants, Mrs L.E.Scotto and P.M.Mikkelsen, who provided immeasurable aid throughout our collective tenure with the Smithsonian Institution, during which time much of the data incorporated herein were collected and synthesized. Several persons have read in entirety, or commented on portions of, the many previous drafts of the manuscript; among them Drs D.L.Felder, J.R.McConaugha and A.J.Provenzano, Jr, A.L.Rice and P.A.Sandifer. Mr J.L. Martin and Dr N.N.Rabalais provided a necessary (and welcome) critique. Dr A.M.Wenner deserves special tribute for his patient and careful editing of this chapter, a task not making his job any easier. To Kim Wilson I owe a debt of gratitude that will be hard to repay. She alone bore the burden of hectic days and long nights resulting from the assembling of this review. Her quiet love and understanding will always be remembered. Finally, Mrs C.M. Bogan carefully typed the manuscript, and Ms A.M.Garbach's perverse sense of humor brightened the musty stacks of the invertebrate reference collections at the Academy throughout the dark days of the winter of my discontent.

7 FOOTNOTES

1. Sollaud (1923) attempted to separate these stages in African freshwater palaemonids by employing such terms as 'hypomysis' and 'mysis', but his terms were not adopted by later authors.

2. Recent work (Bocquet-Vedrine 1982, Bordereau 1982, Drach et al. 1982) indicates that some cuticular expansion and growth may occur even without molting in some decapods and other arthropods.

3. Kurata (1964a) thought it '. . . unreasonable to define a constant 'normal' pattern independent of its ecological condition'.

4. Gore (1979) considered abbreviated development in the Galatheidae and provided a synopsis of development for all genera in which the larvae were known. He suggested that larvae of the Porcellanidae (two instars usually) share many features with first, fourth and fifth instars of the galatheids.

5. Gurney (1924) noted that some 'Discovery' larvae had developing appendices masculinae on pleopod 2.

6. Mueller (1863) held that 'The Porcellanidae are crabs which have remained stationary at the megalops stage'.

7. The appearance of a mandibular palp in many larvae of other genera is not necessarily restricted to the postlarval, or even the ultimate zoeal stage.

8. Kurata (1975) considered Aikawa's double megalopa in *Charybdis* a result of two, instead of one, species being observed.

9. The traumatic effects of eye ablation itself may be manifested in cessation of feeding which, in turn, would lead to a prolongation of a stage.

REFERENCES

Aikawa, H. 1929. On larval forms of some Brachyura. *Rec. Oceanogr. Wks. Japan.* 2:17-55.

Aikawa, H. 1937. Further notes on brachyuran larva. *Rec. Oceanogr. Wks. Japan.* 9:87-162.

Ali Kuhni, K.H. 1967. An account of the postlarval development, molting and growth of the common stomatopods of the Madras coast. *Proc. Symp. Crustacea, Mar. Biol. Ass. India* (2)2:824-939.

Al Kholy, A.A. 1963. The zoeal stage of *Tetralia glaberrima* (Herbst) from the Red Sea. *Publs. Mar. Biol. Stn. Ghardaqa* 12:137-144.

Al Kholy, A.A. & M.F.Mahmoud 1967. Some larval stages of *Sergestes* sp. and *Synalpheus biunguiculatus* (Stimpson). *Publs. Mar. Biol. Stn. Ghardaqa.* 14:167-176.

Ally, J.R.R. 1975. A description of the laboratory reared larvae of *Cancer gracilis* Dana (Decapoda, Brachyura). *Crustaceana.* 28:231-246.

Andrews, E.A. 1907. The young of the crayfishes *Astacus* and *Cambarus*. *Smith. Contr. Knowl.* 35:1-79.

Andryszak, B.L. & R.H.Gore 1981. The complete larval development in the laboratory of *Micropanope sculptipes* (Crustacea, Decapoda, Xanthidae) with a comparison of larval characters in western Atlantic xanthid genera. *Fish. Bull. Fish Wildl. Serv. US* 79:487-506.

Anger, K. & R.R.Dawirs 1981. Influence of starvation on the larval development of *Hyas araneus* (Decapoda, Majidae). *Helgol. wiss. Meer.* 34:287-311.

Anger, K. & K.K.C.Nair 1979. Laboratory experiments on the larval development of *Hyas araneus* (Decapoda, Majidae). *Helgol. wiss. Meer.* 32:36-54.

Anger, K., R.R.Dawirs, V.Anger, J.W.Goy & J.D.Costlow 1981. Starvation resistance in first stage zoeae of brachyuran crabs in relation to temperature. *J. Crust. Biol.* 1:518-525.

Atkins, D. 1954. The post-embryonic development of British *Pinnotheres* (Crustacea). *Proc. Zool. Soc. Lond.* 124:687-715.

Baba, K. & K.Miyata 1971. Larval development of *Sesarma (Holometopus) dehaani* H.Milne Edwards (Crustacea, Brachyura) reared in the laboratory. *Mem. Fac. Educ. Kumamoto Univ.* 19(1):54-64.

Baba, K. & M.Moriyama 1972. Larval development of *Helice tridens wuana* Rathbun and *H.tridens tridens* de Haan (Crustacea, Brachyura) reared in the laboratory. *Mem. Fac. Educ. Kumamota Univ.* 20(1):49-68.

Bate, C.S. 1878a. Report on the present state of our knowledge of the Crustacea – Part IV. On development. *Rep. Brit. Assc. Adv. Sci.* 48:193-209.

Bate, C.S. 1878b. On the nauplius stage of prawns. *Ann. Mag. Nat. Hist.* (5)2:79-85.

Bate, C.S. 1882. *Eryoneicus*, a new genus allied to *Willemoesia*. *Ann. Mag. Nat. Hist.* (5)7:456-458.

Bate, C.S. 1888. Crustacea Macrura. In: *Rept. Voy. Challenger, Zool.* 24:i-xc, 1-942.

Benzie, J.A.H. 1982. The complete larval development of *Caridina mccullochi* Roux, 1926 (Decapoda, Atyidae) reared in the laboratory. *J. Crust. Biol.* 2:493-513.

Bernard, F. 1953. Decapoda Eryonidae (*Eryoneicus* et *Willemoesia*). *Dana Rept.* 7:1-93.

Birge, E.A. 1883. Notes on the development of *Panopeus sayi* (Smith). *Johns Hopkins Univ. Stud. Biol. Lab.* 2:411-426.

Bocquet-Védrine, J. 1982. Role de l'épicuticle dans l'extension du tégument chez un crustacé isopode *Crioniscus equitans* Perez. *Bull. Soc. Zool. Fr.* 107:433-436.

Bonner, J.T. 1965. *Size and cycle*. Princeton: Princeton Univ. Press.

Bookhout, C.G. & J.D.Costlow jr 1974. Larval development of *Portunus spinicarpus* reared in the laboratory. *Bull. Mar. Sci.* 24:20-51.

Bordereau, C. 1982. Extension et croissance cuticulaires en l'absence de mue chez les Arthropodes. *Bull. Soc. Zool. Fr.* 107:427-432.

Bosc, L.A.G. 1802. *Histoire naturelle des crustacés, contenant leur description et leurs moeurs; avec figures dessinées d'après natur. Vol.1.* Paris.

Boschi, E.E. & M.A.Scelzo 1970. Desarrollo larval de cangrejo *Corystoides chilensis* Milne Edwards y Lucas en el laboratorio (Decapoda, Brachyura, Atelecyclidae). *Physis, Buenos Aires* 39:113-124.

Boschi, E.E., M.A.Scelzo & B.Goldstein 1967. Desarrollo larval de dos especies de Crustáceos Decápodes en el laboratorio. *Pachycheles haigae* Rodrigues da Costa (Porcellanidae) y *Chasmagnathus granulata* Dana (Grapsidae). *Boln. Inst. Biol. Mar. Mar del Plata* 12:1-46.

Bott, R. 1969. Die Süsswasserkrabben Süd-Amerikas und ihre Stammegeschichte. *Abh. Senckenb. naturforsch. Ges.* 518:1-94.

Bourdillon-Casanova, L. 1956. Note sur le présence de *Porcellana bluteli* (Risso) Alvarez dans le Golfe de Marseille et sur le développement larvaire de cette espèce. *Rapp. P.-v. Réun. Comm. Explor. Scient. Mer. Mediterr.* 13:225-232.

Bourdillon-Casanova, L. 1960. Le meroplancton du Golfe de Marseille: Les larves de Crustacés Décapodes. *Recl. Trav. Stn mar. Endoume.* Fasc.30, Bull.18:1-286.

Bouvier, E.-L. 1891. Les Glaucothoés sont-elles de larves de Pagures? *Ann. Soc. Nat. Zool.* (7)12:65-82.

Bouvier, E.-L. 1905a. Palinurides et Eryonides recueillis dans l'Atlantique oriental pendant les campagnes de l'Hirondelle et de la Princesse-Alice. *Bull. Mus. Océanogr. Monaco* 28:1-7.

Bouvier, E.-L. 1905b. Nouvelles observations sur les Glaucothoés. *Bull. Mus. Océanogr. Monaco* 51:1-15.

Bouvier, E.-L. 1917. Crustacés Décapodes (Macroures Marcheurs) provenant des campagnes des Yachts Hirondelle et Princesse Alice 1885-1915. *Résult. Camp. Scient. Prince Albert I.* 59:1-140.

Bowman, T.E. & L.G.Abele 1982. Classification of the recent Crustacea. In: L.G.Abele (ed.), *The Biology of Crustacea. Vol.1:* 1-27. New York: Academic Press.

Boyd, C.M. & M.W.Johnson 1963. Variation in the larval stages of a decapod crustacean, *Pleuroncodes planipes* Stimpson (Galatheidae). *Biol. Bull.* 124:141-152.

Broad, A.C. 1957a. Larval development of *Palaemonetes pugio* Holthius. *Biol. Bull.* 112:144-161.

Broad, A.C. 1957b. The relationship between diet and larval development in *Palaemonetes. Biol. Bull.* 112:162-170.

Brooks, W.K. 1882. The metamorphosis of *Penaeus. Johns Hopkins Univ. Circ.* 2:6-7.

Brooks, W.K. & F.H.Herrick 1892. The embryology and metamorphosis of the Macrura. *Mém. Nat. Acad. Sci.* 5:321-576.

Buchanan, D.V. & R.E.Millemann 1969. The prezoeal stage of the Dungeness Crab, *Cancer magister* Dana. *Biol. Bull.* 137:250-255.

Cameron, A.M. 1966. The first zoea of the Soldier Crab *Mictyris longicarpus* (Grapsoidea: Mictyridae). *Proc. Linn. Soc. NSW* 90:222-224.

Campodonico, I. & L.Guzman 1981. Larval development of *Paralomis granulosa* (Jacquinot) under laboratory conditions (Decapoda, Anomura, Lithodidae). *Crustaceana* 40:272-285.

Cano, G. 1891. Sviluppo postembrionale dei Dorippidei, Leucosiadi, Corystoidei e Grapsidi. *Memoire Soc. Ital. Sci. R.Accad. Sci. Fis. Mat.* 8, (3)4:1-14.

Carli, A. 1978. Chiave analitica per l'identificazione degli stadi larvali planctonici e del primo stadio postlarvale di *Palaemon elegans, P.serratus, P.xiphias* (Crustacea Decapoda). *Mem. Biol. Mar. Oceanogr. n.s.* (8)5:115-122.

Chamberlain, N.A. 1961. Studies on the larval development of *Neopanope texana sayi* (Smith) and other crabs of the family Xanthidae. *Johns Hopkins Univ. Chesapeake Bay Inst. Tech. Rept.* 22:1-37.

Christiansen, M.E. 1973. The complete larval development of *Hyas araneus* (Linnaeus) and *Hyas coarctatus* Leach (Decapoda, Brachyura, Majidae) reared in the laboratory. *Norw. J. Zool.* 21:63-89.

Christiansen, M.E., J.D.Costlow Jr & R.J.Monroe 1977. Effects of the juvenile hormone mimic ZR-515 (Altosid) on larval development of the mud crab *Rhithropanopeus harrisii* in various salinities and cyclic temperatures. *Mar. Biol.* 39:269-279.

Churchill, E.P. 1942. The zoeal stages of the blue crab *Callinectes sapidus* Rathbun. *Contr. Chesapeake Biol. Lab. Md. St. Bd. Nat. Resour.* 49:1-26.

Conn, H.W. 1884. The significance of the larval skin of decapods. *Johns Hopkins Univ. Stud. Biol. Lab.* 3:1-26.

Connolly, C.J. 1923. The larval stages and megalops of *Cancer amoenus* (Herbst). *Contr. Can. Biol. Fish.* n.s. 1:337-352.

Cook, H.L. & M.A.Murphy 1971. Early developmental stages of the brown shrimp, *Penaeus aztecus* Ives, reared in the laboratory. *Fishery Bull. Fish Wildl. Serv. US* 69:223-239.

Costlow, J.D.Jr 1962. The effect of eye-stalk extirpation on metamorphosis of megalops of the crab, *Callinectes sapidus* Rathbun. *Am. Zool.* 2:401-402.

Costlow, J.D.Jr 1963a. Regeneration and metamorphosis in larvae of the blue crab, *Callinectes sapidus.* Rathbun. *J. Exp. Zool.* 152:219-227.

Costlow, J.D.Jr 1963b. The effect of eyestalk extirpation on metamorphosis of megalops of the blue crab, *Callinectes sapidus* Rathbun. *Gen. Comp. Endocr.* 3:120-130.

Costlow, J.D. Jr 1965. Variability in larval stages of the blue crab *Callinectes sapidus. Biol. Bull.* 128:58-66.

Costlow, J.D. Jr 1968. Metamorphosis in crustaceans. In: W.Etkin & L.I.Gilbert (eds.), *Metamorphosis. A Problem in Developmental Biology*:3-41. New York: Appleton-Century-Crofts.

Costlow, J.D. Jr & C.G.Bookhout 1959. The larval development of *Callinectes sapidus* reared in the laboratory. *Biol. Bull.* 116:373-396.

Costlow, J.D. Jr & C.G.Bookhout 1960. The complete larval development of *Sesarma cinereum* (Bosc) reared in the laboratory. *Biol. Bull.* 118:203-214.

Costlow, J.D. Jr & C.G.Bookhout 1962. The larval development of *Sesarma reticulatum* Say reared in the laboratory. *Crustaceana* 4:281-294.

Costlow, J.D.Jr & C.G.Bookhout 1968a. The effect of environmental factors on development of the land crab, *Cardisoma guanhumi* Latreille. *Am. Zool.* 8:399-410.

Costlow, J.D.Jr & C.G.Bookhout 1968b. Larval development of the crab *Leptodius agassizii* A.Milne Edwards in the laboratory (Brachyura, Xanthidae). *Crustaceana*, suppl.2:203-213.

Costlow, J.D.Jr & E.Fagetti 1967. The larval development of the crab, *Cyclograpsus cinereus* Dana, under laboratory conditions. *Pacif. Sci.* 21:166-177.

Couch, R.Q. 1843. On the metamorphosis of the crustaceans, including the Decapoda, Entomostraca, and Pycnogonida. *Trans. Cornwall Polytech. Soc.* 12:17-46.

Couret, C.L. & D.C.L.Wong 1978. Larval development of *Halocaridina rubra* Holthius (Decapoda, Atyidae). *Crustaceana* 34:301-309.

Coutiere, H. 1907. Sur quelques formes larvaires énigmatiques d'*Eucyphotes* provenant des collections de SAS le Prince de Monaco. *Bull. Inst. Océanogr. Monaco* 104:1-70.

Crane, J. 1940. Eastern Pacific Expeditions of the New York Zoological Society. XVIII. On the post-embryonic development of brachyuran crabs of the genus *Ocypode. Zoologica* 25(1):65-82.

Crosnier, A. 1972. Naupliosoma, phyllosomes et pseudibacus de *Scyllarides herklotsi* (Herklots) (Crustacea, Decapoda, Scyllaridae) récoltés par l'Ombango dans le sud de Golfe de Guinée. *Cah. ORSTOM Océanogr.* 10:139-149.

Dakin, W.J. 1938. The habits and life-history of a penaeid prawn (*Penaeus plebejus* Hesse). *Proc. Zool. Soc. Lond.* (A)1938:163-183.

Dalley, R. 1980. The survival and development of the shrimp *Crangon crangon* (L.) reared in the laboratory under non-circadian light-dark cycles. *J. Exp. Mar. Biol. Ecol.* 47:101-112.

Davis, C.C. 1964. A study of the hatching process in aquatic invertebrates. XIII. Events of eclosion in the American lobster, *Homarus americanus* Milne Edwards (Astacura, Homaridae). *Am. Midl. Nat.* 72:203-210.

Davis, C.C. 1965. A study of the hatching process in aquatic invertebrates. XX. The blue crab, *Callinectes sapidus* (Rathbun, XXI. The nemertean, *Carcinonemertes carcinophila* (Kolliker). *Chesapeake Sci.* 6:201-208.

Davis, C.C. 1966. A study of the natching process in aquatic invertebrates. XXIII. Eclosion in *Petrolisthes armatus* (Gibbes) (Anomura, Porcellanidae). *Int. Rev. gesammt. Hydrobiol.* 51:791-796.

De Beer, G.R. 1951. *Embryos and Ancestors.* Oxford: Clarendon Press.

Devine, C.E. 1966. Ecology of *Callianassa filholi* Milne Edwards 1878 (Crustacea, Thalassinidea). *Trans. R.Soc. NZ (Zool.)* 8:93-110.

Dechancé, M. 1963. Développement direct chez un paguride, *Paguristes abbreviatus* Dechancé, et remarques sur le développement des *Paguristes. Bull. Mus. Natn. Hist., Nat., Paris* 35:488-495.

Desmukh, S. 1968. On the first phyllosomae of the Bombay spiny lobsters (*Panulirus*) with a note on the unidentified first *Panulirus* phyllosomae from India (Palinuridae). *Crustaceana*, suppl.2:47-58.

Dexter, B.L. 1981. Setogenesis and molting in planktonic crustaceans. *J. Plankt. Res.* 3:1-13.

Diaz, H. & J.D.Costlow jr 1972. Larval development of *Ocypode quadrata* (Brachyura: Crustacea) under laboratory conditions. *Mar. Biol.* 15:120-131.

Dobkin, S. 1961. Early developmental stages of pink shrimp, *Penaeus duorarum* from Florida waters. *Fish. Bull. Fish Wildl. Serv. US* 61:321-349.

Dobkin, S. 1965a. The first post-embryonic stage of *Synalpheus brooksi* Coutiere. *Bull. Mar. Sci.* 15:450-462.

Dobkin, S. 1965b. The early larval stages of *Glyphocrangon spinicauda* A.Milne Edwards. *Bull. Mar. Sci.* 15:872-884.

Dobkin, S. 1968. The larval development of a species of *Thor* (Caridea, Hippolytidae) from south Florida, USA. *Crustaceana*, suppl.2:1-18.

Dobkin, S. 1971. A contribution to the knowledge of the larval development of *Macrobrachium acanthurus* (Wiegmann 1836) (Decapoda, Palaemonidae). *Crustaceana* 21:294-297.

Drach, P., E.Jacques, Ch.Jeuniaux, M.F.Voss-Foucart & J.C.Bussers 1982. Les expansions cuticulaires bordantes, formation nouvelle des décapodes Natantia. Données morphologiques et biochimiques chez *Palaemon serratus* (Pennant 1877) [sic]. *Bull. Soc. Zool. Fr.* 107:437-447.

DuCane, C. 1839. On the metamorphosis of the Crustacea. *Ann. Mag. Nat. Hist.* (1)3:438-440.

Dujardin, F. 1843. Observations sur les métamorphoses de la *Porcellana longicornis* et description de la Zoé, qui est la larvae de ce crustacé. *C.R. hebd. Séanc. Acad. Sci., Paris* 16:1204-1207.

Elofsson, R. 1961. The larvae of *Pasiphaea multidentata* (Esmark) and *Pasiphaea tarda* (Kroyer). *Sarsia* 4:43-53.

Fage, L. & Th.Monod 1936. Biospeologica. LXIII. La faune marine du Jameo de Aqua. Lac souterrain de l'île de Lanzarote (Canaries). *Archs Zool. Exp. Gén.* 78:97-113.

Fagetti, G., & E. & I.Campodonico 1971. Desarrollo larval en el laboratorio de *Taliepus aentatus* (Milne Edwards) (Crustacea Brachyura: Majidae Acanthonychinae). *Revta. Biol. Mar., Valparaiso* 14:1-14.

Faxon, W. 1879. On some young stages in the development of *Hippa*, *Porcellana* and *Pinnixa*. *Bull Mus. Comp. Zool.* 5:253-268.

Faxon, W. 1882. Selections from embryological monographs. I: Crustacea. *Mem. Mus. Comp. Zool.* 9.

Faxon, W. 1895. Reports on an exploration off the west coasts of Mexico, Central and South America, and off the Galapagos Islands . . . by the US Fish Commission Steamer 'Albatross', during 1891 . . . XV. The stalk-eyed Crustacea. *Mem. Mus. Comp. Zool.* 18:1-292.

Fielder, D.R., J.G.Greenwood & J.C.Ryall 1975. Larval development of the Tiger Prawn, *Penaeus esculentus* Haswell, 1879 (Decapoda, Penaeidae), reared in the laboratory. *Aust. J. Mar. Fresh. Res.* 26: 155-175.

Fincham, A.A. 1979. Larval development of British prawns and shrimps (Crustacea: Decapoda: Natantia). 3. *Palaemon (Palaemon) longirostris* H.Milne Edwards 1837 and the effect of antibiotic on morphogenesis. *Bull. BM(NH) (Zool.)* 37:17-46.

Fincham, A.A. & D.I.Williamson 1978. Crustacea Decapoda: Larvae. VI. Caridea Families: Palaemonidae and Processidae. *Fich. Ident. Zooplanct.* (159/160):1-8.

Foxon, G.E.H. 1934. Notes on the swimming methods and habits of certain crustacean larvae. *J. Mar. Biol. Assc. UK* 19:829-850.

Foxon, G.E.H. 1936. A note on recapitulation in the larvae of the Decapoda Crustacea. *Ann. Mag. Nat. Hist.* (10)18:117-123.

Fraser, F.C. 1936. On the development and distribution of the young stages of krill *(Euphausia superba)*. *'Discovery' Rept.* 14:1-192.

Freeman, J.A. & J.D.Costlow 1980. The molt cycle and its hormonal control in *Rhithropanopeus harrisii* larvae. *Dev. Biol.* 74:479-485.

Fukuda, Y. & K.Baba 1976. Complete larval development of the sesarminid crabs, *Chiromantes bidens, Holometopus haematocheir, Parasesarma plicatum,* and *Sesarmops intermedius,* reared in the laboratory. *Mem. Fac. Educ. Kumamoto Univ.* 25:61-75 (in Japanese, English abstract).

Gilchrist, J.D.F. 1913. A free-swimming nauplioid stage in *Palinurus. J. Linn. Soc. (Zool.)* 32:225-231.

Gilchrist, J.D.F. 1916. Larval and postlarval stages of *Jasus lalandii* (Milne Edw.), Ortmann. *J. Linn. Soc. Zool.* 33:101-125.

Glaister, J.P. 1976. Postembryonic growth and development of *Caridina nilotica aruensis* Roux (decapoda: Atyidae) reared in the laboratory. *Aust. J. Mar. Fresh. Res.* 27:263-278.

Goldstein, B. 1971. Développement larvaire de *Macropipus marmoreus* (Leach) en laboratoire (Crustacea, Decapoda, Portunidae). *Bull. Mus. Natn. Hist., Nat., Paris* (2)42:919-943.

Goldstein, B. & C.G.Bookhout 1972. The larval development of *Pagurus prideauxi* Leach 1814, under laboratory conditions (Decapoda, Paguridea). *Crustaceana* 23:263-281.

Gonor, S.L. & J.J.Gonor 1973a. Descriptions of the larvae of four north Pacific Porcellanidae (Crustacea: Anomura). *Fish. Bull. Fish Wildl. Serv. US* 71:189-223.

Gonor, S.L. & J.J.Gonor 1973b. Feeding, cleaning, and swimming behavior in larval stages of porcellanid crabs (Crustacea: Anomura). *Fish. Bull. Fish Wildl. Serv. US* 71:225-234.

Goodbody, I. 1960. Abbreviated development in a pinnotherid crab. *Nature* 185:705-706.

Gore, R.H. 1968. The larval development of the commensal crab *Polyonyx gibbesi* Haig 1956 (Crustacea: Decapoda). *Biol. Bull.* 135:111-129.

Gore, R.H. 1970. *Petrolisthes armatus:* A redescription of larval development under laboratory conditions (Decapoda, Porcellanidae). *Crustaceana* 18:75-89.

Gore, R.H. 1971a. *Megalobrachium poeyi* (Crustacea, Decapoda, Porcellanidae): Comparison between larval development in Atlantic and Pacific specimens reared in the laboratory. *Pacif. Sci.* 25:404-425.

Gore, R.H. 1971b. *Petrolisthes tridentatus:* The development of larvae from a Pacific specimen in laboratory culture with a discussion of larval characters in the genus (Crustacea: Decapoda: Porcellanidae). *Biol. Bull.* 141:485-501.

Gore, R.H. 1979. Larval development of *Galathea rostrata* under laboratory conditions, with a discussion of larval development in the Galatheidae (Crustacea, Anomura). *Fish. Bull. Fish Wildl. Serv. US* 76:781-806.

Gore, R.H. & L.E.Scotto 1982. *Cyclograpsus integer* H.Milne Edwards, 1837 (Brachyura, Grapsidae): The complete larval development in the laboratory with notes on larvae of the genus *Cyclograpsus*. *Fish. Bull. Fish Wildl. Serv. US* 80:501-521.

Gore, R.H. & L.E.Scotto 1983. Studies on decapod Crustacea from the Indian River region of Florida, XXVII. *Phimochirus holthuisi* (Provenzano, 1961) (Anomura; Paguridae): The complete larval development under laboratory conditions and the systematic position of its larvae. *J. Crust. Biol.* 3:93-116.

Gore, R.H. & C.L.Van Dover 1980. Studies on decapod Crustacea from the Indian River region of Florida. XIX. Larval development in the laboratory of *Lepidopa richmondi* Benedict 1903, with notes on larvae of American species in the genus (Anomura: Albuneidae). *Proc. Biol. Soc. Wash.* 93: 1016-1034.

Gore, R.H., C.L.Van Dover, & J.R.Factor 1981a. Studies on decapod Crustacea from the Indian River region of Florida. XVIII. Rediscovery of *Periclimenes (Periclimenes) pandionis* Holthius 1951 (Caridea, Palaemonidae) with notes on the males and zoeal stages. *Crustaceana* 40:253-265.

Gore, R.H., C.L.Van Dover & K.A.Wilson 1981b. Studies on decapod Crustacea from the Indian River region of Florida. XX. *Micropanope barbadensis* (Rathbun 1921): The complete larval development under laboratory conditions (Brachyura, Xanthidae). *J. Crust. Biol.* 1:28-50.

Gould, S.J. 1977. *Ontogeny and Phylogeny.* Cambridge: Belknap Press.

Goy, J.W. & A.J.Provenzano Jr 1978. Larval development of the rare burrowing mud shrimp *Naushonia crangonoides* Kingsley (Decapoda: Thalassinidea; Laomediidae). *Biol. Bull.* 154:241-261.

Green, J. 1963. *A Biology of Crustacea.* London: H.F. & G.Witherby, Ltd.

Greenwood, J.G. 1965. The larval development of *Petrolisthes elongatus* (H.Milne Edwards) and *Petrolisthes novaezelandiae* Filhol (Anomura, Porcellanidae) with notes on breeding. *Crustaceana* 8:285-307.

Greenwood, J.G. & D.R.Fielder 1979. The zoeal stages and megalopa of *Portunus rubromarginatus* (Lanchester) (Decapoda: Portunidae), reared in the laboratory. *J. Plankt. Res.* 1:191-205.

Greenwood, J.G., D.R.Fielder & M.J.Thorne 1976. The larval life history of *Macrobrachium novae-hollandiae* (de Man 1908) (Decapoda, Palaemonidae), reared in the laboratory. *Crustaceana* 30:252-286.

Guaita, E.F. Primer estadio larval de cuatro crustaceos braquiuros de la Bahia de Valparaíso. *Revta Biol. Mar., Valparaiso* 10:143-153.

Guérin-Méneville, F.E. 1830. Crustaces et Arachnides. In: L.I.Duperrey (ed.), *Voyage de la Coquille, Zool.* 2(2):xii + 9-151.

Gurney, R. 1924. Crustacea. Part IX – Decapod larvae. *Br. Antarct. Terra Nova Exped. 1910. Zoology* 8:37-202.

Gurney, R. 1926. The protozoeal stage in decapod development. *Ann. Mag. Nat. Hist.* (9)18:19-27.

Gurney, R. 1927. Larvae of the Crustacea Decapoda. Zoological results of the Cambridge Expedition to the Suez Canal. *Trans. Zool. Soc. Lond.* 22:231-286.

Gurney, R. 1935. Notes on some decapod Crustacea of Bermuda. I. The larvae of *Leptochela* and *Latreutes*. *Proc. Zool. Soc. Lond.* 1935:785-793.

Gurney, R. 1937. Notes on some decapod and stomatopod Crustacea from the Red Sea. III-V. *Proc. Zool. Soc. Lond.* (B)1937:321-336.

Gurney, R. 1938a. Notes on some decapod Crustacea from the Red Sea. VI-VIII. *Proc. Zool. Soc. Lond.* (B)1938:73-84.

Gurney, R. 1938b. The larvae of the decapod Crustacea Palaemonidae and Alpheidae. *Brit. Mus. Great Barrier Reef Exped. 1928-1929, Scient. Rep.* 6:1-60.

Gurney, R. 1942. *The Larvae of Decapod Crustacea*. London: Ray Society.

Gurney, R. & M.V.Lebour 1940. Larvae of decapod Crustacea. Part VI. The genus *Sergestes*. *'Discovery' Rept.* 20:1-68.

Gurney, R. & M.V.Lebour 1941. On the larvae of certain Crustacea Macrura, mainly from Bermuda. *J. Linn. Soc., Zool.* 41:89-181.

Hailstone, S. & J.O.Westwood 1835. Descriptions of some species of crustaceous animals; with illustrations and remarks. *Loudon's Mag. Nat. Hist.* 8:261-276.

Hale, H.M. 1925. The development of two Australian sponge crabs. *Proc. Linn. Soc. NSW* 50:405-413.

Hale, H.M. 1931. The post-embryonic development of an Australian xanthid crab (*Pilumnus vestitus*, Haswell). *Rec. S.Aust. Mus.* 4:321-331.

Hart, J.F.L. 1935. The larval development of British Columbia Brachyura. I. Xanthidae, Pinnotheridae (in part) and Grapsidae. *Can. J. Res.* 12:411-432.

Hartnoll, R.G. & R.Dalley 1981. The control of size variation within instars of a crustacean. *J. Exp. Mar. Biol. Ecol.* 53:235-239.

Haynes, E. 1976. Description of zoeae of coonstripe shrimp, *Pandalus hypsinotus*, reared in the laboratory. *Fish. Bull. Fish Wildl. Serv. US* 74:323-342.

Heegaard, P.E. 1953. Observations on spawning and larval history of the shrimp, *Penaeus setiferus* (L.). *Publs. Inst. Mar. Sci. Univ. Tex.* 3:73-105.

Heegaard, P.E. 1963. Decapod larvae from the Gulf of Napoli hatched in captivity. *Vidensk. Meddr. dansk natuurh. Foren.* 125:449-493.

Herring, P.J. 1974a. Size, density and lipid content of some decapod eggs. *Deep-Sea Res.* 21:91-94.

Herring, P.J. 1974b. Observations on the embryonic development of some deep-living decapod crustaceans, with particular reference to species of *Acanthephyra*. *Mar. Biol.* 5:25-33.

Hiatt, R.W. 1948. The biology of the lined shore crab, *Pachygrapsus crassipes* Randall. *Pacif. Sci.* 2: 135-213.

Hobbs, H.H.Jr 1974. Synopsis of the families and genera of crayfish (Crustacea: Decapoda). *Smith. Contr. Zool.* 164:1-32.

Hong, S.Y. 1974. The larval development of *Pinnaxodes major* Ortmann (Decapoda, Brachyura, Pinnotheridae) under laboratory conditions. *Publs. Mar. Lab. Busan Fish. Coll.* 7:87-99.

Hong, S.Y. 1976. Zoeal stages of *Orithyia sinica* (Linnaeus) (Decapoda, Calappidae) reared in the laboratory. *Publs. Inst. Mar. Sci. Natn. Fish. Univ. Busan* 9:17-23.

Hubschman, J.E. 1963. Development and function of neurosecretory sites in the eyestalks of larval *Palaemonetes*. *Biol. Bull.* 125:96-113.

Hudjinaga, M. 1942. Reproduction, development and rearing of *Penaeus japonicus* Bate. *Jap. J. Zool.* 10:305-393.

Huni, A.A.D. 1979. Larval development of the porcellanid crab, *Petrolisthes galathinus* (Bosc 1802), reared in the laboratory. *Libyan J. Sci.* 9B:21-40.

Hunte, W. 1979a. The complete larval development of the freshwater shrimp *Micratya poeyi* (Guérin-Méneville) reared in the laboratory. *Crustaceana* suppl.5:153-166.

Hunte, W. 1979b. The complete larval development of the freshwater shrimp *Atya innocous* (Herbst) reared in the laboratory (Decapoda, Atyidae). *Crustaceana* suppl.5:231-242.

Hyman, O.W. 1924. Studies on larvae of crabs of the family Grapsidae. *Proc. US Nat. Mus.* 65:1-8.

Hyman, O.W. 1925. Studies on the larvae of crabs of the family Xanthidae. *Proc. US Nat. Mus.* 67: 1-22.

Ingle, R.W. 1981. The larval and postlarval development of the edible crab, *Cancer pagurus* Linnaeus (Decapoda: Brachyura). *Bull. BM(NH) Zool.* 40:211-236.

Ingle, R.W. & P.F.Clark 1980. The larval and postlarval development of Gibb's spider crab, *Pisa armata* (Latreille) family Majidae: subfamily Pisinae, reared in the laboratory. *J. Nat. Hist.* 14:723-735.

Ingle, R.W. & A.L.Rice 1971. The larval development of the masked crab, *Corystes cassivelaunus* (Pennant) (Brachyura, Corystidae), reared in the laboratory. *Crustaceana* 20:271-284.

Iwata, F. & K.Konishi 1981. Larval development in laboratory of *Cancer amphioetus* Rathbun, in comparison with those of seven other species of *Cancer* (Decapoda, Brachyura). *Publs. Seto Mar. Biol. Lab.* 26:369-391.

Johns, D.M. & W.H.Lang 1977. Larval development of the spider crab, *Libinia emarginata* (Majidae). *Fish. Bull. Fish Wildl. Serv. US* 75:831-841.

Johnson, M.W. 1951. A giant phyllosoma larva of a loricate crustacean from the tropical Pacific. *Trans. Am. Microsc. Soc.* 70:274-278.

Johnson, M.W. 1960. Production and distribution of larvae of the spiny lobster, *Panulirus interruptus* (Randall) with records on *P.gracilis* Streets. *Bull. Scripps Instn. Oceanogr.* 7:413-462.

Johnson, M.W. 1968a. Palinurid phyllosoma larvae from the Hawaiian Archipelago (Palinuridae). *Crustaceana* suppl. 2:59-79.

Johnson, M.W. 1968b. The phyllosoma larvae of scyllarid lobsters in the Gulf of California and off Central America with special reference to *Evibacus princeps* (Palinuridea). *Crustaceana* suppl.2: 98-116.

Johnson, M.W. & M.Knight 1966. The phyllosoma larvae of the spiny lobster *Panulirus inflatus* (Bouvier). *Crustaceana* 10:31-47.

Johnson, M.W. & W.M.Lewis 1942. Pelagic larval stages of the sand crabs *Emerita analoga* (Stimpson), *Blepharipoda occidentalis* Randall, and *Lepidopa myops* Stimpson. *Biol. Bull.* 83:67-87.

Jones, M.B. & M.J.Simons 1982. Responses of embryonic stages of the estuarine mud crab, *Macrophthalmus hirtipes* (Jacquinot), to salinity. *Int. J. Invert. Repro.* 4:273-279.

Jorgensen, O.M. 1925. The early stages of *Nephrops norvegicus,* from the Northumberland plankton, together with a note on the post-larval development of *Homarus vulgaris. J. Mar. Biol. Assc. UK* 13: 870-876.

Kakati, V.S. & K.N.Sankolli 1975. Larval development of the pea crab *Pinnotheres gracilis* Burger, under laboratory conditions (Decapoda, Brachyura). *Bull. Dep. Mar. Sci. Univ. Cochin* 7:965-979.

Kemp, S. 1910a. The decapod Natantia of the coasts of Ireland. *Scient. Invest. Fish. Bd. Ireland,* 1908: 1-190.

Kemp, S. 1910b. The Decapoda collected by the 'Huxley' from the north side of the Bay of Biscay in August, 1906. *J. Mar. Biol. Assc. UK* 8:407-420.

Knight, M.D. 1966. The larval development of *Polyonyx quadriunguiculatus* Glassell and *Pachycheles rudis* Stimpson (Decapoda, Porcellanidae) cultured in the laboratory. *Crustaceana* 10:75-97.

Knight, M.D. 1967. The larval development of the sand crab, *Emerita rathbunae* Schmitt (Decapoda, Hippidae). *Pacif. Sci.* 21(1):58-77.

Knight, M.D. 1970. The larval development of *Lepidopa myops* Stimpson, (Decapoda, Albuneidae) reared in the laboratory, and the zoeal stages of another species of the genus from California and the Pacific coast of Baja California, Mexico. *Crustaceana* 19:125-156.

Knowlton, R.E. 1965. Effects of some environmental factors on the larval development of *Palaemonetes vulgaris* (Say). *J. Elisha Mitchell Scient. Soc.* 81:87.

Knowlton, R.E. 1970. Abbreviated larval development in *Alpheus heterochaelis* Say (Crustacea, Decapoda, Caridea) from coastal North Carolina. *Am. Zool.* 10:542-543.

Knowlton, R.E. 1972. Effects of temperature and salinity on the larval development of *Alpheus heterochaelis* Say (Crustacea, Decapoda, Caridea). *Am. Zool.* 12:725.

Knowlton, R.E. 1973. Larval development of the snapping shrimp *Alpheus heterochaelis* Say, reared in the laboratory. *J. Nat. Hist.* 7:273-306.

Knowlton, R.E. 1974. Larval developmental processes and controlling factors in decapod Crustacea, with emphasis on Caridea. *Thalassia Jugosl.* 10:138-158.

Kon, T. 1979. Ecological studies on larvae of the crabs belonging to the genus *Chionoecetes.* I. The influence of starvation on the survival and growth of the Zuwai crab. *Bull. Jap. Soc. Scient. Fish.* 45:7-9.

Konishi, K. 1981a. A description of laboratory-reared larvae of the pinnotherid crab *Sakaina japonica* Serene (Decapoda, Brachyura). *J. Fac. Sci. Hokkaido Univ.* (VI) Zool. 22:165-176.

Konishi, K. 1981b. A description of laboratory-reared larvae of the commensal crab *Pinnaxodes mutuensis* Sakai (Decapoda, Brachyura) from Hokkaido, Japan. *Annotnes Zool. Jap.* 54:213-229.

Kurata, H. 1962. Studies on the age and growth of Crustacea. *Bull. Hokkaido Reg. Fish. Res. Lab.* 24: 1-115.

Kurata, H. 1964a. Larvae of decapod Crustacea of Hokkaido. 3. Pandalidae. *Bull. Hokkaido Reg. Fish. Res. Lab.* 28:23-34.

Kurata, H. 1964b. Larvae of decapod Crustacea of Hokkaido. 7. Porcellanidae. *Bull. Hokkaido Reg. Fish. Res. Lab.* 29:66-70.

Kurata, H. 1968. Larvae of Decapoda Natantia of Arasaki, Sagami Bay. IV. Palaemoninae. *Bull. Tokai Reg. Fish. Res. Lab.* 56:143-159.

Kurata, H. 1969. Larvae of Decapoda Brachyura of Arasaki, Sagami Bay. IV. Majidae. *Bull. Tokai Reg. Fish. Res. Lab.* 57:81-127.

Kurata, H. 1975. Larvae of Decapoda Brachyura of Arasaki, Sagami Bay. V. The swimming crabs of subfamily Portuninae. *Bull. Nansei Reg. Fish. Res. Lab.* 8:39-65.

Kurata, H. & T.Midorikawa 1975. The larval stages of the swimming crabs, *Portunus pelagicus* and *P. sanguinolentus* reared in the laboratory. *Bull. Nansei Reg. Fish. Res. Lab.* 8:29-38.

Kurata, H. & V.Pusadee 1974. Larvae and early postlarvae of a shrimp, *Metapenaeus burkenroadi*, reared in the laboratory. *Bull. Nansei Reg. Fish. Lab.* 7:69-84.

Kwon, C.S. & Y.Uno 1969. The larval development of *Macrobrachium nipponense* (de Haan) reared in the laboratory. *La Mer* 7:278-294.

Lang, W.H. & A.M.Young 1977. The larval development of *Clibanarius vittatus* (Bosc) (Crustacea: Decapoda: Diogenidae) reared in the laboratory. *Biol. Bull.* 152:84-104.

Lebour, M.V. 1928. The larval stages of the Plymouth Brachyura. *Proc. Zool. Soc. Lond.* 1928:473-560.

Lebour, M.V. 1930a. The larval stages of *Caridion*, with a description of a new species, *C.steveni. Proc. Zool. Soc. Lond.* 1930:181-193.

Lebour, M.V. 1930b. The larvae of the Plymouth Galatheidae. I. *Munida banffica, Galathea strigosa* and *Galathea dispersa. J. Mar. Biol. Assc. UK* 17:175-187.

Lebour, M.V. 1940. The larvae of the British species of *Spirontocaris* and their relation to *Thor* (Crustacea, Decapoda). *J. Mar. Biol. Assc. UK* 24:505-514.

Lebour, M.V. 1941. The stenopid larvae of Bermuda. In: R.Gurney & M.V.Lebour, On the larvae of certain Crustacea Macrura, mainly from Bermuda. *J. Linn. Soc. Zool.* (A)277:161-181.

Lebour, M.V. 1943. The larvae of the genus *Porcellana* (Crustacea, Decapoda) and related forms. *J. Mar. Biol. Assc. UK* 25:727-737.

Lebour, M.V. 1944. Larval crabs from Bermuda. *Zoologica, NY* 29:113-128.

Lebour, M.V. 1950. Notes on some larval decapods (Crustacea) from Bermuda. *Proc. Zool. Soc. Lond.* 120:369-379.

Le Roux, A. 1961. Contribution à l'étude du développement larvaire de *Porcellana platycheles* Pennant (Crustacé Décapode). *C.R. Séanc. Acad. Sci., Paris* 253:2146-2148.

Le Roux, A. 1963. Contribution à l'étude du développement larvaire d'*Hippolyte inermis* Leach. (Crustacé Décapode Macroure). *C.R. Séanc. Acad. Sci., Paris* 256:3499-3501.

Le Roux, A. 1966a. Le développement larvaire de *Porcellana longicornis* Pennant (Crustacé Décapode Anomoure Galathéide). *Cah. Biol. Mar.* 7:69-78.

Le Roux, A. 1966b. Contribution à l'étude du développement larvaire de *Clibanarius erythropus* (Latreille) (Crustacé Décapode Anomoure Diogénidé). *Cah. Biol. Mar.* 7:225-230.

Le Roux, A. 1970. Contribution à l'étude du développement larvaire de *Palaemonetes varians* (Leach) (Decapoda, Palaemonidae). *C.R. Séanc. Acad. Sci., Paris* 270:851-854.

Le Roux, A. 1980. Effets de l'ablation des pédoncules oculaires et de quelques conditions d'élevage sur le développement de *Pisidia longicornis* (Linné) (Crustacé, Décapode, Anomoure). *Archs. Zool. Exp. Gén.* 121:97-114.

Little, G. 1968. Induced winter breeding and larval development in the shrimp *Palaemonetes pugio* Holthius (Caridea, Palaemonidae). *Crustaceana* suppl.2:19-26.

Lloyd, A.J. & C.M.Young 1947. The biology of *Crangon vulgaris* L. in the Bristol Channel and Severn Estuary. *J. Mar. Biol. Assc. UK* 26:626-661.

Lockwood, A.P.M. 1967. *Aspects of the physiology of Crustacea.* San Francisco: W.H.Freeman & Co.

Lumare, F. & S.Gozzo 1972. Sviluppo larvale del Crostaceo Xantideo *Eriphia verrucosa* (Forskal, 1775) in condizioni di laboratorio. *Boll. Pesca Piscic. Idrobiol.* 27:185-209.

Lyons, W.G. 1980. The postlarval stage of scyllaridean lobsters. *Fisheries* 5:47-49.

MacDonald, J.D., R.B.Pike & D.I.Williamson 1957. Larvae of the British species of *Diogenes, Pagurus, Anapagurus* and *Lithodes* (Crustacea, Decapoda). *Proc. Zool. Soc. Lond.* 128:209-257.

Makarov, R. 1968a. On the larval development of the genus *Sclerocrangon* G.O.Sars (Caridea, Crango-nidae). *Crustaceana* suppl.2:27-37.

Makarov, R. 1968b. Ob ukorochyenii lichinochnogo razvitiya u desyatinogikh rakoobraznikh (Crusta-cea, Decapoda). [The abbreviation of larval development in decapods (Crustacea, Decapoda).] *Zool. Zh.* 47:348-359. (English abstract)

Marukawa, H. 1933. Biological and fishery research on Japanese King-crab *Paralithodes camtschatica* (Tilesius). *J. Imp. Fish. Exp. Stn.* 4:i-ii, 1-122 [in Japanese], 123-152 [in English].

Matsuda, R. 1979. Abnormal metamorphosis and arthropod evolution. In: A.D.Gupta (ed.), *Arthropod phylogeny:*137-256. New York: Van Nostrand.

Mayer, P. 1877. Zur Entwicklungsgeschichte der Dekapoden.*Jena. Zeit. Naturw.* 11:187-269.

McCann, C. 1937. Notes on the common land crab *Paratelphusa (Barytelphusa) guerini* (M.Edw.) of Salsette Island. *J. Bombay Nat. Hist. Soc.* 39:531-542.

McConaugha, J.R. 1980. Identification of the Y-organ in the larval stages of the crab, *Cancer anthonyi* Rathbun. *J. Morph.* 164:83-88.

McConaugha, J.R. & J.D.Costlow 1981. Ecdysone regulation of larval crustacean molting. *Comp. Bio-chem. Physiol.* 68A:91-93.

Mir, R.D. 1961. The external morphology of the first zoeal stages of the crabs, *Cancer magister* Dana, *Cancer antennarius* Stimpson, and *Cancer anthonyi* Rathbun. *Calif. Fish Game* 47:103-111.

Mirkes, D.Z., W.B.Vernberg & P.J.DeCoursey 1978. Effects of cadmium and mercury on the behavioral responses and development of *Eurypanopeus depressus* larvae. *Mar. Biol.* 47:143-147.

Miyake, S. 1935. Note on the zoeal stages of *Pinnotheres latissimus.* *Bult. Sci. Fak. terk. Kjuśu Univ.* 6:192-200.

Mizue, K. & Y.Iwamoto 1961. On the development and growth of *Neocaridina denticulata* de Haan. *Bull. Fac. Fish. Nagasaki Univ.* 10:15-24.

Mohamed, K.H., P.Vedavyasa Rao & C.Suseelan 1971. The first phyllosoma stage of the Indian deep-sea spiny lobster, *Puerulus sewelli* Ramadan. *Proc. Indian Acad. Sci.* (B)74:208-215.

Montgomery, S.K. 1922. Direct development in a dromiid crab. *Proc. Zool. Soc. Lond.* 922:193-196.

Moreira, C. 1912. Embryologie du *Cardisoma guanhumi* Latr. *Mem. Soc. Zool. Fr.* 25:155-161.

Mueller, F. 1863a. On the transformations of the Porcellanae. *Ann. Mag. Nat. Hist.* (3)11:47-50.

Mueller, F. 1863b. Die Verwandlung der Garneelen. *Arch. Naturgesch.* 29:8-23.

Muraoka, K. & K.Konishi 1977. Note on the first zoea of *Pinnotheres sinensis* Shen (Crustacea, Brachy-ura, Pinnotheridae) from Tokyo Bay. *Res. Crustacea Carcinol. Soc. Jap.* 8:46-50.

Nayak, V.N. 1981. Larval development of the hermit crab *Diogenes planimanus* Henderson (Decapoda, Anomura, Diogenidae) in the laboratory. *Indian J. Mar. Sci.* 10:136-141.

Ngoc-Ho, N. 1976. The larval development of the prawns *Macrobrachium equidens* and *Macrobrachium* sp. (Decapoda: Palaemonidae), reared in the laboratory. *J. Zool.* 178:15-55.

Ngoc-Ho, N. 1977. The larval development of *Upogebia darwini* (Crustacea, Thalassinidea) reared in the laboratory, with a redescription of the adult. *J. Zool.* 181:439-464.

Nichols, J.H. & P.Lawton 1978. The occurrence of the larval stages of the lobster *Homarus gammarus* (Linnaeus 1758) off the northeast coast of England in 1976. *J. Cons. Int. Explor. Mer* 38:234-243.

Noble, A.L. 1974. On the eggs and early larval stages of *Pinnotheres gracilis* Burger and *Pinnotheres modiolicolus* Burger. *J. Mar. Biol. Assc. India* 16:175-181.

Nyblade, C.F. 1970. Larval development of *Pagurus annulipes* (Stimpson 1862) and *Pagurus pollicaris* Say 1817 reared in the laboratory. *Biol. Bull.* 139:557-573.

Ong, K.-S. &J.D.Costlow Jr. The effect of salinity and temperature on the larval development of the stone crab, *Menippe mercenaria* (Say), reared in the laboratory. *Chesapeake Sci.* 11:16-29.

Pace, F.R., R.Harris & V.Jaccarini 1976. The embryonic development of the Mediterranean freshwater crab, *Potamon edulis (= P.fluviatilis)* (Crustacea, Decapoda, Potamonidae). *J. Zool.* 180:93-106.

Pandian, T.J. & S.Katre 1972. Effect of hatching time on larval mortality and survival of the prawn *Macrobrachium idae. Mar. Biol.* 13:330-337.

Passano, L.M. 1960. Molting and its control. In: T.H.Waterman & F.A.Chace, jr (eds.), *The Physiology of the Crustacea. Volume 1:*473-536. New York: Academic Press.

Paul, A.J. & J.M.Paul 1980. The effect of early starvation on later feeding success of King Crab zoeae. *J. Exp. Mar. Biol. Ecol.* 44:247-251.

Pearson, J.C. 1939. The early life histories of some American Penaeidae, chiefly the commercial shrimp, *Penaeus setiferus* (Linn.). *Bull. Bur. Fish., Wash.* 49:1-73.

Perkins, H.C. 1973. The larval stages of the deep sea Red Crab, *Geryon quinquedens* Smith, reared under laboratory conditions (Decapoda: Brachyrhyncha). *Fish. Bull. Fish Wildl. Serv. US* 71:69-82.

Phillips, B.F., P.A.Brown, D.W.Rimmer & D.D.Reid 1979. Distribution and dispersal of the phyllosoma larvae of the western Rock Lobster, *Panulirus cygnus*, in the south-eastern Indian Ocean. *Aust. J. Mar. Fresh. Res.* 30:773-783.

Pike, R.B. & R.G.Wear 1969. Newly hatched larvae of the genera *Gastroptychus* and *Uroptychus* (Crustacea, Decapoda, Galatheidae) from New Zealand waters. *Trans. R.Soc. NZ Zool.* 11:189-195.

Pike, R.B. & D.I.Williamson 1961 The larvae of *Spirontocaris* and related genera (Decapoda, Hippolytidae). *Crustaceana* 2:187-208.

Pike, R.B. & D.I.Williamson 1966. The first zoeal stage of *Campylonotus rathbunae* Schmitt and its bearing on the systematic position of the Campylonotidae (Decapoda, Caridea). *Trans. R.Soc. NZ Zool.* 7:209-213.

Poole, R.L. 1966. A description of laboratory reared zoeas of *Cancer magister* Dana, and megalopae taken under natural conditions. *Crustaceana* 11:83-87.

Powell, C.B. 1979. Suppression of larval development in the African freshwater shrimp *Desmocaris trispinosa* (Decapoda, Palaemonidae). *Crustaceana* suppl.5:185-194.

Provenzano, A.J.Jr 1962a. The larval development of *Calcinus tibicen* (Herbst) (Crustacea, Anomura) in the laboratory. *Biol. Bull.* 123:179-202.

Provenzano, A.J.Jr 1962b. The larval development of the tropical land hermit *Coenobita clypeatus* (Herbst) in the laboratory. *Crustaceana* 4:207-228.

Provenzano, A.J.Jr 1967a. Recent advances in the laboratory culture of decapod larvae. *Proc. Symp. Crustacea Mar. Biol. Assc. India, Part II:*940-945.

Provenzano, A.J.Jr 1967b. The zoeal stages and glaucothoe of the tropical eastern Pacific hermit crab *Trizopagurus magnificus* (Bouvier 1898) (Decapoda; Diogenidae), reared in the laboratory. *Pacif. Sci.* 21:457-473.

Provenzano, A.J.Jr 1968. The complete larval development of the West Indian hermit crab *Petrochirus diogenes* (L.) (Decapoda, Diogenidae) reared in the laboratory. *Bull. Mar Sci.* 18:143-181.

Provenzano, A.J.Jr 1978. Larval development of the hermit crab, *Paguristes spinipes* Milne Edwards 1880 (Decapoda, Diogenidae) reared in the laboratory. *Bull. Mar. Sci.* 28:512-526.

Rabalais, N.N. & J.N.Cameron 1983. An abbreviated pattern of larval development in *Uca subcylindrica* Stimpson 1859 (Crustacea, Decapoda, Ocypodidae) reared in the laboratory. *J. Crust. Biol.* 3:519-541.

Rajabai, K.G. 1959. Studies on the larval development of Brachyura I. The early and post larval development of *Dotilla blanfordi* Alcock. *Ann. Mag. Nat. Hist.* (13)2:129-135.

Raja Bai Naidu, K.G. 1951. Some stages in the development and bionomics of *Ocypoda platytarsis*. *Proc. Indian Acad. Sci.* 33:32-40.

Rathbun, M.J. 1914. Stalk-eyed crustaceans collected at the Monte Bello Islands. *Proc. Zool. Soc. Lond.* 1914:653-664.

Reese, E.S. & R.A.Kinzie 1968. The larval development of the coconut robber crab *Birgus latro* (L.) in the laboratory (Anomura, Paguridea). *Crustaceana* suppl.2:117-144.

Rice, A.L. 1968. Growth 'rules' and the larvae of decapod crustaceans. *J. Nat. Hist.* 2:525-530.

Rice, A.L. 1980a. The first zoeal stage of *Ebalia nux* Milne Edwards, with a discussion of the zoeal characters of the Leucosiidae (Crustacea, Decapoda, Brachyura). *J. Nat. Hist.* 14:331-337.

Rice, A.L. 1980b. Crab zoeal morphology and its bearing on the classification of the Brachyura. *Trans. Zool. Soc. Lond.* 1980:35:271-424.

Rice, A.L. 1981a. The zoea of *Acanthodromia erinacea* A.Milne Edwards; the first description of a dynomenid larva (Decapoda, Dromioidea). *J. Crust. Biol.* 1:174-176.

Rice, A.L. 1981b. The megalopa stage in brachyuran crabs. The Podotremata Guinot. *J. Nat. Hist.* 15:1003-1011.

Rice, A.L. & A.J.Provenzano Jr 1970. The larval stages of *Homola barbata* (Fabricius) (Crustacea, Decapoda, Homolidae) reared in the laboratory. *Bull. Mar. Sci.* 20:446-471.

Richters, F. 1873. Die Phyllosomen. Ein Beitrag zur Entwicklungsgeschichte der Loricaten. *Zeit. wiss. Zool.* 23:623-646.

Roberts, M.H.Jr 1968. Larval development of the decapod *Euceramus praelongus* in laboratory culture. *Chesapeake Sci.* 9:121-130.

Roberts, M.H.Jr 1969. Larval development of *Bathynectes superbus* (Costa) reared in the laboratory. *Biol. Bull.* 137:338-351.

Roberts, M.H.Jr 1970. Larval development of *Pagurus longicarpus* Say reared in the laboratory. I. Description of larval instars. *Biol. Bull.* 139:188-202.

Roberts, M.H.Jr 1971. Larval development of *Pagurus longicarpus* Say reared in the laboratory. IV. Aspects of the ecology of the megalopae. *Biol. Bull.* 141:162-166.

Robertson, P.B. 1968a. The complete larval development of the sand lobster *Scyllarus americanus* (Smith), (Decapoda, Scyllaridae) in the laboratory, with notes on larvae from the plankton. *Bull. Mar. Sci.* 18:294-342.

Robertson, P.B. 1968b. A giant scyllarid phyllosoma larva from the Caribbean Sea, with notes on smaller specimens (Palinuridae). *Crustaceana* suppl.2:83-97.

Robertson, P.B. 1969. The early larval development of the scyllarid lobster *Scyllarides aequinoctialis* (Lund) in the laboratory, with a revision of the larval characters of the genus. *Deep-Sea Res.* 16: 557-586.

Rochanaburanon, T. & D.I.Williamson 1976. Laboratory survival of larvae of *Palaemon elegans* Rathke and other caridean shrimps in relation to their distribution and ecology. *Estuar. Coast. Mar. Sci.* 4: 83-91.

Roesijadi, G. 1976. Descriptions of the prezoeae of *Cancer antennarius* and *Cancer productus* Randall and the larval stages of *Cancer antennarius* Stimpson (Decapoda, Brachyura). *Crustaceana* 31:275-296.

Rothlisberg, P.C. 1979. Combined effects of temperature and salinity on the survival and growth of the larvae of *Pandalus jordani* (Decapoda: Pandalidae). *Mar. Biol.* 54:125-134.

Rothlisberg, P.C. 1980. A complete larval description of *Pandalus jordani* Rathbun (Decapoda, Pandalidae) and its relation to other members of the genus *Pandalus. Crustaceana* 38:19-48.

Saisho, T. 1966. Studies on the phyllosoma larvae with reference to the oceanographical conditions. *Mem. Fac. Fish. Kagoshima Univ.* 15:177-239.

Sandifer, P.A. 1972. Effects of diet on larval development of *Thor floridanus* (Decapoda, Caridea) in the laboratory. *Va. J. Sci.* 23:5-8.

Sandifer, P.A. 1973a. Effects of temperature and salinity on larval development of grass shrimp, *Palaemonetes vulgaris* (Decapoda, Caridea). *Fish. Bull. Fish Wildl. Serv. US* 71:115-123.

Sandifer, P.A. 1973b. Larvae of the burrowing shrimp, *Upogebia affinis*, (Crustacea, Decapoda, Upogebiidae) from Virginia plankton. *Chesapeake Sci.* 14:98-104.

Sandifer, P.A. 1973c. Mud shrimp (*Callianassa*) larvae (Crustacea, Decapoda, Callianassidae) from the Virginia plankton. *Chesapeake Sci.* 14:149-159.

Sandifer, P.A. 1974. Larval stages of the shrimp, *Ogyrides limicola* Williams 1955 (Decapoda, Caridea) obtained in the laboratory. *Crustaceana* 26:37-60.

Sandifer, P.A. & T.I.J.Smith 1979. Possible significance of variation in the larval development of palaemonid shrimp. *J. Exp. Mar. Biol. Ecol.* 39:55-64.

Sandoz, M. & S.H.Hopkins 1944. Zoeal larvae of the blue crab, *Callinectes sapidus* Rathbun. *J. Wash. Acad. Sci.* 34:132-133.

Sandoz, M. & R.Rogers 1944. The effect of environmental factors on hatching, molting, and survival of zoea larvae of the blue crab, *Callinectes sapidus* Rathbun. *Ecology* 25:216-228.

Sankolli, K.N. & S.Shenoy 1968. Larval development of a dromiid crab, *Conchoecetes artificiosus* (Fabr.) (Decapoda, Crustacea) in the laboratory. *J. Mar. Biol. Assc. India* 9:96-110.

Sars, G.O. 1890. Bidrag til Kundskaben om Decapodernes Forvandlinger. II. *Lithodes-Eupagurus-Spiropagurus-Galathodes-Galathea-Munida-Porcellana-(Nephrops). Arch. Math. Naturv.* 3:133-201.

Sato, S. 1958. Studies on larval development and fishery biology of King Crab, *Paralithodes camtschatica* (Tilesius). *Bull. Hokkaido Reg. Fish. Res. Lab.* 17:1-103.

Scelzo, M.A. & E.E.Boschi 1975. Cultivo del langostino *Hymenopenaeus muelleri* (Crustacea, Decapoda, Penaeidae). *Physis, Valparaiso* (A)34:193-197.

Scelzo, M.A. & V.Lichtschein de Bastida 1978. Desarrollo larval y metamorfosis del cangrejo *Cyrtograpsus altimanus* Rathbun 1914 (Brachyura, Grapsidae) en laboratorio, con observaciones sobre la ecologia de la especie. *Physis, Valparaiso* (A)38:103-126.

Schatzlein, F.C. & J.D.Costlow jr 1978. Oxygen consumption of the larvae of the decapod crustaceans, *Emerita talpoida* (Say) and *Libinia emarginata* Leach. *Comp. Biochem. Physiol.* 61A:441-450.

Scheltema, R.S. 1972. Dispersal of larvae as a mean of genetic exchange between widely separated populations of shoal-water benthic invertebrate species. In: B.Battaglia (ed.), *Fifth European Marine Biological Symposium:* 101-114. Padua: Piccin Editore.

Scotto, L.E. 1979. Larval development of the Cuban stone crab, *Menippe nodifrons* (Brachyura, Xanthidae), under laboratory conditions with notes on the status of the family Menippidae. *Fish. Bull. Fish Wildl. Serv. US* 77:359-386.

Selbie, C.M. 1914. The Decapoda Reptantia of the coasts of Ireland. Part I. Palinura, Astacura and Anomura (except Paguridea). *Scient. Invest. Fish. Bd. Ireland* 1914:1-116.

Shen, J.C. 1935. An investigation of the post-larval development of the shore crab *Carcinus maenas*, with special reference to the external secondary sexual characters. *Proc. Zool. Soc. Lond.* 1935:1-33.

Shenoy, S. 1967. Studies on larval development in Anomura (Crustacea, Decapoda). II. *Proc. Symp. Crustacea Mar. Biol. Assc. India, Part II:*777-804.

Shepherd, M.C. 1969. The larval morphology of *Polyonyx transversus* (Haswell) *Pisidia dispar* (Stimpson) and *Pisidia streptochiroides* (de Man) (Decapoda: Porcellanidae). *Proc. R.Soc. Queensld.* 80: 97-124.

Shokita, S. 1970a. Studies on the multiplication of the freshwater prawn *Macrobrachium formosense* Bate. I. The larval development reared in the laboratory. *Biol. Mag. Okinawa* 6:1-12.

Shokita, S. 1970b. A note on the development of eggs and larvae of *Penaeus latisulcatus* Kishinouye reared in an aquarium. *Biol. Mag. Okinawa* 6:34-36.

Shokita, S. 1973a. Abbreviated larval development of fresh-water atyid shrimp, *Caridina brevirostris* Stimpson from Iriomote Island of the Ryukyus (Decapoda, Atyidae). *Bull. Sci. Engng Div. Univ. Ryukyus* 16:222-231.

Shokita, S. 1973b. Abbreviated larval development of the fresh-water prawn *Macrobrachium shokitai* Fujino et Baba (Decapoda, Palaemonidae) from Iriomote Island of the Ryukyus. *Annotnes Zool. Jap.* 46:111-126.

Shokita, S. 1976. Early life-history of the land-locked atyid shrimp, *Caridina denticulata ishigakiensis* Fujino et Shokita, from the Ryukyu Islands. *Res. Crustacea Carcinol. Soc., Jap.* 6:1-10.

Shokita, S. 1977a. Larval development of palaemonid prawn, *Palaemon (Palaemon) debilis* Dana from the Ryukyu Islands. *Bull. Sci. Engng. Div. Univ. Ryukyus* 23:57-76.

Shokita, S. 1977b. Abbreviated metamorphosis of land-locked fresh-water prawn, *Macrobrachium asperulum* (Von Martens) from Taiwan. *Annotnes Zool. Jap.* 50:110-122.

Silberbauer, B.I. 1971. The biology of the South African Rock Lobster *Jasus lalandii* (H.Milne Edwards). I. Development. *Investl. Rep. Div. Sea Fish. Repub. S.Afr.* 92:1-70.

Simon, M. 1977. *Guide to Biological Terms in Melanesian Pidgin.* WAU Ecology Institute, Handbook No.3. Joint Publ. WAU Ecol. Inst., Papua Nenguime, Univ. Technology, N.Guinea.

Sims, H.W. Jr 1964. Four giant scyllarid phyllosoma larvae from the Florida Straits with notes on smaller similar specimens. *Crustaceana* 7:259-266.

Sims, H.W. Jr 1965a. Notes on the occurrence of prenaupliosoma larvae of spiny lobsters in the plankton. *Bull. Mar. Sci.* 15:223-227.

Sims, H.W. Jr 1965b. The phyllosoma larvae of *Parribacus. Quart. J. Fla. Acad. Sci.* 28:142-172.

Sims, H.W. Jr & C.L.Brown Jr 1968. A giant scyllarid phyllosoma taken north of Bermuda (Palinuridea). *Crustaceana* suppl.2:78-82.

Smith, S.I. 1880. On the species of *Pinnixa* inhabiting the New England coast, with remarks on their early stages. *Trans. Conn. Acad. Arts Sci.* 4:247-253.

Snodgrass, R.E. 1956. Crustacean metamorphoses. *Smith. Misc. Coll.* 131(10):1-78.

Snodgrass, R.E. 1961. The caterpillar and the butterfly. *Smith. Misc. Coll.* 143(6):1-51.

Soh, C.L. 1969. Abbreviated development of a non-marine crab, *Sesarma (Geosesarma) perracae* (Brachyura; Grapsidae), from Singapore. *J. Zool.* 158:357-370.

Sollaud, E. 1923. Le développement larvaire des 'Palaemoninae'. I. Partie descriptive. La condensation progressive de l'ontogénèse. *Bull. Biol. Fr. Belg.* 57:509-603.

Srinivasagam, S. & S.Natarajan 1976. Early development of *Podophthalmus vigil* (Fabricius) in the laboratory & its fishery off Porto Novo. *Indian J. Mar. Sci.* 5:137-140.

Sund, O. 1925. The Challenger Eryonidea. *Ann. Mag. Nat. Hist.* (9)6:220-226.

Templeman, W. 1936. Fourth stage larvae of *Homarus americanus* intermediate in form between normal third and fourth stages. *J. Biol. Bd. Cand.* 2:349-354.

Thiriot, A. 1973. Stades larvaires de Parthenopidae meditérranèens: *Heterocrypta maltzani* Miers et *Parthenope massena* (H.Milne Edwards). *Cah. Biol. Mar.* 14:111-134.

Thomas, M.M., K.V. George & M.Kathirvel 1974a. On the spawning and early development of the marine prawn, *Parapenaeopsis stylifera* (H.Milne Edwards) in the laboratory. *Indian J. Fish.* 21:266-271.

Thomas, M.M., M.Kathirvel & N.N.Pillai 1974b. Laboratory spawning and early development of *Parapenaeopsis acclivirostris* (Alcock) (Decapoda: Penaeidae). *J. Mar. Biol. Assoc. India* 16:731-740.

Thompson, J.V. 1828. On the metamorphoses of the Crustacea, and on Zoea, exposing their singular structure, and demonstrating that they are not, as has been supposed, a peculiar genus, but the larva of Crustacea!! *Zoological Researches and Illustrations of nondescript or imperfectly known animals.* Vol.1, part 1:1-11, pls.1-2; plus Addenda to Memoir 1, pp.63-67, pl.8.

Thompson, J.V. 1836. Memoir on the metamorphosis in *Porcellana* and *Portunus*. *Ent. Mag. Lond.* 3: 275-280.

Thompson, M.T. 1903. The metamorphosis of the hermit-crab. *Proc. Boston Soc. Nat. Hist.* 31:147-209.

Thompson, W. 1836. Memoir on the metamorphosis and natural history of the *Pinnotheres*, or pea-crabs. *Ent. Mag. Lond.* 3:85-90.

Thorson, G. 1950. Reproductive and larval ecology of marine bottom invertebrates. *Biol. Rev.* 25:1-45.

Travis, D.F. 1954. The molting cycle of the spiny lobster, *Panulirus argus* Latreille. I. Molting and growth in laboratory-maintained individuals. *Biol. Bull.* 107:433-450.

Valentine, J.W. 1981. Emergence and radiation of multicellular organisms. In: J.Billingham (ed.), *Life in the Universe:* 229-257. Cambridge: MIT Press.

Von Bonde, C. 1936. The reproduction, embryology and metamorphosis of the Cape crawfish (*Jasus lalandii*) (Milne Edwards) Ortmann. *Investl. Rep. Fish. Mar. Biol. Surv. Div. Un. S.Afr.* 6:1-25.

Warner, G.F. 1967. The life history of the mangrove tree crab, *Aratus pisoni*. *J. Zool.* 153:321-335.

Warner, G.F. 1968. The larval development of the mangrove tree crab, *Aratus pisonii* (H.Milne Edwards), reared in the laboratory (Brachyura, Grapsidae). *Crustaceana* suppl.2:249-258.

Wear, R.G. 1964a. Larvae of *Petrolisthes novaezelandiae* Filhol, 1885 (Crustacea, Decapoda, Anomura). *Trans. R.Soc. NZ Zool.* 4:229-244.

Wear, R.G. 1964b. Larvae of *Petrolisthes elongatus* (Milne Edwards 1837). (Crustacea, Decapoda, Anomura). *Trans. R.Soc. NZ Zool.* 5:39-53.

Wear, R.G. 1965a. Larvae of *Petrocheles spinosus* Miers 1876 (Crustacea, Decapoda, Anomura) with keys to New Zealand porcellanid larvae. *Trans. R.Soc. NZ Zool.* 5:147-168.

Wear, R.G. 1965b. Breeding cycles and pre-zoea larvae of *Petrolisthes elongatus* (Milne Edwards 1837). (Crustacea, Decapoda). *Trans. R.Soc. NZ Zool.* 5:169-175.

Wear, R.G. 1965c. Pre-zoea larvae of *Petrolisthes novaezealandiae* Filhol 1885 (Crustacea, Decapoda, Anomura). *Trans. R.Soc. NZ Zool.* 6:127-132.

Wear, R.G. 1966. Pre-zoea larvae of *Petrocheles spinosus* Miers 1876 (Crustacea, Decapoda Anomura). *Trans. R.Soc. NZ Zool.* 8:119-124.

Wear, R.G. 1967. Life-history studies on New Zealand Brachyura. I. Embryonic and post-embryonic development of *Pilumnus novaezealandiae* Filhol, 1886, and of *P.lumpinus* Bennett 1964 (Xanthidae, Pilumninae). *NZ J. Mar. Fresh. Res.* 1:482-535.

Wear, R.G. 1968. Life-history studies on New Zealand Brachyura. 2. Family Xanthidae. Larvae of *Heterozius rotundifrons* A.Milne Edwards 1867, *Ozius truncatus* H.Milne Edwards 1834, and *Heteropanope (Pilumnopeus) serratifrons* (Kinahan 1856). *NZ J Mar. Fresh. Res.* 2:293-332.

Wear, R.G. 1970a. Notes and bibliography on the larvae of xanthid crabs. *Pacif. Sci.* 24:84-89.

Wear, R.G. 1970b. Some larval stages of *Petalomera wilsoni* (Fulton & Grant 1902) (Decapoda, Dromiidae). *Crustaceana* 18:1-12.

Wear, R.G. 1974. Incubation in British decapod Crustacea, and the effects of temperature on the rate and success of embryonic development. *J. Mar. Biol. Ass. UK* 54:745-762.

Wear, R.G. 1976. Studies on the larval development of *Metanephrops challengeri* (Balss 1914) (Decapoda, Nephropidae). *Crustaceana* 30:113-122.

Wear, R.G. & E.J.Batham 1975. Larvae of the deep sea crab *Cymonomus bathamae* Dell 1971 (Decapoda, Dorippidae) with observations on larval affinities of the Tymolinae. *Crustaceana* 28:113-120.

Webb, G.E. 1919. The development of the species of *Upogebia* from Plymouth Sound. *J. Mar. Biol. Assc. UK* 12:81-134.

Webb, G.E. 1921. The larvae of the Decapoda Macrura and Anomura of Plymouth. *J. Mar. Biol. Assc. UK* 12:385-417.

Webber, W.R. & R.G.Wear 1981. Life history studies on New Zealand Brachyura. 5. Larvae of the family Majidae. *NZ J. Mar. Fresh. Res.* 15:331-383.

Williams, B.G. 1968. Laboratory rearing of the larval stages of *Carcinus maenas* (L.). Crustacea: Decapoda. *J. Nat. Hist.* 2:121-126.

Williamson, D.I. 1965. Some larval stages of three Australian crabs belonging to the families Homolidae and Raninidae, and observations on the affinities of these families (Crustacea: Decapoda). *Aust. J. Mar. Fresh. Res.* 16:369-398.

Williamson, D.I. 1967. On a collection of planktonic Decapoda and Stomatopoda (Crustacea) from the Mediterranean coast of Israel. *Bull. Sea Fish. Res. Stn. Israel* 45:1-64.

Williamson, D.I. 1969. Names of larvae in the Decapoda and Euphausiacea. *Crustaceana* 16:210-213.

Williamson, D.I. 1970. On a collection of planktonic Decapoda and *Stomatopoda* (Crustacea) from the east coast of the Sinai Peninsula, Northern Red Sea. *No.45. Bull. Sea Fish. Res. Stn. Israel* 56:1-48.

Williamson, D.I. 1976. Larvae of Stenopodidea (Crustacea, Decapoda) from the Indian Ocean. *J. Nat. Hist.* 10:497-509.

Williamson, D.I. 1983. Larval morphology and diversity. In: L.G.Abele (ed.), *Biology of the Crustacea, Vol.2:*43-110. New York: Academic Press.

Williamson, H.C. 1911. Report on larval and later stages of certain decapod Crustacea. *Scient. Invest. Fish. Bd. Scot.* 1910:1-20.

Williamson, H.C. 1915. Crustacea Decapoda. Larven. *Nord. Plankt. 1927, Zool. Teil.* Bd.3, Lief.18: 315-588.

Wilson, K.A. 1980. Studies on decapod Crustacea from the Indian River region of Florida. XV. The larval development under laboratory conditions of *Euchirograpsus americanus* A.Milne Edwards 1880 (Crustacea, Decapoda: Grapsidae) with notes on grapsid subfamilial larval characters. *Bull. Mar. Sci.* 30:756-775.

Wilson, K.A. & R.H.Gore 1980. Studies on decapod Crustacea from the Indian River region of Florida. XVII. Larval stages of *Plagusia depressa* (Fabricius 1775) cultured under laboratory conditions (Brachyura: Grapsidae). *Bull. Mar. Sci.* 30:776-789.

Wilson, K.A., L.E.Scotto & R.H.Gore 1979. Studies on decapod Crustacea from the Indian River region of Florida. XIII. Larval development under laboratory conditions of the spider crab *Mithrax forceps* (A.Milne Edwards 1875) (Brachyura: Majidae). *Proc. Biol. Soc. Wash.* 92:307-327.

Yang, W.T. 1967. A study of zoeal, megalopal, and early crab stages of some oxyrhynchous crabs (Crustacea: Decapoda). Unpubl. Doctoral Diss. Univ. Miami, 459 pp.

Yang, W.T. 1968. The zoeae, megalopa, and first crab of *Epialtus dilatatus* (Brachyura, Majidae) reared in the laboratory. *Crustaceana* suppl.2:181-202.

Yang, W.T. 1971. The larval and postlarval development of *Parthenope serrata* reared in the laboratory and the systematic position of the Parthenopinae (Crustacea, Brachyura). *Biol. Bull.* 140:166-189.

Yaqoob, M. 1974. Larval development of *Petrolisthes rufescens* (Heller, 1861) (Decapoda: Porcellanidae) under laboratory conditions. *Pakist. J. Zool.* 6:47-61.

Yaqoob, M. 1977. The development of larvae of *Petrolisthes ornatus* Paulson 1875 (Decapoda, Porcellanidae) under laboratory conditions. *Crustaceana* 32:241-255.

Yaqoob, M. 1979. Larval development of *Pisidia dehaani* (Krauss 1843) under laboratory conditions (Decapoda, Porcellanidae). *Crustaceana* suppl.5:69-76.

Yatsuzuka, K. 1957. Study of brachyuran zoea, artificial rearing and development. *Suisan Gaku Shusei* 1957:571-590.

Yatsuzuka, K. 1962. Studies on the artificial rearing of the larval Brachyura, especially of the larval Blue-Crab, *Neptunus pelagicus* Linnaeus. *Rep. Usa Mar. Biol. Stn. Kochi Univ.* 9:1-4.

NANCY N.RABALAIS* / ROBERT H.GORE**

* Port Aransas Marine Laboratory, Marine Science Institute, University of Texas, USA (Present address: Louisiana Universities, Marine Consortium, Chauvin, USA)

** The Academy of Natural Sciences of Philadelphia, Pennsylvania, USA

ABBREVIATED DEVELOPMENT IN DECAPODS

ABSTRACT

Although most decapod crustaceans possess early developmental stages that are normally long in developing, the incidence of abbreviated development within the group is frequent and widespread across phylogenetic lines. Abbreviated development occupies the far end of a continuum of decapod developmental patterns, whereby the developmental sequence is of shorter duration than that normally seen in the majority of related species in a taxon, which results in fewer morphologically discrete instars and/or a reduced ontogenetic duration. Examples span the continuum from those in which the young hatch in the form of the adult, often with retention of some larval characteristics, to those which hatch as nonfunctional zoeal-type form retained on the pleopods of the parental female (i.e., direct development). In abbreviated development of the advanced type, the young hatch as zoeae but in a state considerably more developed than that seen in their congeneric relatives. Abbreviated development may also involve postlarval stages which are eliminated or are morphologically advanced and more adult-like in form as a result of metamorphosis from the larva. Developmental sequences may also be shortened compared to their regular counterparts, with the elimination of one or more otherwise regularly occurring stage(s) within the ontogenetic sequence (i.e., accelerated development). Features associated with abbreviated development are often related to size and number of eggs and the lecithotrophic nature of early developmental stages. These trends, however, are not consistent across the continuum of developmental patterns. Differences in survivorship and maturation rates of those species with abbreviated development, as opposed to those with longer planktonic existence, are most obvious among closely related taxa where the developmental schemes differ significantly. Increased survivorship may be related to decreased nutritional vulnerability, reduced number of critical premetamorphic molt periods, avoidance of suboptimal conditions for growth and development, antipredation adaptations, and reduced dispersal from specialized and restrictive adult habitats. Some general trends are apparent in the occurrence of abbreviated development in species with respect to habitat. The elimination or partial suppression of pelagic larval stages would seem advantageous in several environs, yet other species occupy similar habitats but do not possess similar developmental modes. Those characteristics of abbreviated development which would enhance survivorship in a particular ecological situation are stressed. The evidence reveals an array of developmental patterns. Shortened sequences, while numerically less frequent, are not aberrant, and the evolutionary implications are far-reaching.

1 INTRODUCTION

The developmental patterns of decapod crustaceans generally follow a scheme of anamorphic changes in a series of larval stages which occur over a prolonged planktonic existence (Snodgrass 1956, Knowlton 1974). The larval sequence usually ends when a metamorphic molt transforms the larva to a form substantially different from those preceding and more similar to the eventual adult form. In some instances, the free-swimming larval phases are suppressed or eliminated altogether. This epimorphic type of development (Knowlton 1974) is seen in species in normally long-developing families such as the Atyidae, Alpheidae, or Hippolytidae that have little or no pelagic zoeal phase (e.g. Dobkin 1965a, Makarov 1968, Benzie 1982). In the freshwater crabs or crayfishes, the young hatch more or less directly into the parental form (Gurney 1942, Smart 1962). In still others, such as some pinnotherid or ocypodid crabs, an extremely short pelagic stage exists, lasting in some cases no more than 24 to 48 hours before attaining the postlarval stage (Goodbody 1960, Rabalais & Cameron 1983) (see Gore in this volume, for a treatment of larval development patterns).

Morphological and durational variations also occur in the postlarval phase. This phase may produce an abrupt and morphologically distinct form which undergoes an additional molt (also often considered metamorphic). Some morphologically distinct postlarval forms, however, may be more juvenile in form, resulting in suppression of this transitionary phase. On the other hand, the postlarval phase may consist of gradual change into the adult form which is attained through a series of successive molts (see Felder et al., this volume, for a treatment of postlarval development).

The pattern that emerges then is a continuum of developmental types in decapod crustaceans in which there are often combinations of larval and postlarval variability. This continuum differs somewhat, but not significantly, from that discussed by Gore (this volume) for larval development, in that we are here considering both larval and postlarval phases. Much of the variability seen in these developmental patterns is associated with a shortening of the time interval required to attain a juvenile form that possesses the overall morphological characteristics of the adult. This, of course, is relative to the greater temporal extent of other development patterns. A shortening, or abbreviation, of developmental sequences is characteristic of peracaridans but among other crustacean groups is sporadic (Knowlton 1974). However, a review of the literature shows that among the decapod Crustacea, which in general undergo a prolonged developmental sequence, there are many instances of abbreviated development. In this chapter we review these many reports and use them as examples to support the categories that we develop. We then summarize the overall trends of abbreviated development and discuss the adaptive significance of different developmental patterns.

2 HISTORICAL USAGE OF TERMS

A brief consideration of historical usage of terms related to abbreviated development will prove useful to our later treatment of definitions and categories. Throughout the literature are numerous references to 'abbreviated development', with different meaning imparted by different authors. Such inconsistent application has unfortunately muddied the concepts associated with these terms. Moreover, evolutionary implications often become attached to such wording so that an abbreviated, or shortened, development may imply a less primitive or more derived state in a taxa than longer, or extended, development.

Hale (1931) used the terms 'direct development' or 'abbreviated metamorphosis'. In describing other species, he said that although development was not direct, it was 'curtailed'. Gurney (1938a) described some European upogebiids as having development that is 'slightly abbreviated'. Gurney (1942) then summarized abbreviated development and, while not defining the term, per se, included forms which had partial or complete suppression of larval stages. A similar approach was used by Williamson (1982a) in his brief treatment of abbreviated development. The development pattern of *Pinnotheres moseri* was termed 'abbreviated' by Goodbody (1960) who did not consider *P.pinnotheres* to be in the same category, even though it develops through two zoeal stages (Lebour 1928a) compared

Table 1. Comparison of number of zoeal stages and habitat types in pinnotherid crabs.

Number of zoeal stages	Species	Habitat and reference
0	*Pinnotheres* sp.	Offshore sublittoral sponge reef, precise microhabitat not given (Kurata 1970)
1	*Pinnotheres moseri*	Obligate commensal in atrial cavity of *Ascidia nigra* (Goodbody 1960)
2	*Pinnotheres pinnotheres*	Mantle cavity of *Mytilus, Modiolus, Pinna* and *Ascidia* (Lebour 1928a, Atkins 1955)
	Pinnotheres taylori	Transparent ascidian (Hart 1935)
3	*Pinnotheres boninensis*	Not stated in English (Muraoka 1977b)
	Pinnotheres aff. *sinensis*	Mantle cavity of clam, *Tapes japonica* (Yatsuzuka & Iwasaki 1979)
	Sakaina japonica	Free-living or commensal with polychaete worms, *Loimia medusa* (Konishi 1981a)
	Pinnotheres chamae	Commensal with *Chama congregata* (Roberts 1975)
	Dissodactylus crinitichelis	Obligate parasite on *Mellita sexiesperforata* (Pohle & Telford 1981)
	Dissodactylus primitivus	Symbiotic with *Meoma ventricosa* (Pohle & Telford 1983)
4	*Pinnotheres ostreum*	Mantle cavity of *Ostrea virginica* (Sandoz & Hopkins 1947)
	Pinnotheres pisum	Mantle cavity of *Mytilus edulis* (Atkins 1955)
	Pinnotheres gracilis	Commensal with *Katelysia opina* (Kakati & Sankolli 1975)
	Ostracotheres tridacnae	Mantle cavity of clam, never seen to leave (Gohar & Al-Kholy 1957)
	Fabia subquadrata	Mantle cavity of *Modiolus modiolus* (Irvine & Coffin 1960)
	Pinnaxodes mutuensis	Mantle cavity of several bivalves (Konishi 1981b)
5	*Pinnaxodes major*	Mantle cavity of *Mytilus edulis* (Hong 1974)
	Pinnotheres maculatus	Mantle cavity of *Atrina* and *Aequipecten* (Costlow & Bookhout 1966)
	Pinnixa longipes	Commensal with tube-building polychaete, *Axiothella rubrocincta* (Bousquette 1980)
	Pinnixa chaetopterana	Commensal in burrows of polychaete worms and callianassid shrimp (Sandifer 1972)
	Pinnixa sayana	Commensal in burrows of polychaetes, *Arenicola cristata* (Sandifer 1972)
	Pinnixa rathbuni	Free-living (Sekiguchi 1978)

to the three, four, or five zoeal stages of the majority of the family (Table 1). Hartnoll (1964a) did not hesitate to call the development of freshwater grapsid crabs he studied as 'abbreviated'. Other authors have been more hesitant. For example, Wear (1968) did not consider the developmental series of *Heterozius rotundifrons* to be 'abbreviated' because there was a series of two planktonic larval stages. He termed the development sequence of this species 'incompletely abbreviated', thus implying that only 'direct' development is abbreviated. The development of *Pilumnus vestitus, P.lumpinus* and *P.novaezealandiae* was labeled by him (Wear 1967) as 'direct or abbreviated'. Rice et al. (1970) called the larval development of *Conchoecetes artificiosus* 'relatively abbreviated', yet this species was deleted by Lang & Young (1980) from the category of 'abbreviated development' in their treatment of developmental patterns in the Dromiacea. Ngoc-Ho (1977) recognized three types of developmental sequences – 'an extremely abbreviated development' in *Upogebia savignyi* (Gurney 1937a), a 'slightly abbreviated larval development' in his description of *U.darwinii*, and the others with three, four or five zoeal stages, with four or five most commonly occurring.

Other authors have placed 'abbreviated development' within a series of developmental patterns. Sollaud (1923) outlined three basic types of larval development for the Palaemonidae. In the first were marine and brackish water species characterized by numerous small eggs and a larval life of many stages. The second were freshwater species which produce fewer larger eggs and have an abbreviated larval life, usually of three stages. Lastly, were the freshwater species in which the newly hatched forms were postlarval. These groups correspond to those outlined by Knowlton (1973) in his comparisons of alpheid larval development. Without classifying upon the basis of habitat types, he defined 'extended, or prolonged' larval development as that in which the larva hatched as a typical zoea and molted many times with larval development lasting several weeks, 'abbreviated development' as hatching as an advanced zoea completing larval development in four or fewer molts, and 'direct development' as hatching in the adult form. Dobkin (1969) noted the extreme variability of developmental patterns in the Caridea and Stenopodidea. He defined 'abbreviated development' as a decrease in the number of planktonic larval stages and 'direct development' as the complete suppression of planktonic larvae, with the newly hatched individual being similar to the adult in form and undergoing no metamorphosis. He dealt with variations in these categories by adding that others 'appeared to be evolving toward abbreviated development'.

Soh (1969) recognized four types of abbreviated development, to which he added a fifth. These categories were based on the form in which the individual hatches from the egg: 1) direct – hatches as a juvenile crab (Calman 1911, Gurney 1942), 2) a short-lived zoea followed by a megalopa (Goodbody 1960, Wear 1967), 3) a megalopa larva without natatory pleopods molting to a megalopa with natatory pleopods (Hale 1931), 4) a single 'true' megalopa with natatory pleopods (Wear 1967), and 5) two zoeal stages of advanced characters but highly modified, then a megalopa. In other words, he devised a category for almost every type of variation in brachyuran development known up to that time.

Gore (1979) in a review of galatheid developmental types categorized them as 'direct' – those larvae which hatch from the egg in a form morphologically similar to the adult and undergo no further metamorphosis, 'advanced' – usually hatch in the ultimate or penultimate zoeal stage and thus may undergo additional ecdyses prior to metamorphosis, and 'abbreviated' – hatch as early zoeae but may dispense with one or more intermediate stages in completing their larval development. He later discussed the 'accelerated morphological development' of Atlantic galatheids collected from the plankton. Readdressing these concepts, he has both redefined and restricted these definitions into a more workable and uniform scheme (Gore, this volume).

The many usages of terminology for abbreviated development has been confusing and promises to continue so, as more and more decapod developmental patterns are documented from laboratory-reared specimens and supplemented by plankton studies. The addition of more detailed descriptions will certainly reveal that decapod crustacean development patterns cannot be considered normal, because variability is the all encompassing characteristic of the group (see Gore, this volume, section 3). However, a critical analysis of developmental patterns, with recognition of those exhibiting abbreviation, remains long overdue and will pinpoint comparable life history tactics and associated characteristics.

3 WHAT IS ABBREVIATED DEVELOPMENT?

We envisage a definition of abbreviated development that is consistent with the idea of a continuum of developmental types as stated above and in Gore (this volume). In this sense abbreviated development with its many variations occupies the far end of a continuum that proceeds through regular development and into extended development, at the other extreme. The point at which any one developmental sequence fails to be 'abbreviated' and becomes 'regular' is not precisely definable. The terms are not mutually exclusive, and each is relative to the other forms of development. The futility in trying to categorize decapod development into discrete units and attempting to provide a proper place to put everything convinced us that a continuum concept is more realistic and desirable, especially since biological entities, whether species, classes, communities, or populations, defy strict classification. As such, we have developed a broad definition of abbreviated development within which we propose subheadings. Supporting evidence for these categories is a seemingly limitless store of descriptions of decapod crustacean developmental patterns.

Gore (this volume) provided definitions for all of the larval molting modes presently recognized within the literature. But before proceeding to an in-depth consideration of abbreviated development, the definition for regular development should be restated; viz. 'the numerically predominant type of ontogenetic growth and developmental staging that consistently recurs in a particular decapod group'. As Gore noted, this definition encompasses the subheadings of direct development (with complete suppression of larval stages), as well as development incorporating a consistent number of stages within a taxon (e.g. two zoeae in Majidae; four zoeae in most Xanthidae, etc.). The distinction made here, however, is between regular development as a consistently similar ontogeny, and irregular development such as that occurring as a consequence of environmental or genetic factors (e.g. Sandifer & Smith 1979). Within the latter category is placed the type of development best classified as sporadic, as exhibited by some caridean shrimp and portunid crabs.

The category of abbreviated development has often been ambiguously applied, and many authors have placed both regular and irregular development modes within this framework. It is helpful, therefore, to re-emphasize Gore's definition, which considers abbreviated development as a developmental sequence of shorter duration than that normally seen in the majority of related species in a taxon, and which results in fewer morphologically discrete instars and/or a reduced ontogenetic duration. These shortened developmental times can occur across the larval and postlarval continuum, thereby more rapidly producing a juvenile form possessing the general morphological characteristics of the adult. Contrasted with regular development, the concept of abbreviated development rests on the assumption that longer ontogeny is 'normal', primarily based on the preponderance of longer developmental sequences appearing in a taxon. For example, the abbreviated development of freshwater potamoid crabs (Gurney 1942) is a regular developmental mode for this group but differs greatly from the general developmental patterns of other decapods. Other examples occur in the Grapsidae (Table 2).

In addition to these examples, variations in postlarval stages may abbreviate the developmental patterns. For example, the megalopal stage may be skipped (Lucas 1971) or may be morphologically advanced so that a more crab-like form metamorphoses from the larva (Soh 1969, Rabalais & Cameron 1983).

With these concepts in mind, we now direct attention to their refinement with supportive examples. These examples will both substantiate these subheadings as well as show the array of variability in the continuum. Even though much of the information in the recorded liter-

Table 2. Comparison of number of zoeal stages and habitat preferences of grapsid crabs

Number of zoeal stages	Species	Hatitat and reference
2	*Metopaulius depressus*	Terrestrial, arboreal, tank bromeliads (Hartnoll 1964a)
(2?)	*Sesarma bidentatum*	Terrestrial, damp habitats, forest periphery, wet and muddy (Hartnoll 1964a)
	Sesarma jarvisi	Terrestrial, montane, limestone talus and rock rubble (Abele & Means 1977)
2	*Geosesarma perracae*	Freshwater, terrestrial (Soh 1969)
3	*Sesarma reticulatum*	Estuarine, intertidal (Costlow & Bookhout 1962)
4	*Sesarma cinereum*	Estuarine, intertidal (Costlow & Bookhout 1960)
	Aratus pisonii	Semi-terrestrial, mangroves, estuarine (Hartnoll 1965, Warner 1968)
	Chiromantes bidens	Estuarine (Fukuda & Baba 1976)
	Parasesarma plicatum	Estuarine (Fukuda & Baba 1976)
	Chasmagnathus convexus	Estuarine, reed marshes (Baba & Fukuda 1972)
	Holometopus haematocheir	Estuarine (Fukuda & Baba 1976)
5	*Sesarmops intermedius*	Estuarine (Fukuda & Baba 1976)
	Grapsus strigosus	Marine, rocky intertidal (Gohar & Al-Kholy 1957)
	Cyrtograpsus altimanus	Marine, rocky intertidal and subtidal (Scelzo & Lichtschein de Bastida 1978)
	Pachygrapsus marmoratus	Marine, intertidal (Hyman 1924)
	Leptograpsus variegatus	Marine, rocky intertidal, upper shore (Wear 1970c)
	Planes marinus	Oceanic on debris and floating weed (Wear 1970c)
	Hemigrapsus crenulatus	Estuarine, intertidal or semi-terrestrial (Wear 1970c)
	Hemigrapsus edwardsi	Marine, intertidal mud, gravel, rocks, sheltered shores (Wear 1970c)
	Cyclograpsus lauvauxi (= punctatus)	Littoral fringe, among stones and pebbles (Wear 1970c, Fagetti & Campodonico 1971)
	Helice crassa	Estuarine, intertidal, semi-terrestrial (Wear 1970c)
	Plagusia chabras	Marine, exposed rocky coasts and wave-washed reefs (Wear 1970c)
	Plagusia depressa	Marine, rocky intertidal (Wilson & Gore 1980)
	Euchirograpsus americanus	Marine, subtidal oculinid coral reefs (Wilson 1980)

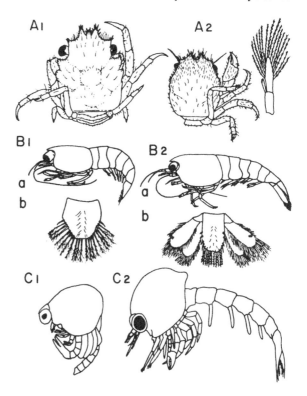

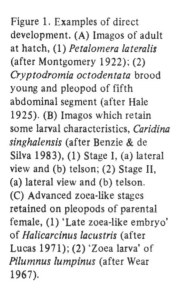

Figure 1. Examples of direct development. (A) Imagos of adult at hatch, (1) *Petalomera lateralis* (after Montgomery 1922); (2) *Cryptodromia octodentata* brood young and pleopod of fifth abdominal segment (after Hale 1925). (B) Imagos which retain some larval characteristics, *Caridina singhalensis* (after Benzie & de Silva 1983), (1) Stage I, (a) lateral view and (b) telson; (2) Stage II, (a) lateral view and (b) telson. (C) Advanced zoea-like stages retained on pleopods of parental female, (1) 'Late zoea-like embryo' of *Halicarcinus lacustris* (after Lucas 1971); (2) 'Zoea larva' of *Pilumnus lumpinus* (after Wear 1967).

ature is vague and does not provide the details required to properly analyze the abbreviated nature of a developmental sequence and its adaptive value, the evidence reveals an array of developmental patterns and points to far reaching evolutionary implications.

3.1 *Direct development*

In direct development no free-swimming larval stages exist. The young hatch more or less in the form of the adult, possessing distinct juvenile morphology. Imagos of the adult are usually produced at hatching (Fig.1A). In some, morphological equivalents are often held on the female's abdomen, or are non-swimming if they fall to the bottom. The young never, or rarely, leave the pleopods of the maternal parent or else remain in her immediate vicinity.

Some examples of species with direct development are straight-forward, while in others the distinction is less easily made; although the young hatch from the egg essentially in the adult form, certain larval characteristics are retained and drop out as molting continues (Fig.1B) (see Gore, this volume, section 3.1.2). In other cases, a zoea- or megalopa-type young hatches from the egg (Fig.1C); although it is a nonfunctional, nonswimming, non-pelagic stage usually retained on the pleopods of the females, it nevertheless undergoes a discrete molting series before becoming a juvenile crab with a full complement of adult characters. In some cases, it is difficult to draw the line between what is 'direct' and what is 'advanced'. Often a late embryo, zoea or megalopa, is held under the abdomen of the adult female and does not swim, although it possesses some typical larval characteristics. If these forms are present, they are usually soft, flaccid, amorphous, and natatorily nonfunctional.

3.1.1 *Phylogenetic treatment*

Direct development appears to be the rule among the freshwater and terrestrial group Potamoidea, where the young hatch in the form of the parent and are carried for a time under the abdomen of the female (Lanchester 1901, Calman 1909, 1911, Moreira 1912, Koba 1936, McCann 1937, Gurney 1942, Fernando 1960, Bott 1969, Pace et al. 1976). There is no free swimming phase requiring a metamorphic molt, and both the zoea and the megalopa stages have been eliminated. The female crabs carry up to several hundred fully formed young on their pleopods (Lanchester 1901, McCann 1937, Chace & Hobbs 1969). The developmental type is unrecorded for more than half the families in this superfamily (Gore, this volume, Table 1). Whether more variability may surface as our knowledge is expanded, or whether the developmental pattern proves to be consistent in the group remains unknown. Those patterns recorded, especially the actual hatching sequence and early development, are not well described. Pesta (1930) reported a female *Geosesarma noduliferum* de Man with young crabs in the brood pouch, and Johnson (1965) suggested that amphibious and terrestrial species of *Geosesarma* might perhaps have direct development. However, the developmental sequence of other non-marine grapsids, although abbreviated, is advanced rather than direct as described above (see section 3.2).

The situation is similar in the freshwater crayfishes (astacids and parastacids) which regularly hatch young with all adult appendages except the first pleopods and uropods (Andrews 1907, Hale 1927, Gurney 1942, Smart 1962), and they remain attached to the pleopods of the adult. Interestingly, uropods appear in stage 3 (perhaps analogous to other normal larval series?), which shows a reminiscence of three free larval stages (as in *Homarus*) retained in ontogeny. Pleopods on the first abdominal somite appear in stage 4.

Members of the Aeglidae which inhabit freshwater streams of South America (Schmitt 1942) are reported to have shortened development with young that hatch in the form of the adult; however, no descriptions of the embryos or young have been published (Müller 1880, 1892, Mouchet 1932 in Gurney 1942). Burns (1972) reported female *Aegla* maintained eggs and brood under their abdomens for five months.

Examples of direct development occur in some marine caridean shrimps such as *Synalpheus brooksi* (Dobkin 1965a), *S.longicarpus* (Brooks & Herrick 1892), many more freshwater Caridea, especially in the Palaemonidae (Boone 1935, Sollaud 1923, Gurney 1938b, Powell 1979), Atyidae (*Caridina* Shen 1939, Shokita 1973a, 1976, Benzie & de Silva 1983; *Neocaridina*, Mizue & Iwamoto 1961), and some stenopodidean shrimps (*Richardina*, *Spongicoloides*, Kemp 1910a,b). Direct development has been reported by Hale (1927) for an Australian pagurid hermit crab which inhabits tunnels of soft stone. He described females with broods of 30 young clinging to the abdominal appendages. He suggested that they hatch at a very advanced stage of development, since their forms were of the adult. Dechancé (1963) also described direct development in a paguridean crab.

Direct development has been described in some marine brachyuran crabs. Hale (1925) described the adult females of *Cryptodromia octodentata* and *Petalomera lateralis* with young crabs beneath the abdomen similar in aspect to the adult crabs, but with pleopods having natatory setae as in the megalopae of some other crabs. A female *C.octodentata* carried approximately 530 young crabs. The early stages of young crabs in the brood pouch of *P.lateralis* were soft and ready to molt. Both the early and later stages possessed natatory pleopods. Montgomery (1922) also described young crabs on the abdomen of a female *P.lateralis,* which bore four pairs of pleopods similar in form to those of the adult and not like those of other megalopae. These, however, were smaller than those described by Hale (1925). It is possible that Montgomery observed later stages than Hale, based on the lower larval number contained on the female (20) (i.e., some dispersed). Size differences could be attributed to methodology or the independent nature of growth and morphogenesis (Fincham 1979a, and others). Rathbun (1914) reported young crabs carried beneath the abdomen of the spider crab *Paranaxia (= Naxoides) serpulifera,* presumably with direct development. The 160 young crabs were of two stages, probably contiguous molts, and it was impossible to tell whether the younger of the two stages hatched directly from the egg or not.

Wear (1967) described the development of *Pilumnus novaezealandiae* which hatches as a megalopa in a soft, flaccid condition and remains inside a type of brood pouch formed by the pleopods on the parental female. The megalopae eventually become more robust and swim away from the female but always return either to her carapace, pereopods or abdomen. Although the megalopae possess pleopods fringed with 12-15 natatory setae and are active swimmers, they are otherwise advanced morphologically and more similar to the first crab, even though they remain associated with the parent.

Kurata (1970) reported an instance of direct development in a pinnotherid crab, *Pinnotheres* sp. The larvae emerged from the brood pouch of the female and were found to have 'characters of the adult in every respect' (Pl.90A, B). The female crab was not identified.

3.1.2 *Inferences*
Direct development has been inferred for several species. Fage & Monod (1936) hypothesized that the young *Munidopsis polymorpha* were well advanced in development inside the eggs, which were extremely large (1.5 to 1.8 mm, diameter) and probably hatched into a form nearly like the adult. They felt that such a development would not be surprising considering the large egg size and the restrictive habitat of the adult (see section 5.5), i.e., which is a littoral cave in the Canary Islands formed by lava tunnels connecting to the sea. However, abbreviated development of other than 'direct' type is also a possibility, and egg size is not necessarily an accurate determinant of developmental type (section 5.1).

Hale (1925, 1927) hypothesized direct development for several Australian decapods, based on egg size, but these have not been verified: an undescribed pagurid with eggs protected by a brood pouch on the side of the female's abdomen, two dromiids, two thalassinids, and a crangonid shrimp. Dobkin (1969) also compiled literature on caridean shrimp that were reported to undergo direct development. Much of the information on these species is limited and often based on sparse data such as presence of large eggs, or a late embryo dissected from the egg. For example, Boone (1935) reported direct development in *Conchodytes biunguiculatus,* an inquiline pontoniid shrimp, based on the presence of many young in the brood pouch of an adult female. Bruce (1972), on the other hand, indicated that this species, correctly identified as *C.meleagrinae* Peters, most probably hatched as normal stage I larvae, based on the presence of very numerous and small ova. Bruce (1972) then dissected the late embryo from an egg of *Pontonia minuta* Baker and found it to be in an advanced state of development compared to normal stage I pontoniid larvae; however, the exact development sequence was not known.

3.1.3 *Those which defy categorization*
A group of brachyurans, which almost defy categorization, hatch as a first zoea of some

form but advanced in nature. This zoeal stage undergoes an extremely rapid development, lasting usually no longer than a few hours or less before metamorphosing to the megalopal stage. Because in some species, it does not swim and is basically nonfunctional, these cases are considered examples of direct development in the sense that the pre-juvenile phase usually remains associated with (but not always attached to) the parent. In others, the imago sheds one or two larval characters.

The xanthid crab, *Pilumnus lumpinus* hatches as an advanced zoea which possesses rudimentary natatory maxillipedal setae. These are not functional, and the zoeae do not swim. At eclosion, the zoeae are not retained on the pleon of the female crab, but lie on the bottom, rapidly flexing and extending the abdomen, while molting both the prezoeal and zoeal skins. Both are shed upon molt to the megalopa 30 minutes later. The megalopae of *P. lumpinus* are more active swimmers than those of *P. novaezealandiae* (Wear 1967) and, although also not attached to the parent female, remain in contact with her. The developmental sequence of *P. vestitus* described by Hale (1931) is strikingly similar to that described above for *P. lumpinus,* although he categorized the stages differently. Hale described *P. vestitus* as hatching in a soft, flaccid megalopa stage, which is sedentary among the pleopods of the female. The large pleopods lack natatory setae. This stage was not studied in detail by Hale, but it appears that the stage figured by him as a megalopa is actually an advanced zoea still encased in a prezoeal cuticle similar to one described by Wear (1967) for *P. lumpinus.* Hale labeled the next stage as the first crab, but noted that it possesses pleopods with natatory setae. He did not comment on the swimming behavior, nor the association with the female crab. The abdomen of this so-called first crab stage (Hale 1925: 320, Fig.10) is more megalopal than crab-like in general form. That of his second crab stage has assumed a more adult form. Furthermore, the ontogenetic changes in the form of the carapaces of Hale's first to fifth young crab series more closely parallel changes in the carapaces of megalopa through fourth crab stage in *P. novaezealandiae.* (Wear 1967) and several xanthids described by Martin et al. (1984). Wear (1967) commented on these stages, as well as on a megalopa without natatory setae and a first crab stage with natatory setae, and proposed that they were all megalopae, the first one without natatory setae and a second one with them.

We suggest, from comparing this development sequence with that of *P. lumpinus* and others, as well as with the ontogeny of morphological characters in megalopa through early juvenile crab stages, that the first two stages of Hale's developmental sequence for *P. vestitus* are a first zoea (advanced but nonfunctional) and a megalopa with natatory setae, followed by a first crab stage. The stage at which the juvenile crabs leave the pleon of the adult female was not given by Hale (1931). *P. vestitus,* therefore, parallels closely the developmental sequence of *P. lumpinus.*

The young of *Halicarcinus lacustris* (Chilton), an Australian freshwater hymenosomatid, hatch on the female pleopods in a form that Lucas (1971) terms a 'late embryo'. This stage is 'zoea-like', with striking similarities to the advanced zoea of other species with advanced development (see section 3.2). The four natatory setae are present but rudimentary, and the zoeae, if lost off the female's pleopods, fall to the bottom and lie there until they subsequently molt. This stage is advanced morphologically, without pleopods, but is characteristic of late zoea of other *Halicarcinus* with three zoeal stages. The stage molts directly to the juvenile crab, as is characteristic of those Hymenosomatidae with no megalopa (see section 3.3). Lucas (1971) alternately calls this first stage 'late embryo', 'zoea-like', 'zoeal', and 'prezoeal'. It most certainly is not pre-zoeal sensu Gore (this volume). We suggest it is an advanced zoea (albeit with rudimentary setae and natatorily non-functional) and classify this developmental pattern as 'direct'.

The development of *Upogebia savignyi* progresses through a simple but very short larval stage (Gurney 1937a). The appendages of this zoea, except for the antenna, antennule, and the vestigial exopods on the first three pairs of pereopods, are all generally similar to those of the adult and quite different from those of all other *Upogebia* larvae so far described (Ngoc-Ho 1977). Gurney noted that there is complete suppression of the free-swimming stages and that the inert larva hatches from the egg and immediately molts to the adult form.

In some cases the imago is nearly complete, and only adds or refines some adult limbs or mouthparts or completes the telson morphology (Dobkin 1968, Shokita 1973b, Benzie & de Silva 1983). For example, Shokita (1977) distinguishes a 'megalopal phase' from a second zoeal phase in *Macrobrachium*, but the gradation between the two was so gradual that his terminology seems more one of convenience than actual delineation of instar. Otherwise, his 'megalopa' is a combinatorial stage exhibiting some few zoeal characters (e.g., incompletely formed telson), with many more postlarval characters (e.g., presence of uropods, functional pleopods with appendices internae, and ventral rostral teeth). In another case (Shokita 1973b), the larvae of *Macrobrachium shokitai* Fujino & Baba had all the appendages of the adult except the uropods, yet two zoeal stages were distinguishable, with development including appearance of uropods, eyes becoming mobile, and

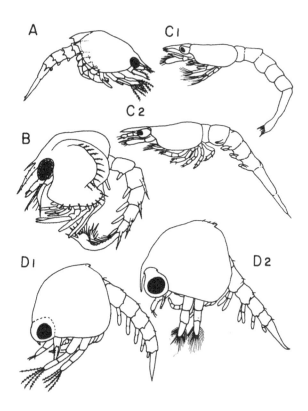

Figure 2. Examples of advanced development: (A) *Paralomis granulosa* (after Campodonico & Guzman 1981); (B) *Gastroptychus* n.sp. (after Pike & Wear 1969); (C) *Callianassa kraussi*, (1) First stage larva and (2) Second stage larva (after Forbes 1973); (D) *Uca subcylindrica*, (1) Zoea I, and (2) Zoea II (after Rabalais & Cameron 1983).

appendices internae on pleopods. The postlarval stage was attained in about 30 hours after hatching.

3.2 *Advanced development*

In this type of abbreviated development, the young hatch as zoeae, but in a state considerably more developed than that seen in their congeneric relatives (Fig.2). These are functional, swimming zoeae compared to those nonfunctional late embryos, zoeae, or megalopae of direct development. They often hatch in the ultimate or penultimate zoeal stage and thus may undergo additional ecdyses prior to metamorphosis. The larval development is thus shortened in comparison, both durationally and ecdysially, so that a reduced number of morphologically advanced instars obtains before the metamorphic molt to the postzoeal stage. As with direct development, the young may remain in the vicinity of the parent owing to the short time spent in the plankton.

In these species, although a few characteristics consistent with those of longer-developing forms are retained in the first zoeal stage, the reduced number of zoeal stages often have varying degrees of maturational development, exhibiting morphological features that would normally appear in later stages of their congeners (e.g., Table 3). Often, in addition to characteristics of maturationally advanced stages, there are characteristics probably related to the highly modified nature of these zoeae (Table 3). For example, there are usually, but not always, reduced carapacial spines associated with a globular carapace and sometimes modification of the feeding appendages, so that accessory setae are rudimentary or missing (section 5.3.1). The second zoeal stage of many brachyurans exhibiting advanced development often have more natatory setae than the last stage of closely related species (e.g. 14 in *Metopaulius depressus*, Hartnoll 1964a; 24 to 26 in *Uca subcylindrica*, Rabalais & Cameron 1983). There is an absence of lateral projections on the abdominal somites where they normally occur and, in one species (*Uca subcylindrica*), a greater number of telsonal processes compared to other related crabs. These highly modified features may be related to several factors, including a more benthic existence, better swimming ability and related requirements of dispersal and recruitment, lecithotrophic nature and non-feeding behavior, lack of predators, or an unknown adaptive significance upon which we can only speculate.

3.2.1 *Phylogenetic treatment*

Examples of advanced development are numerous, especially in the caridean shrimp. In the Palaemonidae, a few *Macrobrachium* hatch as advanced zoeae (Fielder 1970, Shokita 1973b, 1977, Siqueira Bueno 1980), compared to the 9 to 12 larval stages of most other members of the genus (Kwon & Uno 1969, Uno & Kwon 1969, Choudhury 1970, 1971a). In zoea I of these species, pleopods are present, although rudimentary and nonsegmented in some. All pereopods except the fifth are biramous, and the first two have non-functional chelae. Pleopods exist as small buds. In zoea II the antennal flagellum is multi-articulated and the pereopods bear well-developed exopods (except the fifth which remains uniramous), and the pleopods are now elongated and biramous. Zoea III is easily recognized by the appearance of uropods, which were visible beneath the telsonal cuticle in zoea II. Abbreviation is carried to the extreme in *M.hendersodayanum*, which passes through a single zoeal stage (Jalihal & Sankolli 1975). Abbreviated developmental sequences are also found in *Palaemonetes*, where zoeae hatch in an advanced condition and complete larval development occurs in shorter time and passes through fewer stages than most other members of the

Table 3. Comparison of morphological characteristics between advanced zoeae of an abbreviated development sequence and typical zoeae of longer-developing species (after Rabalais & Cameron 1983). Roman numerals represent zoeal stages.

Characteristics	Advanced zoeae abbreviated development (e.g., *Uca subcylindrica*)	Typical zoeae longer-developing species (e.g., *Uca* species)
Eyes	Sessile in I, stalked in II	Sessile in I, stalked in II
Carapacial spines	Dorsal and rostral reduced	Dorsal and rostral elongate
Antenna: endopod	2x protopodal spine in I and II	Present as bud in III, cf. ¾ of protopodal spine in IV, 1.5x protopodal spine in V
Mandible: mandibular palp	Fully-formed in I and II	As bud in megalopae
Maxilla: scaphognathite	Later stage of type 8 (Van Dover et al. 1982); 25-30 setae in I and II	Type 7 series (Van Dover et al. 1982); 4 setae in I, 23-27 setae in V
Natatory setae on maxillipedal exopods	4 in I, 24-26 in II	4, 6, 8, 10, 10-12 in I-V
Third maxilliped	Well formed in I and II	Absent until II as bud, finger-shaped in V
Accessory setae on feeding appendages	Reduced or absent in I and II	Numerous and fully-formed in I-V
Pereopods	Fully-formed and segmented but non-functional in I and II	Buds of 4 pairs in III, buds of 5 pairs in IV, finger-shaped in V
Gills	4 pairs well formed but undifferentiated and buds of 2 pairs in I and II	2 or 3 pairs of gill buds in V
Pleopods	Fully-formed and segmented with endopods in I and II	Buds in III, more prominent in IV, finger-shaped with indistinct segments in V
Abdomen: somites	6 in I and II	5 in I-III, 6 in IV and V
Lateral projections	Absent	Present on 2 and 3
Telsonal processes	Most often 6 + 6 in I and II	Most often 3 + 3 in I and II, 4 + 4 in III, 4 + 4 in IV, 4 + 4 or 6 + 6 in V

genus (e.g., *P.paludosus* in Dobkin (1963), *P.sinensis* described by Shen (1939), and *P.antennarius* as described by Mayer (1881)). These species have three zoeal stages with development of morphological features similar to that described above for *Macrobrachium*.

Advanced development also occurs in the Alpheidae, and several authors have reported advanced larvae in this group (e.g. *Racilius compressus*, Bruce 1974; *Synalpheus goodei*, Gurney 1949, as well as *Alpheus heterochaelis* and others reviewed by Knowlton 1973). Here again the zoea I hatches in an advanced condition with all the appendages of the adult in rudimentary form, similar to those described for the *Macrobrachium* with advanced development, followed by two more zoeal stages before molt to the postlarva. On the other hand, Bourdillon-Casanova (1960) demonstrated that *Synalpheus laevimanus* hatched with all its appendages present, but the larvae passed through four zoeal stages to complete development. This number of stages is still shortened compared to most alpheids, which develop through a series of 9 to 13 zoeal stages (Knowlton 1973).

The Pandalidae also have advanced development in some species (e.g. Haynes 1976, 1980). In *Pandalus hypsinotus* advanced characters are seen in the loss of thoracic exopods on the last three pairs of pereopods and in the presence of a proximal extension in the maxilla. In other less advanced larvae, thoracic exopods are usually found on all pereopods and the proximal maxillary extension is not seen in stage I. Yet *P.hypsinotus* retains six zoeal stages in its development, although the Japanese and northeastern Pacific forms differ in the heterochronic appearance of some of the larval features. Pike & Williamson (1961, 1964) and Rothlisberg (1980) noted the tendency of fewer larval stages in the Pandalidae to be accompanied by an increase in the number of telsonal spines and a reduction in the number of thoracic exopods, but this is not necessarily the case with some pandalids (e.g. *Chlorotocella* with a normal telson, Pike & Williamson 1964, and a full complement of thoracic exopods in some pandalids, Haynes 1980). In the Hippolytidae, *Lebbeus* was dissected from the egg, described by Krøyer (1842) and later reported by Pike & Williamson (1961) to have larvae in a very advanced condition, with all appendages except the uropods present (and segmented). The larvae also lack exopods on any of the pereopods, and have 8 + 8 telson spines. Larval development consists of two, or perhaps three, stages. Such a complete suppression of exopods on the pereopods of zoeal stages is rare in the Caridea but is known in *Cryptocheles* and *Bythocaris* (Hippolytidae) and in *Sclerocrangon* (Crangonidae) (Sars 1890, 1912 in Gurney 1942). These differ from *Lebbeus* in that 'larvae' are not pelagically free-swimming.

Makarov (1968) considered the question of advanced versus direct development in deep-sea species, especially the Crangonidae. He made the point that the production of highly advanced larvae in *Sclerocrangon*, for example, is not direct development as it has sometimes been called, because the stages that hatch are still unquestionably larval, even though they cling to the parental pleopods. They exhibit no direct developmental stage equivalent to that seen in the freshwater potamonid crabs. Makarov showed that several genera (*Sabinea, Nectocrangon, Lebbeus, Chorismus,* and *Notocrangon*) all exhibited abbreviated cycles with advanced zoeal stages. In *Notocrangron* (Makarov 1973), for example, pleopods were present (advanced) but the eyes otherwise remained fused, a mandibular palp was absent, the pereopods were more or less embryonic, and the uropod was not yet separated from the telson in stage I (regular characteristics). With subsequent stages the more advanced characters appeared. Makarov concluded that two stages were eliminated (not skipped) in comparison to other crangonids during the evolutionary shortening of the life cycle. In another crangonid, *Pontocaris pennata,* three zoeal stages were described (Sankolli & Shenoy 1976), with the first hatching in an advanced stage, bearing more than the usual four natatory setae on the maxillipedal exopods, but with sessile eyes as is usually the case in stage I (see section 3.2). Five abdominal somites were retained throughout development (an advanced character according to Rice 1980), and the uropods and antennal flagella developed quickly in the later stages; however, the pereopods remained relatively undeveloped.

Dobkin (1965b) described the larvae of two forms of *Glyphocrangon spinicauda,* which produce larvae in a very advanced condition. The two forms appear to be similarly advanced, except that form B is larger, darker blue-green with more orange chromatophores, without spines on the posterolateral edge of abdominal somites 2-5 and with 11 natatory setae instead of 9 as in form A. Both forms are vigorous swimmers (i.e. planktonic) but appear demersal. It is probable that these and larval stages of other deep-sea species exist in close proximity to the bottom. Several other caridean shrimp have been reported by Dobkin

(1969) to have abbreviated development, based on large egg size and a reported low number of zoeal stages; many of these developmental sequences are not fully detailed and require further analyses.

In the Nephropidae, *Homarus americanus* and *Nephrops norvegicus* regularly hatch in an advanced stage but still possess a total of three zoeal stages (Gurney 1942). In *Nephrops norvegicus* stage I there is an unsegmented mandibular palp, five pairs of biramous pereopods present, and legs 1-3 chelate, but no pleopods or uropods until stage II. This development is strikingly similar to those carideans described above with advanced development. In addition, Wear (1976) described the first zoeal larva of *Metanephrops challengeri,* which is even more advanced than that of *Nephrops norvegicus* but similar to its zoea III, with biramous pleopods on abdominal segments 2-5. The mouthparts have accessory setae which are short, fine, or few in number. All were weak swimmers in this hatching and died before molt to the next stage. They were presumed postlarvae, based on natatory pleopods and loss of thoracic setae. The zoeal stage was estimated to last four days or less.

The Thalassinidea display a wide range of developmental sequences. Most upogebiids and callianassids develop through four or five zoeal stages (Webb 1919, Sandifer 1973a,b, Rodrigues 1976, Ngoc-Ho 1977). The abbreviated development of some species described by Sandifer (1973a), Ngoc-Ho (1977) and Gurney (1938a) fall into the category of 'advanced', as opposed to those of 'regular', with four or five zoeal stages. For example, Gurney (1938a) stated that the zoea I of *Upogebia* D.I. was as advanced as the last stage IV of *U.deltaura* and that it was probable that the first molt would give rise to the postlarva, as it does in *U.savignyi.* In zoea I of *Upogebia darwinii* (Ngoc-Ho 1977) the pereopods are present but not as well formed as in zoea III. Exopods are present as buds on legs 1-3, and pleopods are present as buds on abdominal somites 2 and 3 and subcuticular on somites 4 and 5. In zoea II pereopods 1-5 are more developed and the pleopod buds are more elongate. In zoea III segmentation of pereopods and pleopods is more distinct, and the mandibular palp is present as a bud. There is a regular progression of development of mouthparts and sensory appendages. In this developmental sequence there is actually a compression into three stages that which is normally four or five stages. Similarly, *Callianassa* sp. B (= *C.atlantica* Rathbun?) develops through three larval stages (Sandifer 1973a). The larvae of *Callianassa kraussi* Stebbing have lost the planktonic phase, even though passing through two zoeal stages (Forbes 1973). The larvae do not swim, but lie on their backs in the parent's burrow water until metamorphosing 3-5 days after hatch, whereupon they dig into the wall of the parent burrow and eventually emerge at the surface 2-3 months later. This is another instance of a situation where development is advanced (but non-planktonic) to nearly direct.

Many axiids exhibit abbreviated development of the advanced type. *Axius stirhynchus* (Leach) passes through two zoeal stages (Gurney 1938b). Gurney also examined larvae of *A.plectrorhynchus* dissected from the egg and found them to be similar to those of *A.stirhynchus* in developmental state. The North Atlantic *Calocaris macandreae* hatches in an even more advanced state than *Axius stirhynchus* (Gurney 1938b) but still is suspected to have only two zoeal stages. In *Eiconaxius parvus,* the young leave the egg in almost the adult form. Gurney (1938b) proposed that most axiids and their allies would have abbreviated development but cited cases (e.g. *Calocaris alcocki*) where development may be regular, with the young hatching as 'normal larva', presumably molting through more than two stages.

The anomuran crustaceans are well known for having abbreviated larval development.

In the lithodids, *Paralomis granulosa,* the zoea I hatches in an advanced condition and passes through a second zoeal stage with slight refinement of the characters. It then metamorphoses to the megalopa (Campodonico & Guzman 1981). *Lithodes maja* is the only other lithodid known with two zoeal stages (MacDonald et al. 1957). Although it does not hatch in as advanced a condition as *Paralomis granulosa,* the zoea I does possess mobile eyes, a mandibular palp, and pleopods and pereopods which are slightly less developed; the latter are not segmented. Other lithodids develop through three or four zoeal stages (Campodonico & Guzman 1981).

Several hermit crabs exhibit advanced development. In *Paguristes sericeus,* there are only two zoeal stages (instead of three as in other members of the genus). Rice & Provenzano (1965) described the zoea I in which the anterior pereopods are present as unsegmented buds. Similarly, with *Lithopagurus yucatanicus* (Provenzano 1968), the zoea I is advanced, since pleopods and pereopods are present as buds visible through the very delicate cuticle. Nayak & Kakati (1977) and Nayak (1981) found three zoeal stages in the hermit crabs *Diogenes diogenes* and *D.planimanus* as compared to the four or five zoeal stages of other *Diogenes* (see review in Nayak 1981). The zoea III is advanced in antennular, mandibular, pleopodal and telsonal characters over those of zoea III of other *Diogenes* with longer development.

In the galatheids, advanced development occurs in *Munidopsis tridentata,* and although it hatches in an advanced condition (Sars 1889, Samuelsen 1972), it is not in the ultimate or penultimate stage since only three zoeal stages occur (versus four or five for other galatheids (Gore 1979). The morphologically advanced characteristics of the first zoea are refined in the two successive stages, with some advanced characters not appearing until zoea III (e.g. six abdominal somites and marginally setose pleopods). Larvae in the related family Chirostylidae may be even more advanced than the galatheids (Pike & Wear 1969). For example, an undescribed *Gastroptychus,* and *Uroptychus* cf. *politus* hatch as zoea I with a full complement of pereopods, including the first chelae, biramous pleopods, six abdominal somites, and a mandibular palp.

In the 'anomurous Branchyura' (i.e. Dromiacea), larval development, as in some caridean families, is sporadic, ranging from direct (see section 3.1) to at least five or six larval stages. Advanced development has been noted in *Conchoecetes artificiosus* by Sankolli & Shenoy (1968) with the two zoeal stages and megalopa having many advanced features, such as the presence of pleopods in zoea I, absence of exopods on the legs, and uniramous uropods. *Hypoconcha sabulosa* (Lang & Young 1980) and *H.arcuata* (Kircher 1970) hatch in a more maturationally developed condition than other dromiids, but not to the degree seen in *Conchoecetes,* and pass through three zoeal stages. Rice et al (1970) noted intermediate conditions of *H.arcuata* between *Conchoecetes artificiosus* (see section 3.2) and *Dromidia* (six zoeal stages, Rice & Provenzano 1966) and *Dromia* (four or five zoeal stages, Rice et al. 1970) and stated that the 'advanced' larvae of the first two point to the fact that the Dromiacea bridge the gap between the more primitive dromiids and the anomurans. (We caution the use of 'primitive' and 'advanced' characters when dealing with abbreviated development, where morphology of larvae is modified and not related to phylogenies (see also Rice 1981a).) In *Petalomera wilsoni* (Fulton & Grant), the third zoea seen by Wear (1970a) beneath the culticle of zoea II near molting almost parallels that of *H.arcuata* in the degree of development (Wear 1977). Thus, there are zoeal stages in *P.wilsoni* similar to the *Hypoconcha* described above, and zoea III is probably the final stage (Wear 1977).

In the Branchyura, advanced development has been noted in several families. The larvae of *Cymonomus bathamae* (Wear & Batham 1975) emerge as typically advanced zoea similar to others of advanced type (i.e., mandibular palp, maxilla with scaphognathite broad and well developed with a fringe of 35 marginal setae; pereopods 1-5 present, well developed and segmented; first pereopods chelate; gill buds present, but not differentiated into lamellae; only five abdominal somites; with long pleopods on abdominal somites 2-4; setae of next stage of development visible beneath zoeal cuticle). Based on such characters seen in a zoea I close to molting, Wear & Batham suggested that the next stage was a megalopa, although one of the zoea I molted to a zoea II that was substantially unchanged except for six natatory setae and more protruding (but not stalked) eyes.

Some portunid crabs develop through three or four zoeal stages rather than the five, six or seven which is more common in the family (Costlow & Bookhout 1959, Kurata & Omi 1969, Roberts 1969, Bookhout & Costlow 1974, 1977, Kurata & Midorikowa 1975, Greenwood & Fielder 1980). Examples are three in *Portunus rubromarginata* (Greenwood & Fielder 1979), four in *P.pelagicus* (Kurata & Midorikawa 1975, Shinkarenko 1979) and three in *Thalamita danae* (Fielder & Greenwood 1979). There is an apparent compression of zoeal characters, such as a change in the number of natatory setae on the maxillipedal exopods (see section 4.2). Differing degrees of development occur at intermediate points, depending on the number of zoeal molts. Fielder & Greenwood (1979) termed these 'a more abbreviated life history'; the distinction according to placement in our continuum of abbreviated developmental sequences is 'advanced'.

A few xanthids, other than those with direct development (see sections 3.1.1 and 3.1.3), have less than the usual four zoeal stages (see Wear 1970b and Andryszak & Gore 1981 for reviews). In *Heterozius rotundifrons* (if it remains classified as a xanthid) and *Epixanthus dentatus* (White), there are two zoeal stages (Wear 1968, Saba et al. 1978). There are some morphological attributes in zoea I usually seen in a later zoea. The third maxilliped is biramous but short and unsegmented and without setae, the pereopods are present as short uniramous unsegmented buds (first pereopod with a rudimentary chela at its tip), and pleopod buds are present on the abdominal somites. In *Pilodius nigrocrinitus* Stimpson, there are three zoeal stages with some characteristics normally appearing in zoea III in closely related species appearing in zoea II (Terada 1982). Similar examples are found in the Calappidae (Hong 1976).

The larval history of *Pinnotheres moseri,* a commensal with the ascidian, *Ascidia nigra,* is also abbreviated (Goodbody 1960). The female liberates an advanced first zoea which possesses well-developed natatory setae. The swimming behavior of this short-lived zoea, however, is not described. Within a day, there is metamorphosis to the megalopa. *P.pinnotheres* developes through two zoeal stages (Lebour 1928a). This is also the case with *P.taylori* (Hart 1935) with two zoeal stages, compared to the three, four or five zoeal stages in most other described pinnotherids (Table 1), with the exception of those with direct development (section 3.1.1).

In Jamaica, there are five endemic species of grapsid crabs which never enter sea water (Hartnoll 1964a, 1971). At least three of these species exhibit abbreviated development of the advanced type (Hartnoll 1964a, Abele & Means 1977). *Metopaulius depressus,* an inhabitant of tanks (leaf axils) of large bromeliads, hatches as an advanced zoea lasting one day before molt to a second zoea which is very similar. Within two more days the second zoea has metamorphosed to the megalopa and by day nine or ten the first crab is present. There is evidence that the eggs hatch and develop within the water of the bromeliad. The megalopa exhibits advanced features, but with non-natatory pleopods and apparent confinement to a benthic existence. *Sesarma bidentatum,* an inhabitant of mountain freshwater streams and rivers, produces an advanced zoea I similar to that of *M.depressus* (Hartnoll 1964a). *S.jarvisi* inhabits limestone talus and rock rubble substrates on mountain slopes (Abele & Means 1977). The first zoea of this species (Abele & Means 1977) are also similar to those of *M.depressus* (Hartnoll 1964a). The subsequent stages of *S.bidentatum* and *S.jarvisi* were not obtained, but, based on the known characters, the abbreviated development is probably similar to *M.depressus.* Another non-marine grapsid with advanced development is *Geosesarma perracae* which hatches as an advanced zoea I, followed by a zoea II exhibiting insignificant changes other than in the number of natatory setae on the maxillipedal exopods, and then followed by an advanced megalopa with undeveloped pleopods. Other terrestrial and amphibious *Geosesarma* are suspected of having abbreviated development (Soh 1969). *Sesarma reticulatum* which has three zoeal stages (Costlow & Bookhout 1962), as compared to *S.cinereum* with four zoeal stages (Costlow & Bookhout 1960), as well as most other grapsids with four or five zoeal stages (Table 2), except for those described above, exhibits advanced development. In *S.reticulatum,* the eggs and three zoeal stages are larger than those of *S.cinereum.* The second zoea of *S.reticulatum* is similar to the third zoea of *S.cinereum,* and the third zoea is similar to the fourth zoea, thus indicating an eliminated stage. There is no variability in the number of stages (Costlow & Bookhout 1962), so that there are consistently fewer stages and not a periodically skipped stage as in *Micropanope barbadensis* (Gore et al. 1981).

Only one member of the Ocypodidae has been definitely recorded with abbreviated development (Thurman 1979; Rabalais & Cameron 1981). (Although Bate (1879) described what was obviously an advanced zoea I and attributed it to *Gelassimus (= Uca),* the identification of the specimen was not accompanied by sufficient information to validate it.) Rabalais & Cameron (1983) described the development of *Uca subcyclindrica,* an inhabitant of semi-arid habitats in the extreme western coastal Gulf of Mexico. This species hatches as an advanced zoea; within a half day the zoea I molts to an equally advanced zoea II which differs substantially only in the number of natatory setae and in refinement of some of the appendages. Within two more days, advanced megalopae (especially in regard to the feeding appendages) are present, and within seven to nine days of hatch, first crabs are present.

Raja Bai Naidu (1951) inferred that the semi-terrestrial ocypodid crab, *Ocypode platytarsis* produces larval stages directly in the burrows where sufficient water occurs to allow development. She noted that no berried females were ever seen liberating zoeae near the water's edge, that the first zoeal stages were advanced because of shortened carapacial spines, and that a reduced planktonic period would be natural in a form that is becoming progressively terrestrial. Pursuing this line of reasoning, she recorded finding a megalopal stage of another ocypodid crab, *Dotilla blandorfi,* in a burrow exposed at low tide (Rajabai 1960). But, without actual evidence, abbreviated development for these ocypodids must remain unsubstantiated. For example, the burrow was also a suspected nursery area for *Uca subcylindrica* (Rabalais & Cameron 1981), but, to date, no larval stages have been found there (Rabalais unpublished data). *Uca subcylindrica* larvae are sometimes found in conjunction with burrow openings filled with standing water, when the burrows were associated with depressions in the sediment which collected rainfall. Very dry weather in South Texas in 1982 (76 days without substantial rainfall between early June and mid-August) provided a perfect opportunity to examine burrows for larval stages, but none were found. Indications are that *Uca subcylindrica* will forestall hatching until it rains. Females with eggs were seldom seen out of the burrow until then, remaining under a plugged entrance.

Rabalais (unpublished data) never found evidence of larval release except immediately following a rain. In a severe drought, the burrow water remains a possible nursery area, but there is almost always some rainfall throughout the year, although infrequent and sporadic at best. We would predict that the casual observations of Rajabai may not have fully documented the precise conditions present and that the zoeal form described was an aberrant, prematurely hatched stage. A discussion of terrestrial and semi-terrestrial species and abbreviated development is given in section 5.5.

There are two families which regularly undergo a development of two zoeal and one megalopal stage − 1) all described Majidae (most recently reviewed by Ingle 1979 and Rice 1980), except *Paranaxia* (see section 3.1.1); and 2) all described Porcellanidae (many authors), except *Petrocheles spinosus* with five zoeal stages (Wear 1965). In comparison to other brachyurans and anomurans with a more extended development, these two families obviously exhibit abbreviated development; however, any attempts to place them within a specific category is fruitless because of the variability exhibited in the many species represented. Rather, the two families, especially the Majidae, are good examples of the continuum concept of decapod developmental sequences. Zoea I of majids show a precocious development of certain characteristics compared with the first two zoea of other Brachyura (Hartnoll 1964b). In fact some, for example *Chlorinoides longispinus,* hatch as an advanced zoea (Terada 1981) with most, if not all, of the characteristics ascribed to them. Often majids are lecithotrophic with large, yolky eggs, non-feeding behavior, and rudimentary mouthparts (Kurata 1969, Provenzano & Brownell 1977, Terada 1981) (see section 5.3.1). Other majid zoea I do not possess as many developmentally mature characteristics at hatch. For instance, some zoea I lack a mandibular palp while others have it, pereopods may be present only as buds rather than fully developed as in others, pleopods may not be present even as buds, the scaphognathite blade of the maxilla progresses through a developmental change (Van Dover et al. 1982), rather than being of the form seen in most advanced zoeae, and usually definition of the sixth abdominal somite does not occur until zoea II. In the porcellanids more maturationally advanced characteristics, such as pereopods, pleopods, and mandibular palps, are usually absent until zoea II, although some of these morphological features may be present as primordia.

3.3 *One more category (= postlarva)*

There is yet another type of abbreviated developmental sequence, based on the postlarva. The postlarval stage may appear postembryonically as an imago of the adult and has already been discussed in the section on direct development (3.1). When the megalopa is advanced, a metamorphic stage closer to the first juvenile crab occurs. For example, the megalopae of *Metopaulius depressus* described by Hartnoll (1964a) are advanced with rudimentary pleopods and confined to a benthic existence. Those of *Geosesarma perracae* also have rudimentary pleopods (Soh 1969), but this form was considered larval although it has a crab-like form, due to the 'larval' type of telson and the nature of antennule and antenna (no longer anterior in the first crab). In *Corystes cassivelaunus,* Lebour (1928b) recognized a megalopal stage that was very large and far more advanced and crab-like than megalopa-like. The megalopa of *Uca subcylindrica* is also more advanced than others of the genus described, based on the advanced feeding appendages which are more similar to the first crabs of other *Uca* (Rabalais & Cameron 1983). Thus the megalopa exists in a condition capable of benthic existence, with well-equipped feeding appendages, even though the pleo-

pods bear more natatory setae than those of other *Uca,* a feature probably related to flooding typical of its habitat. In fact, *Uca subcylindrica* megalopae have been found in field
populations on moist sediments devoid of any standing water; when these were placed in
water, they actively swam (Rabalais & Cameron 1983). This points to the advantage of
retainment of natatory pleopods in megalopae that remain associated with parental females
(e.g. *Cryptodromia octodentata* and *Petalomera lateralis* in section 3.1.1), when that swimming ability may be necessary. The typical megalopal stage is abbreviated in *Trigonoplax
unguiformis* (Fukuda 1981), and natatory pleopods are absent. However, the mouthparts
bear many progressive and retrogressive characters typical of the megalopae, and the abdomen is shown to be intermediate between the last zoea and the first crab. Williams (1980)
described the 'crablike megalopae' of *Bythograea thermydron,* which have megalopal eyes
and abdominal characters but resemble crab stages in general body shape.

The Hymenosomatidae have an abbreviated developmental sequence by virtue of the
aspect of complete elimination of the megalopal stage (Al-Kholy 1959b, Broekhuysen 1955,
Lucas 1971, Muraoka 1977a, Rice 1981b). A first crab instar follows the last zoeal stage,
except in *Halicarcinus lacustris,* which has direct development. Although Lucas stated that
this may be the modification of a megalopa to a crab-like condition, rather than the omission of a developmental stage, the telson is crab-like, along with other morphological characteristics. Rice (1981b) conjectured that the necessary morphological and behavioral changes
between the zoea and juvenile crab stages may have been too extensive to be accomplished
initially through a somewhat unsatisfactory intermediate phase. In the members of this
family, absence of the megalopa is accompanied by zoeal stages without pleopods or buds
even in the last stage. In any event, crab stages or crab-like stages are reached more quickly.
Other authors have postulated an eliminated megalopal stage, because the resultant morphotype from the ultimate zoeal molt bore such a close resemblance to the adult of the
species; however, most of these cases have been refuted (see Gore, this volume). Further
treatment of postlarval development will be given by Felder et al. (this volume).

4 ACCELERATED DEVELOPMENT

Gore (this volume) classified accelerated development under the main heading of irregular
development because it appears in response to operating environmental influences and is
not necessarily a consistent part of any given development. The definition is worth repeat-
ing. In accelerated development the larval series in individuals within a species, or species
among closely related forms, becomes shorter than their regular counterparts, owing to
the direct elimination, probably as a consequence of environmental adaptation, of one or
more otherwise regularly occurring stages within the ontogenetic sequence. In this respect,
reduction and stage elimination is intermittent and does not always occur within all individuals in a series. Thus, the elimination of, say, a postlarval stage such as the megalopa in
the Hymenosomatidae is a form of abbreviated development under the main heading of
regular development, whereas the occasional elimination of intermediate zoeal stages (as in
some Palaemonidae) is a type of accelerated development if the resultant larval duration is
shortened.

As Gore noted, there is often a fine line between advanced and accelerated development,
but the criterion of intermittency employed as defined above should allow distinction. For
example, if most grapsid crabs have four or five zoeal stages, but *Sesarma reticulatum* has

only three, consistently (Costlow & Bookhout 1962), then its development is advanced and not accelerated. On the other hand, if a larval series in *Micropanope barbadensis* occasionally skips a stage, for whatever genetic or environmental reason, thus differing from a normal sequence of four zoeal stages as exhibited by its specific counterparts, then individuals displaying this skipped staging are undergoing accelerated development (Gore et al. 1981). It is easily visualized that the skipping or acceleration of development will produce an advanced type of later zoea, but the one does not necessarily follow the other. Gore (this volume) has provided several subcategories within the framework of accelerated development including skipped, combinatorial, and precocious staging. For our purposes, we will consider some additional examples.

4.1 *Acceleration in the larval sequence within a species*

The presence of variation in the number of larval stages and duration of the larval sequence within a species due to environmental factors, laboratory-induced factors and inherent variability have been well documented in the literature and elsewhere in this volume. However, certain points relative to shortening of developmental sequences are pertinent to our discussions and are included here.

The once widely held assumption that decapods generally pass through a fixed number of larval stages was replaced by the idea that instances of variability are less frequently encountered in Macrura and Anomura and rarely in Brachyura (Knowlton 1974). This idea was based on the fact that the shorter larval phase of these groups, compared to the Natantia, produced a fixed number of molts which were usually morphologically distinct, comparable to the early stages in natant development, and that, in general, the degree of larval variation is an index of primitiveness. Yet, with more and more exhaustive studies, usually based on laboratory-reared larvae, we have seen variability in larval sequencing in numerous species. Fincham (1979a) suggested that the size and morphology in early larval stages are largely predictable; however, in late larval development there is some independence of the developmental processes of molting, growth, and morphogenesis (see also Broad & Hubschman 1962, 1963, Broad 1957a, Gore, this volume). Situations such as these create variability in both plankton and laboratory-reared material. Provenzano et al. (1978) and Sandifer & Smith (1979) stated that genetic or other quality variations between individual broods of larvae may also account for significant variation in larval duration and/or in survival rates.

Dobkin (1963) and Fielder (1970) noted the lack of variability in the larval stages of caridean shrimp with abbreviated development, whereas those with prolonged larval life exhibit considerable variation in the number of stages, size, and morphological complexity (Broad 1957a,b, Knowlton 1965, Ewald 1969, Hubschman & Rose 1969, Choudhury 1971a, Fincham 1977, 1979b). Gore, citing several supportive examples, suggests (this volume) that variation may be stage-restricted and unable to occur before a given maturational state. Other factors, however, may also be operating. For example, Dobkin (1963) attributed the lack of variation in abbreviated development sequences in caridean shrimp to the presence of large internal food reserves and lack of variability in diet.

One species with abbreviated development and related large yolk reserves has been found to exhibit variability, although rarely, in the development sequence (Rabalais & Cameron 1983). An additional molt (which did not differ in morphology) at the end of the larval sequence between zoea II and the megalopa was present in a few cultures of laboratory-reared *Uca subcylindrica*. There were also indications from larvae collected in the field that this may have also been the case in natural populations (Rabalais, unpublished data). This additional molt is not a consistently occurring stage and certainly does not represent an abbreviation, either ecdysially or durationally, but does point out that variation can occur within a species with an abbreviated development sequence of two zoeal stages and associated large yolk reserves.

That the larval development time decreases as temperature increases has been consistently observed for different species (e.g. Rice & Provenzano 1966, Sandifer 1973c, Andryszak & Gore 1981, Gore et al. 1981). In *Galathea rostrata*, where the days required to attain the next stage were less in 20°C than in 15°C, overall mortality was higher; on the other hand, more megalopae were attained at 20°C and succeeded to first crab (Gore 1979). Even though larval duration is already abbreviated, *Uca subcylindrica* metamorphoses faster at higher temperatures (Rabalais & Cameron 1982).

Duration of larval development may also be reduced by skipped staging, in which one or more instars may be bypassed by a species as it proceeds towards metamorphosis. Skipped staging is probably the most common type of accelerated development. Often there is little consistency in whether a stage will be deleted or not, and environmental factors seem to play an important role in this respect. For example, extremes of salinity, if not lethal, sometimes prolong duration of larval stages (e.g. Ong & Costlow 1970, Sandifer 1973c, Huni 1979), even in abbreviated developmental series (Rabalais unpublished data); however, in one instance, seven individuals of *Hypoconcha arcuata* developed through only two zoeal stages instead of the normal three when raised in 40‰ (Kircher 1970). The results of all the developmental sequences indicated response to some form of stress; those raised in 20 and 40‰ only went to megalopa, whereas those raised in 25, 30 and 35‰ went to first crab.

In *Galathea rostrata* some zoea III molted to an advanced stage IV with some but not all of the features of stage V (Gore 1979). These molted directly to megalopa and eventually to first crab. This resulted in a shortening of the duration of the larval period, allowing earlier postlarval metamorphosis. In *Micropanope barbadensis,* there may be an elimination of the terminal stage with a developmental sequence of three zoeae instead of four (Gore et al. 1981). However, in this case, there was not a shortened duration associated with the skipping of a stage. In fact, the duration was slightly longer than that of the four zoeae. Skipped staging occurs most often in those groups with prolonged larval duration of many stages such as is seen in the Caridea and Scyllaridea; however, larval variability in the Macrura, Anomura, and Brachyura is not as uncommon as it was once considered (see Gore, section 3.3.1.1). In another type of staging, combinatorial, an instar exhibits a combination of characters from the succeeding (or preceding) instar, so that the resulting stage may appear more (or less) advanced than the normally numeric stage. Costlow (1965) provided good examples of combinatorial staging.

4.2 *Accelerated development and evolution*

The evolutionary evidence supporting accelerated development in decapod larvae is widespread and consists primarily of comparison of general ontogeny within a genus, variation of characters intraspecifically and intragenerically, and non-sequential appearance of characters in the instar series. Each of these comparisons relate to what is considered regular development as it occurs today. Obviously, what is regular at present might be considered advanced if the ontogeny for the particular species was known, say, during the Pliocene.

Comparison of general ontogeny within a genus is usually the easiest and most obvious means of obtaining evidence regarding acceleration. The differences in the number of naupliar stages in the Penaeidea suggest that some genera have dropped one or more naupliar instars, although the possibility that additional instars have been added cannot be completely excluded (see Fielder et al. 1975). Lebour (1934) found five zoeal stages in a *Dromia* from the North Sea, whereas Al-Kholy (1959a) found only two in a congener from the Red Sea. In the genus *Galathea*, Lebour (1930) noted four or five stages (as did Gore 1979) but Al-Kholy (1959a) listed another having three. In the Pinnotheridae, the genus *Pinnotheres* may exhibit 0-5 zoeal stages, and *Pinnaxodes* can have four or five (Table 1). As before, comparison with what transpires in the majority of taxa suggests that the abbreviated forms have evolved via accelerated development and evolutionary fixation of the shortened cycle.

Variation in appearance of characters has already been discussed (Gore, this volume) within larvae of a species, but this approach has value across generic lines as well. In *Cancer*, for example, the sixth abdominal somite appears by stage V in all species except *C.productus* and *C.magister*, and the retention of five somites throughout development is considered an advanced character (Rice 1980). In the Anomura, Nayak & Kakati (1977) found three, four or five zoeal stages in species of the hermit crab *Diogenes*, and considered the third stage in *D.diogenes* advanced in antennular, mandibular, pleopodal and telsonal characters over those of *D.bicristimanus, D.pugilator* and *D.avarus*. If it can be assumed that four zoeal stages is the regular developmental mode in *Diogenes*, then those individuals exhibiting three or five stages are examples of either a skipped, or a terminally additive type of staging. A similar type of response seems to occur in the hippolytid generic complex *Spirontocaris* (sensu lato), where the larval development ranges from two to nine stages among the species. Some forms hatch with nearly all appendages well developed, but others have no trace of appendages posterior to the maxillipeds (Haynes 1981).

A third line of evidence is seen in the retention of 'out of sequence' characters within species of a genus. Greenwood & Fielder (1979) showed that *Portunus rubromarginatus* has a relatively short development comprising only three zoeal stages and a megalopa, compared to other portunids (see section 3.2.1). The number of maxillipedal setae in the respective stages of *P.rubromarginatus* suggests that intermediate stages have been evolutionarily compressed or dropped. Setal progression does not follow the usual sequence of '2's' seen on the maxillipedal exopodites of other brachyurans where, in a development of five zoeal stages, the progression is 4→6→8→10→12 natatory setae. Instead, in *P.rubromarginatus* the formula is 4→8→12 in the larval instars, implying that a previously existing stage II and IV have been eliminated, leaving what at one time was stage III and V shifted backward toward zoea I. Kurata & Midorikawa (1975) found a similar occurrence in *Portunus pelagicus* which has four zoeal stages. Perkins (1973) stated that in *Geryon quinquedens* the setal progression on the maxillipedal exopods from stages I-IV was 4→10-11→14→17-19, thus suggesting that an earlier zoea II and III may have been condensed so that the numerically present second stage seems to be morphologically equivalent to a previously existing third or fourth stage; in this case the last zoeal stage (IV) may have been compressed backward from a zoea VIII. In yet another interesting case Shenoy & Sankolli (1967) showed that the maxillipedal exopod setae in the two zoeal stages of the porcellanid crab *Petrolisthes boscii* progressed from four to 16 in one molt, suggesting that at least five and perhaps even seven zoeal stages were at one time present in the ancestors, and that the intermediate stages II-VI were subsequently condensed into the ontogenetic sequence seen today. Wear

(1965) had previously demonstrated that another porcellanid crab, *Petrocheles spinosus,* has five instead of the usual two zoeal stages and bore 22 or 24 exopodal setae in the fifth instar. This evidence is admittedly circumstantial and is based on an allegedly general 'rule' first formulated by Lebour (1928b). There are sufficient exceptions (e.g. in the two developmental stages of most majid crabs) to show that what is occurring is neither simple nor even analogous across taxonomic lines. As further information becomes available, especially from those taxa which may skip, or undergo combinatorial staging, the picture may become clearer.

5 CORRELATES (OR NOT) OF ABBREVIATED DEVELOPMENT

As seen in preceding sections, certain characteristics associated with abbreviated development keep surfacing. These features should not, however, be considered fail safe clues that abbreviation occurs in the development of an individual, or, perhaps worse, indicate cause-and-effect relationships. For instance, a misconception replete in the literature is that larger eggs point to a reduction in free larval life. While most species which exhibit abbreviated development produce large eggs, the converse is not necessarily true. Similarly, statements along the lines of 'abbreviated development is a response to . . .' or 'a need for . . . explains the abbreviation of larval life in . . .', while appealing, are teleological at best. In the following sections, we attack some of the myths perpetuated in the literature and stress instead those characteristics of abbreviated development which would be advantageous given a particular ecological situation.

5.1 *Number and size of eggs*

It was long a truism that species with large and few eggs could be expected to have shortened development, whereas those bearing numerous, small eggs underwent an extended period of ontogeny (Gurney 1942, Dobkin 1969, Herring 1974a). The relationship between abbreviated development and the number and size of eggs produced by decapod crustaceans was described by Gurney (1942) as a ratio between the diameter of the egg and the length of the adult. In species with abbreviated development, he found ratios of one-seventeenth or one-ninth compared to a ratio of one-hundredth in species where larval development pursues its normal course. Pike & Wear (1969) noted similar ratios of one-ninth to one-fifth for *Uroptychus* cf. *politus* and *Gastroptychus* n.sp. Such ratios, however, are not always appropriate in comparison of morphologically diverse taxa.

The fact that larger eggs may signify abbreviated development in a species is most evident among closely related species where some exhibit abbreviated development and others do not. For example, most members of the genus *Uca* produce small eggs (0.21-0.28, or 0.34-0.38 mm, diameter) (Gibbs 1974, Rabalais & Cameron 1983) and exhibit a regular developmental sequence of five zoeal stages where recorded (reviewed by Rabalais & Cameron 1983), compared to *Uca subcylindrica* with an abbreviated developmental sequence of two advanced zoeal stages and eggs that average 1.06 mm diameter (Figs.3, 4). Often related to larger egg size is the feature that females produce fewer of these eggs. Thus the number of eggs carried by any one female is quite variable, both at the individual and at the species level. Egg numbers in different species with abbreviated development patterns range from ten or less large-sized ova to several hundred (Wear 1967, 1968, Rabalais &

Figure 3. Comparative measurements of carapace width, egg size, and fecundity per brood for 4 fiddler crabs (*Uca*), 3 with longer developing larvae (*U.longisignalis, U.rapax,* and *U. panacea*) and one with abbreviated development (*U.subcylindrica*). Fecundity = total of unhatched eggs and/or live hatched zoeae and/or dead hatched zoeae. Bars on histograms represent ±1 s.e. n for each sample given at base of histogram (from Rabalais & Cameron 1983).

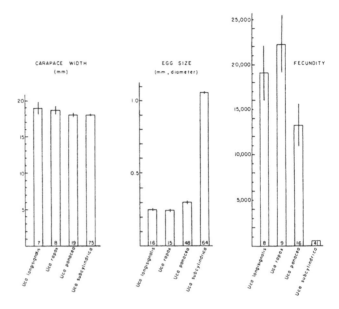

Figure 4. Log number of eggs per brood versus carapace length (mm) for several species of brachyuran crabs. Egg size in mm, diameter, given next to species, where known. Initials correspond to Hr – *Heterozius rotundifrons* (Wear 1968, Jones 1978); Mh – *Macrophthalmus hirtipes* (Simons & Jones 1981); Md – *Metopaulius depressus* (Hartnoll 1964a); Pl and Pn – *Pilumnus lumpinus* and *P.novaezealandiae* (Wear 1967); Ua, Uta and Utr – *Uca annulipes, U.tangeri* and *U.triangularis* (Feest 1969); Ub – *U.burgersi* (Gibbs 1974); Ul, Up, Ur and Us – *U.longisignalis, U. panacea, U.rapax* and *U.subcylindrica* (Rabalais & Cameron 1983); Um – *U.minax* (derived from Gray 1942, Hines 1982). Egg size for *Uca* given by Crane (1941) as 0.22-0.27 mm, diameter.

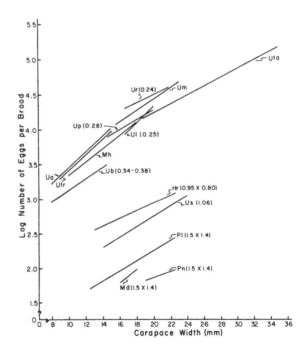

Cameron 1983), compared to the many eggs in species with longer periods of ontogeny (e.g. Graham & Beaven 1942, Yatsuzuka 1962). These differences in egg numbers are illustrated in Figures 3 and 4. (Bearing in mind the allometric constraints of female body size on reproductive 'effort' (Hines 1982), we have included species within a similar size range and general body form.)

As with the size of eggs, the difference between number of eggs in relation to abbreviation of development is most obvious in closely related species (e.g., *Uca subcylindrica* and *Uca* spp.). There is also a trend seen in these figures with decreasing number of eggs with increasing egg size. *Uca burgersi* with 0.34-0.38 mm eggs produces fewer eggs than the majority of *Uca* with 0.22-0.28 mm eggs (but note also the smaller size of the ovigerous females). Larger egg size in this case and as discussed below may be related to modifications which prevent desiccation. In another study, Wear (1968) suggested that the xanthid crab *Heterozius rotundifrons* was a good example of the correspondence between relatively large-sized (0.95 x 0.80 mm, at hatch), few (400-1 600) eggs and abbreviated development (see section 3.2.1), when it is compared with *Heteropanope serratifrons.* The latter xanthid species produces smaller (0.45 x 0.40 mm) and more numerous (2 000 or more) eggs and passes through four zoeal stages. However, the adaptive difference in egg numbers and sizes of either species (i.e., 1 600 vs. 2 000; 0.4 vs. 0.7 mm) seems insignificant, and as Elofsson (1961) has shown, egg size is not necessarily a consistent indicator of any developmental or maturational condition. Similarly, grapsids with abbreviated development produce large eggs (Hartnoll 1964a) compared to other members of the family which produce smaller eggs and pass through four or five zoeal stages (Hartnoll 1965).

Although the large egg size of some species gives a useful hint to the probability of abbreviated development, it is not necessarily a consistent predictor of any developmental or maturational condition. Often assumptions are made concerning abbreviated development based on egg sizes and are found to be incorrect; others remain unverified. For example, Barnard (1950) indicated that direct development without any free-swimming larval stages was associated with few and large eggs carried by a relatively small female *Cymonomus.* Wear & Batham (1970) showed that the developmental sequence of *Cymonomus bathamae,* which possesses few (11-26) eggs, 1.4 mm in diameter at hatch, is abbreviated of the advanced type rather than direct since there are one and probably two larval stages. In another instance, Gurney (1938b) examined larvae of *Axius plectrorhynchus* dissected from the egg and found them not to be any more abbreviated than *A. stirhynchus* which passes through two zoeal stages, thus negating Hale's (1927), report, based on egg size, that *A. plectrorhynchus* hatched in the form of the adult. Milne Edwards & Bouvier (1897) suggested that the large eggs of *Uroptychus nitidus* indicated hatching in the adult form, but Pike & Wear (1969) showed that the eggs of *Gastroptychus* n.sp. (1.8 x 1.7 mm) and *Uroptychus* cf. *politus* (1.4 x 1.2 mm) hatch as advanced zoeae and doubted that the relatively smaller eggs of *U. nitidus* (1.75 mm) and *U. concolor* hatch in the adult form. They felt rather that development is comparable to that described by them of at least one free-swimming larval stage. In the cases listed above, large eggs indicate some type of abbreviated development, although not always direct as suspected.

Throughout the literature are cases of suspected, but not substantiated, abbreviated development of some type based on species with large eggs (e.g. *Munidopsis polymorpha,* 1.8 mm eggs, Fage & Monod 1936; *Geosesarma noduliferum,* 2 mm eggs, Pesta 1930; and two dromiid crabs, 2 mm eggs, Hale 1925, 1927). Hartnoll (1964a, 1971) predicted abbreviated development similar to that of *Metopaulius depressus* (egg size 1.5 x 1.4 mm) and *Sesarma bidentatum* (egg size 1.8 x 1.7 mm) for *S. jarvisi, S. cookei* and *S. verleyi* based on 1+ mm eggs and/or enlarged gonopores of female crabs. Abele & Means (1977) verified that *S. jarvisi* hatched as a large advanced zoea from 1.32 mm eggs. They were unable to find ovigerous females of *S. cookei* but did measure 1.25 mm eggs in the ovaries of a mature female. Hartnoll (1964a) also predicted fewer number of stages for *S. curacaoense* with slightly larger eggs (0.6 x 0.55 mm), perhaps similar to the abbreviated larval development of *S.reticulatum* (see section 3.2.1) but this is not suggested by data on *Uca* given below.

Similar sized eggs does not always indicate similar developmental patterns; and vice versa, larger eggs do not always indicate abbreviation of developmental sequences. For example, *Petalomera wilsoni* (egg size 1.12 x 0.88 mm at hatch) passes through three zoeal stages in an advanced developmental pattern (Wear 1970a, 1977) compared to direct development in *P.lateralis* (egg size 1.14 x 1.00 mm) (Hale 1925). *Ozius truncatus,* a xanthid crab with eggs 0.7 x 0.65 mm, passes through four zoeal stages, compared to *Heteropanopeus serratifrons* with 0.45 x 0.40 mm eggs and four zoeal stages (Wear 1968). Many carideans with large eggs have abbreviated development, yet others, e.g. *Pandalus jordani* with eggs 1.4 x 0.88 mm at hatch, develop through 13 zoeal stages (Rothlisberg 1980). Gurney (1942) pointed out that the egg of *Chlorotocella* is relatively but little smaller (0.6 x 0.5 mm) than that of *Synalpheus laevimanus* (0.75 mm) but the latter has its development greatly abbreviated, and there is a complete series of larvae in a species of *Chlorotocella* described by Gurney (1937b). Elofsson (1961) showed that large eggs are not indicative of shortened development in the oceanic shrimp *Pasiphea. P.multidentata* has large eggs (2.4 x 1.8 mm), those of *P.tarda* are even larger, while those of *P.sivado* are the smallest of the three species (2.0 x 1.5 mm). Yet, larvae of all four pass through four zoeal stages compared to the five zoeal stages of other known pasiphaeid larvae (Williamson 1960). However, the zoeae do hatch in a relatively advanced condition, possess a mandibular palp, pleopods, segmented pereopods bearing exopods, and a relatively well-developed telson, with a large number of telsonal processes. They thus display features shown by many species in which development is abbreviated (Gurney 1942). Abele (1970) suggested abbreviated development in the semi-terrestrial shrimp, *Merguia rhizophorae,* based on large (0.9 mm) eggs, but this was later found not to be the case (Gilchrist et al. 1983).

Larger egg size may be related to factors other than abbreviation of development. For example, the eggs of *Uca panacea* are larger (0.28 mm) than those of *U.longisignalis* (0.24 mm) and *U.rapax* (0.25 mm) (Fig.3), and those of *U.burgersi* are even larger (0.34-0.38 mm). These size differences are likely to be related to the terrestrial and inland habitats of these two species (Powers 1975, Gibbs 1974) compared to more intertidally associated *Uca.* Increased volume and subsequent reduction of the surface to volume ratio would be adaptive as a deterrant to desiccation of the eggs. This is possibly a factor in *U. subcylindrica* as well (at least partially), where eggs average 1.06 mm and where the crabs occupy habitats not only inland and supratidal, but also climatically semiarid. Large eggs may also be advantageous to the semiferrestrial shrimp, *Merguia rhizophorae,* which seems to avoid water (Abele 1970). Storage of yolk for energy reserves results in larger eggs but is not necessarily accompanied by an abbreviation of development. For example, *Macrobrachium rosenbergii,* which hatches from medium-sized eggs, 0.55 x 0.57 mm (Uno & Kwon 1969), will molt through the first three zoeal stages without feeding, but eventually will complete 11 zoeal stages. Salinity may also affect egg size. Late stage eggs of *Macrophthalmus hirtipes* were significantly larger at 18 ‰ than at 36 ‰ salinity (Jones & Simons 1982) during the earlier parts of the breeding season. However, just the opposite was found by Jones (1980) in the crab *Helice crassa* Dana, where egg volume did not increase during development in those crabs from estuarine habitats. Those from marine habitats did increase, so that late stage eggs were significantly smaller under estuarine conditions than marine. Egg size may vary temporally so that eggs of *Macrophthalmus hirtipes* spawned during the latter part of the breeding season were significantly smaller than those laid earlier (Simons & Jones 1982). These size differences may be related to temperature, food resources available to female crabs, salinity, or a combination of any of these factors.

5.2 *Embryonic development*

Large eggs, as is the case with most species with abbreviated development, contain more yolk and thus a greater internal food supply. Larval life is concomitantly shortened and embryonic development is extended, especially where a more advanced stage is produced at hatching. These differences are especially notable in closely related species. For example, *Uca* spp. incubate eggs for 12-15 days (Feest 1969, von Hagen 1970, Ringold 1979, Christy 1982) compared to *Uca subcylindrica,* which is estimated to incubate eggs for one to one and a half months (Rabalais & Cameron 1983). However, the period of incubation for eggs for *U.subcylindrica* is considerably less than for other species with abbreviated development. For example, approximately four months are required for *Pilumnus novaezealandiae* and *P.lumpinus* (Wear 1967), 10-12 weeks for *Metopaulius depressus* (Hartnoll 1964a), five months for *Heterozius rotundifrons* (Wear 1968), 25-29 days for *Alpheus heterochaelis* (Knowlton 1973), and approximately 11 months for *Glyphocrangon* (Provenzano 1967). Yet, *Cambarus l.longulus* with direct development requires only three weeks for embryonic development (Smart 1962). On the other hand, species with smaller eggs such as *Lepidopa myops* (Knight 1970) and *Helice crassa* (Jones 1980) require fairly long incubation periods, 73-77 days and up to 90 days, respectively. Temperature is one parameter responsible for differential embryonic development. Knowlton (1970) stated that temperature acclimation may occur in eggs of *Alpheus heterochaelis* that could alter development time, lengthening it at 20°C or shortening it at 30°C. In *Uca subcylindrica* there is an indication, from collections of ovigerous females with various stages of developed eggs and from time to hatch of eggs on captive females in the laboratory, that the incubation period is longer in the spring (approximately 1.5 months) during low air temperatures than in late summer (approximately 1 month) during high air temperatures (Rabalais unpublished data).

Wear (1967) followed the embryonic development of two xanthid crabs with abbreviated development. He demonstrated that *Pilumnus novaezealandiae* passed through a nauplius, a metanauplius, and four zoeal stages embryonically without discrete ecdyses, but with clearly separate stages definable. The ultimate embryonic stage was a megalopa surrounded by five separate envelopes – the megalopal, zoeal, and prezoeal cuticles, plus the inner and outer egg membranes. Upon hatching, these were all shed by the megalopa, which was retained in the brood pouch of the female. *Pilumnus lumpinus* passed through a discrete nauplius and a two-phased metanauplius within the egg. The latter developed a prezoeal type of cuticle lacking the process sheath usually seen in other prezoeal stages. The egg hatched usually as an advanced zoea surrounded by a prezoeal cuticle, which was shed 20 minutes later with zoeal exoskeleton upon molt to the megalopa.

In species with abbreviated development and large eggs (where measurements are available), the rate of growth in the egg between deposition and eclosion is slightly greater than in eggs from species with longer developmental sequences (see section 4.1 and Table 4 in Gore, this volume). For example, the average increases in egg diameter, area, and volume for six families based on data listed in Table 4 (Gore) are 16, 43 and 65 % respectively. Comparable figures for five species with abbreviated development (Wear 1970a, 1967, 1968, Wear & Batham 1975) are 24, 56 and 84 %. However, when we consider that the egg size (as measured in mm, diameter) between those considered here with abbreviated development is two times greater than those with a longer developmental sequence, then the relative rate of growth for those with abbreviated development is not as great.

5.3 *Lecithotrophic larvae*

Associated with the larger eggs, longer incubation period, and advanced morphology of the larvae is the lecithotrophic nature of the larvae (i.e., hatched from large, yolky eggs upon which they depend for nutritional reserves, versus planktotrophic larvae which hatch from small eggs without an energy reserve of yolk and which depend on planktonic prey for energy). Larger eggs contain more yolk and thus a greater internal food supply. Herring (1974a) noted that the size/density/lipid relationship is an inevitable consequence of the development pattern of the species, but the converse may be equally true. Since lipid provides the main energy source for the embryonic development of decapod crustaceans (Pandian 1967, 1970, Pandian & Schumann 1967), an extension of embryonic development, as occurs in species with abbreviated development, requires increased lipid reserves in the egg. The larvae at hatch usually have substantial amounts of yolk still available to the early (or only) larval stages and may be characterized by relatively undeveloped mouthparts, especially in setation and reduced feeding ability. These characteristics have been noted in several species with abbreviated development, as well as in others without abbreviated development.

5.3.1 *Feeding behavior and related morphologies*

Samuelson (1972) noted that the large amount of yolk in all the zoeal stages of *Munidopsis tridentata* seems to be the sole source of food during development. The larvae lack setae on the feeding appendages and endopodites of natatory appendages and do not eat newly hatched *Artemia*. A similar situation was observed in *Metopaulius depressus* and *Uca subcylindrica* (Hartnoll 1964a, Rabalais & Cameron 1983) in which a large reserve of yolk is present in both advanced zoeal stages, part of which still remains in the megalopa. The zoeal stages do not feed and the mouthparts, although well formed and segmented, are not fully setose (Fig.5B). The megalopae of both species, on the other hand, do feed actively and possess mouthparts with setae associated with feeding (Rabalais & Cameron 1983).

The first zoeae of *Sesarma bidentatum* and *S.jarvisi* contain large quantities of yolk, and the feeding appendages are similar to species described above (Hartnoll 1964a, Abele & Means 1977). The amount of yolk and condition of the mouthparts in subsequent stages is not known. Wear & Batham (1970) described the large, yolky eggs of *Cymonomus bathamae* and the subsequent larvae with relatively non-setose mouthparts. Those setae which are present are sparse or barely protrude through the cuticle. The larval feeding behavior was not recorded.

Larvae of *Lithopagurus yucatanicus* hatch from large 2 mm eggs, apparently subsist on their very obvious yolk reserves, and pass through three zoeal stages without feeding (Provenzano 1968). Mouthparts are rudimentary in structure (the mandible was almost overlooked in zoea I) and non-setose. In *Geosesarma perracae* there is a large quantity of yolk present through the megalopal stage, and none of these stages feed (Soh 1969). The mouthparts of zoea I and zoea II are without setae; those of the megalopa possess some setation, but these setae are mostly short and rudimentary and not as in a normal megalopa. *Geosesarma perracae* will develop to the first crab instar without feeding. Yolk reserves in *Uca subcylindrica* are adequate for development to proceed to the megalopa without feeding,

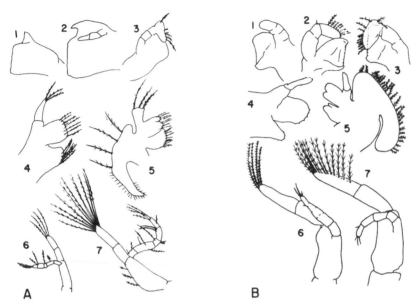

Figure 5. Mouthparts of (A) *Uca* spp. (after Hyman 1920), (1) mandible, zoea IV; (2) mandible, mega-lopa; (3) mandible, crab I; (4) maxillule, zoea V; (5) maxilla, zoea V; (6) first maxilliped, zoea I; (7) first maxilliped, zoea V. (B) *Uca subcylindrica* (from Rabalais & Cameron 1983), (1) mandible, zoea II; (2) mandible, megalopa; (3) mandible, crab I; (4) maxillule, zoea II; (5) maxilla, zoea II; (6) first maxilliped, zoea I; (7) first maxilliped, zoea II.

but remaining reserves in the megalopa are insufficient for molt to the first crab without food (Rabalais & Cameron 1983).

Degrees of lecithotrophy occur in many species and are often unrelated to abbreviation of development or morphological variations. Yolk storage in large eggs forms sufficient energy reserves in some species which have functional mouthparts and which do feed. Uno & Kwon (1969) found that development of *Macrobrachium rosenbergii* reaches zoea III without food. In several species with abbreviated sequences, development under starvation conditions will proceed, yet mouthparts are setose and the larvae will feed (e.g., *Alpheus heterochaelis,* Knowlton 1970, 1973, *Palaemonetes paludosus,* Dobkin 1969, *M.australiense,* Fielder 1970). In *P.paludosus* and *M.australiense,* mouthpart setation is somewhat reduced. In *A.heterochaelis* metamorphosis will proceed in 20 days with food as opposed to 25 days without food. In other species with a prolonged development (where there are non-feeding stages), the feeding appendages have reduced setation. For example, in *Micratya poeyi* with a larval series of ten stages, zoea I does not feed (non-setose mouthparts) and zoea II and successive stages do feed (setose mouthparts) (Hunte 1979).

Reduced setation on the feeding appendages of the glaucothoe in anomurans, following zoeal stages in which the mouthparts are fully setose (MacDonald et al. 1957, Hart 1965, Hoffman 1968, Campodonico & Guzman 1981, Haynes 1982), varies from most of the above examples where non-feeding and non-setose features occur in the initial stages of development. (Many of the species of concern here are lecithotrophic and produce large eggs and larvae with yolk reserves, but only some of them exhibit abbreviated development.) The zoeal stages of *Cryptolithodes typicus* Brandt and *Paralithodes granulosus* are active feeders (Hart 1965, Campodonico & Guzman 1981); those of *Pagurus bernhardus* are not (Dawirs 1980). Development to the first crab stage in *Pagurus bernhardus* will occur without feeding. Development through the zoeal stages and to the glaucothoe in *C.typicus* and *P.granulosus* proceeds normally, although survivorship in cultures of *P.granulosus,* where food is offered, is greater than under starvation conditions. Both Hart (1965) and Campodonico & Guzman (1981) noted that the glaucothoe of *C.typicus* and *P.granulosus* were very inactive, did not feed well, were not strong swimmers and frequently remained immobile on the bottom of the culture containers. Mortalities were greatest in the megalopal intermolt period in *C.typicus,* and only four first crab stages were obtained from a large number of megalopae; no first crabs were obtained for *P.granulosus.* It appears that stored yolk reserves in these two lithodids are insufficient for molt to the first crab, as opposed to the hermit crab *P.bernhardus.*

5.3.2 *Adaptive significance*

As lecithotrophy becomes more prominent than planktotrophy in the larvae, egg size (and consequently the contained yolk) is increased. Development then becomes increasingly transferred from the plankton into the egg, and eventually, for some species, total larval development takes place inside the egg (e.g., Wear 1967). As a consequence substantial quantities of protein and lipid are required. During ovarian development, these substrates are synthesized by the ovigerous females from ingested food. But during peak breeding activity, or when successive broods are produced, stored lipid reserves may be borrowed from the hepatopancreas and transferred to the ovaries, and protein may be mobilized from muscle (Wear 1974). Both activities may severely weaken the female. The fact that successive batches of eggs become smaller in number suggested to Wear that the female may have increasing difficulty in mobilizing sufficient protein-lipid reserves. At the same time, however, the lipid content of the egg is also decreasing over the span of embryonic development, as the growing larvae use those lipids (Herring 1974a). But even though abbreviated development produces a greater drain on the energy reserves of the parental female, the benefit is that embryonic development is longer, during which time the female may replenish her lipid reserves in preparation for the next deposition of eggs. In this sense, parental investment directly aids the larvae, but at the same time indirectly aids the female by providing a resting period of sorts. However, the total reproductive effort for females producing lecithotrophic larvae may be more energetically expensive. The volume of an average egg clutch of *Uca subcylindrica* is twice the average volume of three other species studied in southern Texas (Table 4). If two broods per year are produced by temperate zone *Uca* (as stated in Williams 1965 and Sandifer 1973d) and only one by *U.subcylindrica,* then reproductive effort per year (in terms of volume of eggs) would be equivalent. However, it is possible that *U.subcylindrica* may produce two broods per year [based on observations of peaks in breeding activity and occurrence of ovigerous females spaced temporally to allow for mating, egg deposition, and incubation of the large eggs (Rabalais, unpublished data)]. Other southern Texas *Uca* may produce three or four broods per year, so that seasonal and lifetime reproductive output for comparable species is not known. In any event, it is evident that *U.subcylindrica* produces a larger egg mass in a single reproductive output and probably a greater volume on a seasonal and lifetime basis.

As larval development becomes more and more abbreviated and embryonic periods lengthened, larger energy reserves must be included in the egg. Because the source of

Table 4. Meristics for egg masses of four species of *Uca* (original data from Rabalais & Cameron 1983). Volume = $\frac{4}{3} \pi r^3$, where r = ½ average egg diameter (assuming spherical, but some noted to be slightly elliptical).

Species	Average egg diameter (mm)	Average egg volume (mm³)	Average number of eggs per single egg mass	Average volume of clutch (mm³/clutch)
Uca longisignalis	0.25	8.18×10^{-3}	19,100	156
Uca rapax	0.24	7.24×10^{-3}	22,300	161
Uca panacea	0.28	1.15×10^{-2}	13,300	153
Uca subcylindrica	1.06	6.24×10^{-1}	580	341

energy necessary for embryonic development is limited in female crabs, there may be a certain limit for the energy budget of abbreviated developmental sequences. Once the embryonic energy source is depleted, the developing larvae must either hatch or cease development. We can thus postulate a reserve saturation point (RSP) in the embryo similar to that postulated for the zoea (cf. Anger & Dawirs 1981) (Gore, this volume). The point of no return (PNR) has already been predicted by Sulkin (1978) who stated that 'The point in transition (in crustaceans) between lecithotrophy and planktotrophy may be defined as the point of hatching'.

Part of Crisp's (1974) model of energy relations of marine invertebrate larvae suggested that the parent provides each offspring with energy not greatly in excess of the minimum for survival so that the provision of energy for successful metamorphosis is adequate. As such, energy reserves for planktotrophic larvae which feed are minimal, while those of lecithotrophic larvae are maximal, so that larval stages can survive through planktonic existence without feeding where food supplies are low. Sulkin & Norman (1976) also held that the adaptive significance of abbreviated larval development in brachyuran crabs lay in the fact that larvae were no longer dependent upon favorable prey after hatching, providing sufficient yolk could be stored prior to eclosion.

Where suitable planktonic prey may be in short supply, stored yolk reserves in the larvae for growth and development would be a decided advantage. For example, plankters may be limited in the deep sea, where larvae of species, such as *Lithopagurus yucatanicus* (Provenzano 1968), *Metanephrops challengeri* (Wear 1976), and *Cymonomus bathamae* (Wear & Batham 1970), are expected to develop. For *Metopaulius depressus,* where the nursery areas appear to be limited to collected water in large bromeliad tanks, the quantities of planktonic prey were not discussed (Hartnoll 1964a). For *Uca subcylindrica,* nursery areas are supratidal ephemeral puddles which may last a couple of days to weeks, depending on temperature, winds, and presence or absence of subsequent rainfall (Rabalais & Cameron 1983). Often the duration of these nursery areas is only the few days (two to three) required for metamorphosis to megalopae. There is thus limited time for any other species (unless similarly adapted with short-lived, water-dependent stages) to produce pelagic stages upon which larval *U.subcylindrica* could prey. Lecithotrophic larvae would be advantageous in the above situations if food were limiting.

Shokita (1973a,b) suggested that little food would be present for free-swimming larvae of freshwater shrimps living in landlocked pools at river heads in the mountains of Taiwan, thus accounting for the abbreviated development of *Caridinia brevirostris* Stimpson and

Macrobrachium shokitai which hatch as advanced zoeae of benthonic habit. Although the ecological aspects of these landlocked freshwater nursery areas were not described, one would suspect that there may be plankters available as prey. But this suggestion raises the curious possibility that abbreviated larval development may have developed in some cases simply because favorable prey was scarce at some point in the evolutionary history of a species.

Sulkin (1978) presented a similar viewpoint when he stated that if larvae do not encounter sufficient prey at specific points in their ontogeny, the evolutionary response may be to reduce the number of ontogenetic stages. Yet one would not expect food in the natural environment, especially relatively permanent bodies of freshwater, to be so limiting, given the great variety of plankters available. Lebour (1922) examined marine planktonic brachyuran zoeae and concluded that they fed on diatoms for the most part, but with molluscan, echinoderm and other larvae also being eaten.

Sulkin & van Heukelem (1980) proposed an increased nutritional flexibility exhibited by the planktotrophic larvae of the deep-sea red crab *Geryon quinquedens* as an alternative to the lecithotrophic larval strategy as opposed to the larvae of shallow water stone crab, *Menippe mercenaria,* which are nutritionally vulnerable on a non-variable diet but can reach nearshore waters likely to contain the kinds of prey required to satisfy nutritional needs. The larvae of *G.quinquedens,* which hatch at depths exceeding 200 m, have increased nutritional flexibility and thus reduced dependence on locating and capturing specific prey which satisfy qualitative nutritional requirements. It is important to note, however, that studies such as these, while providing interesting data and possibilities, are using non-natural food supplies in larval cultures. It remains a distinct possibility, for example, that *G.quinquedens* may not require any nutritional flexibility, because sufficient food stuffs of various sizes and qualities would be available in the natural environment. This study and others do show, however, that diet is an important influence in larval development and that nutritional deficiencies are known to affect development.

Of evolutionary importance in all these examples is that species retaining large, lecithotrophic eggs for long periods may be under a selective disadvantage as opposed to species that carry and release their planktotrophic eggs in a shorter period. Should the former suffer predation, both the producing female and the entire late-developing egg supply is lost to the population, whereas in the latter forms there is a better chance that the rapidly developing eggs will be quickly shed and that some larvae may survive even if the parent female does not escape predation. In addition, Ringold (1979) showed that the cost to a semi-terrestrial crab of carrying a large egg mass (20 % of the female's body weight) may be responsible for an increased respiration rate and energetic expenditure. Since this cost is proportional to the length of time over which the eggs are carried, he predicted this factor selects for shorter periods of egg bearing (and against the abbreviation of larval development and the longer periods of embryonic incubation associated with it). Hyman (1920) suggested that ovigerous females run greater risk of predation because of decreased speed and agility while carrying an egg mass. This would also be a factor selecting for shorter incubation periods and smaller egg volumes.

Comparison of closely related species, one with abbreviated development (*Uca subcylindrica*) and others without (western temperate Atlantic *Uca* spp.) shows that behavioral adaptations of *Uca subcylindrica* (and especially ovigerous females) to semi-arid habitats prevent increased respiration and energetic expenditure as well as increased risk of predation. Ovigerous *Uca subcylindrica* females have seldom been found venturing from burrows (usually plugged) except following a rainfall, when they were searching for suitable nursery areas (Rabalais, unpublished data). They thus do not run similar risks as those typically intertidal ovigerous females of *Uca* spp., which may migrate to or congregate in areas of water or move to the water at a nocturnal low tide and bathe their egg masses (Montague 1980). The lecithotrophic nature and lengthy incubation of large egg masses of many species with abbreviated development is obviously advantageous in certain ecological settings, and disadvantages are outweighed by net reproductive success of species with this developmental mode. Some of these other advantageous features of abbreviated development follow.

5.4 *Aspects of development and growth*

In those species with lengthened embryonic development and associated advanced larvae which often have stored energy reserves, there are obvious differences seen in survivorship and maturation rates of those species with abbreviated development and those with regular development of shorter embryonic development and longer planktonic existence. Larval development in *Uca subcylindrica* is considerably shortened compared to those of other *Uca* (Fig.6). In *U.subcylindrica,* megalopae are present within two days, and within five to seven days later first crabs are present. Differences in these developmental times vary with temperature, salinity, and presence or absence of food (Rabalais & Cameron 1982). Other species of *Uca* require an average of 17-28 days to reach megalopa and 27-40 days to reach the first crab stage (Hyman 1920, Herrnkind 1968, Feest 1969, Kurata 1970, Montague 1980). These values may also vary with temperature, salinity, and nutrition.

In addition to a decreased maturation time to megalopa and first crab in laboratory cultures, *Uca subcylindrica* benefits from higher survivorship to these stages. In *U.subcylindrica,* 94% will reach megalopa and first crap in 27.6°C and 15‰ compared to 90% of *U.pugilator* on a similar diet of *Artemia* nauplii (Christiansen & Yang 1976). Although there was better survival and shorter larval duration for *U.pugilator* on a combined diet of *Artemia-*

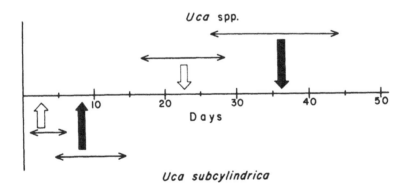

Figure 6. Average development time to megalopa (open vertical arrows) and to first crab stage (dark vertical arrows) of closely related species – one with abbreviated development (*Uca subcylindrica*) and others with longer-developing larvae (*Uca* spp.). Horizontal arrows associated with each heavier vertical arrow indicate ranges of development times. Data for *Uca* spp. from Hyman (1920), Kurata (1970), Feest (1969), Herrnkind (1968), and Montague (1980); data for *Uca subcylindrica* from Rabalais & Cameron (1983).

rotifers-ciliates, there was 82 % survival to the first crab on the *Artemia* diet with a duration of 17 days to first crab (extrapolated from two separate experiments). Considerably lower survivorship to the megalopal stage (76 % at 25°C, 59 % at 28°C) for *U.pugilator* in laboratory cultures was reported by Christy (1982) and even less for *U.longisignalis* (46.3 % at 25°C, 20 ‰) by M.Flynn (personal communication).

In other closely related species, the time to metamorphosis is reduced considerably in abbreviated development compared to regular development. In *Macrobrachium australiense* about six days are required to reach the postlarval stage (Fielder 1970), compared to the 35-45 days required in *M.acanthurus* Wiegman and *M.carcinus* with 10-12 larval stages (Choudhury 1970, 1971a). *Palaemonetes paludosus* will metamorphose to the postlarva after three nonfeeding zoeal stages, with a survival rate of 65 % in 5-10 days, in comparison to *P.vulgaris* which requires 16-30 days to metamorphose after an average of seven zoeal stages (Sandifer 1973c); survivorship was not reported. Broad (1957b) reported that 65 % of *P.pugio* fed *Artemia* reached the postlarval stage and 35 % of *P.vulgaris* fed on the same food reached the postlarval stage. Dobkin (1968, 1969) studied *Thor floridanus* and *Thor dobkini,* the former with a long pelagic life and the latter with abbreviated development. Specimens of *Thor dobkini* generally attained the postlarval stage after two days without being fed, and mortality was low (3 %). On the other hand, only 17 % of *Thor floridanus* reared became postlarvae. *Paralomis granulosus* with advanced development has a 'very abbreviated' duration of larval life (Campodonico & Guzman 1981). Within ten days metamorphosis to megalopa will occur. This is the shortest larval period recorded to date of the lithodids, as compared to 16 days to megalopa in *Lithodes antarcticus* (Campodonico 1971).

5.4.1 *Factors promoting greater survivorship*
As discussed above (section 5.3) on lecithotrophic larvae and in McConaugha (this volume), food availability is often not a limiting factor for many species with abbreviated development. Planktonic larvae must often depend on a patchy and variable source of planktonic prey for the energy resources necessary for growth and survival. Most of the above comparisons of survivorship between species with abbreviated development and those without are based on laboratory cultures where optimal nutritional supplies are provided. Actual survivorship values for these developmental types may be even more disparate in natural environments. Difficulties in following survivorship in natural populations and in ascribing mortalities to nutritional deficiencies prevent us from making more concrete statements along these lines.

Since ecdyses are critical periods in larval life and highest mortality of cultured decapod larvae often occur then (Knudsen 1960, Ong 1966, Knowlton 1970, Roberts 1971), reduction in the number of premetamorphic molts may thus increase larval survivorship (Sandifer 1973c). Larval decapods, like adults, increase the volume of their bodies rapidly and reliably during the period immediately after molting. The mortality among larvae of several species during this critical period may be related to a pre-ecdysial rise in osmotic pressure or to an inability to maintain an osmotic gradient at the time of molt (see Foskett 1977 for a review of osmoregulation in larval crabs). A reduction in the number of molts and its probable related physiological stress would be advantageous to species which inhabit more dilute media, are exposed to extremes or unpredictable fluctuations in salinities, or are incapable of appropriate behavioral responses necessary to avoid suboptimal conditions. Lucas (1971) predicted that abbreviation of planktonic larval development may be expected in species restricted to estuaries, where their chances of encountering adverse conditions in a characteristically heterogeneous environment are increased. This viewpoint is obviously not supported by examples such as that of *Rhithropanopeus harrisii* in which the four larval stages are retained while females are in the estuaries (Cronin 1982, Lambert & Epifanio 1982) and develop normally in salinities of 5 to 40‰ (Kalber & Costlow 1966, Costlow et al. 1966).

Molting places high demands on limited energy reserves and places the larvae at a higher risk to predators and water currents (Benzie 1982). Also, during the process of molt, larvae may possibly become encumbered by an exuvial cast that may prevent swimming, feeding, or proper respiration. Such dangerous conditions could possibly lead to death either from the mechanical problems involved or from the actions of a pelagic predator. Campodonico & Guzman (1981) invoked a 'less time in the plankton' argument by stating that the shorter free larval life of *Paralomis granulosus* might be of survival value when confronted with a species of higher fecundity, such as *Lithodes antarcticus* with longer larval life. Similar arguments were fostered by Forbes (1977) to explain the high recruitment of *Callianassa kraussi.* Benzie (1982) suggested that a reduction in larval molts under selection for faster development would be strongly selected for up to a point (four molts according to him), beyond which further changes of the developmental plan are more fundamental and more difficult to achieve. Yet, one wonders (if these periods are so 'critical') how planktonic larval development continues in the majority of decapod crustaceans.

Problems of physiological stress confronting larval decapod crustaceans may be circumvented by many species with abbreviated development. The resulting advantage would, of course, be an increased survivorship in early developmental life. The molting process, as mentioned above, is a critical period for larval decapods when osmotic stress may prove intolerable. Kalber & Costlow (1966) found that larvae of *Rhithropanopeus harrisii* actively hyperregulate just prior to the molt and for several hours afterwards, during which period there is a rapid, inward diffusion of water when permeability of the body wall to water apparently increases. Other authors have demonstrated that such rises are insufficient to account for water uptake at the molt (cited in Foskett 1977). The larvae of *R.harrisii* are apparently able to develop normally in a wide range of test solutions, but many other larvae succumb to variations of salinity outside a rather limited range (Ong & Costlow 1970, Sastry & McCarthy 1973, Dawirs 1979, Rothlisberg 1979, Huni 1979, Dietrich 1979, Johns 1981). The larvae of many euryhaline species are unable to survive the salinity extremes tolerated by the adults (e.g. Lucas 1972, Dietrich 1979). These larvae exhibit a complex interaction of behavioral adaptations (Sulkin et al. 1980, Harges & Forward 1982, Jacoby 1982), rely on physical export and recruitment from and to the estuaries (Sandifer 1975, Christy 1982, Lambert & Epifanio 1982) but retain a prolonged larval developmental period.

Several characteristics of some species with abbreviated developmental sequences prove advantageous in reducing osmotic stress in environments of dilute media, such as estuaries or more brackish and freshwater environments. Lucas (1972) pointed out the osmoregulatory problems in low salinity waters for small larval stages with high surface-to-volume ratios and thin cuticles. He suggested that osmoregulatory 'work' in decapod larvae could be reduced by 1) lowered body fluid concentrations, 2) decreased surface permeability, and 3) increased volume and thus reduced surface to volume ratio.

Direct development in the freshwater hymenosomatid crab *Halicarcinus lacustris* has at least two physiological advantages over its longer larval developing, estuarine congeners, *H.australis* (Haswell) and *H.paralacustris* Lucas; 1) embryos of *H.lacustris* are twice the diameter (with half the surface to volume ratio), and 2) the period of probable protection from osmosis within the egg membrane is greatly prolonged (i.e., decreased permeability). Observations suggesting that the egg membrane protects the embryo from osmosis were made on two hymenosomatid species in which the eggs develop normally in low salinities, but the emergent larvae soon died (Forbes & Hill 1969, Lucas & Hodgkin 1970). Also, Pandian (1970) showed that there is a period of embryonic development in *Homarus americanus,* in which the egg membrane is variably permeable to salts and water or impermeable to water.

Studies of the internal osmotic concentrations in late stage embryos of *Uca subcylindrica* showed that a fairly constant internal environment (~950 mOsm/kg) was maintained in two incubation conditions – high humidity but not standing water and ca. half seawater (20 ‰) (Rabalais, unpublished data). A reduced surface to volume ratio is seen in many species with abbreviated development, owing to greater volume of the developing embryo. As stated earlier in section 4.2, the egg size, of those considered for determination of embryonic growth rates, is two times greater (diameter, as measured in mm) for those with abbreviated development than for those with a longer developmental sequence. For the species listed in Table 4, the egg diameter of the species with abbreviated development is four times greater and the surface to volume ratio is reduced by one-fourth.

One point to be made on abbreviated development and osmoregulatory capabilities concerns the morphological features of larvae which hatch in an advanced condition or as an imago of the adult crab. Relatively well-developed gills in advanced larvae, which although lamellar are not fully differentiated (for example, *Uca subcylindrica* (Rabalais & Cameron 1983) and *Metopaulius depressus* (Hartnoll 1964a)), may be the sites of osmoregulation in a species (*Uca subcylindrica*) which actively regulates against low external media (Rabalais, unpublished data). We can note the localization of osmoregulatory tissue in the salt absorbing patches of the lamellae of the posterior gills of the adult blue crab and land crabs (Copeland 1968, Copeland & Fitzjarrell 1968) but no such tissues are known for larval or other early developmental forms.

Most crabs do not develop gills until the penultimate or ultimate zoeal stages, and, even in these stages, they are usually present only as gill buds. In majid and porcellanid crabs which develop through two zoeal stages, gill buds are present in both of these instars (Gore 1977, Yang & McLaughlin 1979); however, only in the latter zoeal stage are they to a slight degree lamellar as in other brachyurans (Yang & McLaughlin 1979). Subsequent development of the gills in majids does not approach the general adult form until the sixth crab instar. Other species examined by Yang & McLaughlin (1979) and Hyman (1920, 1924, 1925) in the families Xanthidae, Grapsidae, Ocypodidae, and Portunidae possess well-developed gills in the megalopal stage. Where eggs hatch in pools of rainwater on creek banks (*Uca subcylindrica*) or in the tanks of bromeliads (*Metopaulius depressus*), the presence of these potential salt absorbing tissues would be advantageous for the survival of larval and early postlarval forms.

The inability of most larval *Uca* spp. to osmoregulate in less than 20 ‰ (Dietrich 1979) makes them less tolerant of extremes and fluctuations in salinities that may be encountered in the natural environment. By comparison, larvae with advanced morphological features (i.e., gills) could experience greater survivorship and develop in these salinities (Fig.7).

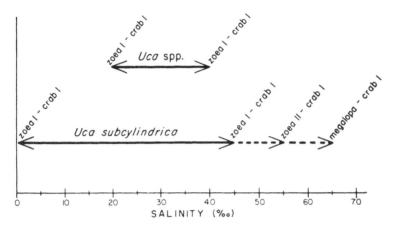

Figure 7. Salinity tolerance of larvae and early postlarvae of closely related species – one with abbreviated development, *Uca subcylindrica* (Rabalais, unpublished data), and others with longer-developing larvae, *Uca* spp. (Dietrich 1979, and Rabalais, unpublished data). Horizontal arrows represent limits for growth and development at various salinities for the stages shown. Lower limit for *U.subcylindrica* is Dietz's pond water mixture of 0.08 ‰ salinity.

Those species which hatch as young crabs would not only benefit from the presence of these potential osmoregulatory tissues but also from an integument less permeable to salts and water than delicate larval stages. Young crabs of Potamoidea live at the bottom of flooded fields or are deposited in or find their way to flowing water (Fernando 1960, Koba 1936, McCann 1937) and must be capable of withstanding osmotic stress.

In other freshwater species in which the larvae survive at low salinities, for example, *Macrobrachium australiense, M.shokitai,* and *Caridina brevirostris* (Lee & Fielder 1981, Shokita 1973a,b), the larvae hatch in an advanced condition. The development of osmoregulatory tissue in these species was not included in their larval descriptions, but the possibility exists that salt absorbing epithelia similar to that observed in larval brown shrimp (*Penaeus aztecus*) and developing *M.rosenbergii* (Talbot et al. 1972, Harrison et al. 1981) is present in these advanced larvae. In another instance, the branchial formula of the first stage of *Astacus leniusculus* Dana was found by Andrews (1907) to be the same as in the adult of the species.

In other species in which the adults occupy freshwater habitats (e.g. many *Macrobrachium* spp., atyid and hippolytid shrimp and some grapsid crabs) larvae develop through a prolonged series, but these stages seem unable to tolerate entirely fresh waters and develop in more brackish waters (Choudhury 1971b,c, Guest & Durocher 1979, Gilchrist et al. 1983). They are swept downstream, where they develop into postlarval stages and then move back into fresher waters. An example is offered by *Micratya poeyi* (Hunte 1979), in which adults inhabit fast flowing, high gradient streams. Attempts to rear them in freshwater were unsuccessful but were successful in 32 ‰ salinity. Hunte observed postlarval forms migrating back to parental populations, thus validating the suggestion by Chace & Hobbs (1969) that the early stages could not avoid being swept down to lower reaches of the streams and eventually out to sea.

Adults of *Sesarma americanum* and *Platychirograpsus typicus* inhabit freshwater streams, often great distances from coastal areas but in other situations only a few hundred meters from the sea (Rathbun 1918, Chace & Hobbs 1969). The larvae, on the other hand, appear to undergo prolonged development (Rabalais, unpublished data). The first three zoeal stages obtained in the laboratory had no advanced larval features and did not tolerate freshwater (Dietz's pond water mixture of 0.08 ‰ salinity). Better survival was obtained in 15 ‰ salinity, but higher salinities were not tested. The larvae of these semiterrestrial grapsids are probably also swept downstream in currents, as suggested by Hartnoll (1965) and Chace & Hobbs (1969). In all the above examples of longer developing species, the necessary osmoregulatory tissues and/or regulatory mechanisms may not be present as is suggested, thus differing from others with abbreviated development; their salinity tolerance is more restrictive and survivorship lower outside of a limited range.

5.4.2 *Larval growth and metamorphosis*

Rice (1968) pointed out that species with abbreviated development and non-feeding larvae do not grow much, if at all, during the molt cycle, and thus that growth is not an essential prerequisite to molting. In contrast to alpheid larvae with a prolonged developmental series of planktonic stages, *Alpheus heterochaelis* zoeae are large at hatching but experience no linear growth during the larval phase (Knowlton 1973). Subsequently, the postlarva is smaller than those of other species. A similar situation exists in *Uca subcylindrica,* in which growth during the larval phase (two stages) is minimal. Size increases 6.5 % (see Table 5) as compared to an average increase of 17.8 % in all larval stages of other *Uca* spp. Growth factor values for many other decapods are given in Table 7 (Gore, this volume). However, the size at hatch of *Uca subcylindrica* is considerably larger than those of other *Uca* (Table 6), so that even with a minimal growth rate, size at metamorphosis is about two times greater for *Uca subcylindrica.* (It is interesting to note that *Uca panacea,* which hatches from a larger egg, is larger at zoea I but does not grow at the rate of other *Uca,* so that size at metamorphosis is similar to the other species which grew at a greater rate from a considerably smaller first zoea.)

Some of the size advantage is minimized with molt to the first crab stage, where *Uca subcylindrica* increases only 15.6 % compared to others at 37.2 %. In any event, size at metamorphosis to megalopa and at the first crab instar is greater in this species with abbre-

Table 5. Increase in size as a percentage of previous larval instar (from Rabalais & Cameron 1983). a – Determined from carapace length and length from dorsal spine to rostral spine in this study. b – Determined from total lengths given in Feest (1969). c – Determined from carapace lengths given in Terada (1979).

Species	Increase in molt from zoea I to II	Increase in molt from zoea II to III	Increase in molt from zoea III to IV	Increase in molt from zoeal IV to V	Mean intermolt increase
Uca subcylindrica[a]	6.5 %	–	–	–	6.5 %
Uca annulipes[b]	25.4 %	14.5 %	13.8 %	16.7 %	17.6 %
Uca triangularis[b]	26.2 %	16.4 %	14.7 %	16.7 %	18.5 %
Uca pugilator[b]	14.5 %	26.9 %	20.0 %	11.1 %	18.1 %
Uca tangeri[b]	25.4 %	9.2 %	22.4 %	11.3 %	17.1 %
Uca lactea[c]	20.5 %	6.4 %	28.0 %	15.6 %	17.6 %

Table 6. Meristics of stages in early development of *Uca* species. All measurements are in millimeters (from Rabalais & Cameron 1983). a – This study. b – From Feest (1969). c – Determined from Novak & Salmon (1974). d – Larvae determined from Terada (1979): postlarvae from Muraoka (1976). e – Determined from Kurata (1970).

Species	First zoea	Last zoea	Megalopa			First crab		
	Carapace length	Carapace length	Carapace width	x	Carapace length	Carapace width	x	Carapace length
Uca subcylindrica[a]	1.22	1.30	1.24	x	1.42	1.47	x	1.50
Uca annulipes[b]	0.35	0.85	0.65	x	0.80	–		
Uca triangularis[b]	0.32	0.60	0.59	x	1.00	–		
Uca tangeri[b]	–	–	1.03	x	1.47	–		
Uca panacea[c]	0.70	1.04	0.72	x	1.04	1.10	x	–
Uca lactea[d]	0.39	0.74	1.02	x	1.35	1.50	x	1.26
Uca pugnax[e]	0.42	0.83	0.68	x	1.00	1.14	x	1.09

viated development (advanced type) than the regular development of other *Uca* spp. Similar larger size at first crab instar was noted for *Halicarcinus lacustris* with direct development, which was the equivalent size of a third crab instar of *H.paralacustris* with regular larval development (Lucas 1972). Where measurements are available for other species with direct development (Wear 1967, Hale 1931), the size of retained megalopae and first crabs of these xanthids are greater than those of other xanthids with a longer larval sequence (Gore et al. 1981, Andryszak & Gore 1981, Kurata et al. 1981).

5.4.3 *General life history patterns*
The general scheme of the reproductive ecology of many species with abbreviated development is clear, but many questions have been raised concerning overall life history patterns. Advanced larvae or imagos maintain an apparent advantage at the point of maternal release based on previously discussed lecithotrophic nature and physiological aspects. Many of these species are larger when they metamorphose to the benthonic stage. Are there constraints on the number of imagos or of large, well-developed larvae with stored energy reserves that a female decapod crustacean can produce per season or per lifetime? What differences are there in survivorship and recruitment of lecithotrophic and planktotrophic larvae? Does a larger size at metamorphosis prove advantageous in recruitment or in decreased time, if at all, before reaching sexual maturity? Are such females forced into a reproductive stage earlier and thereby offset the inability to produce large numbers of offspring per brood or season? Are there differences in the growth rates to sexual maturity or differences in life span among closely related species with such obviously different early developmental sequences? Or do the growth rates converge at some point? Most of these questions must remain unanswered until comparable life history information is available on many more species. Existing data are limited as well. A good example of divergent life history patterns is provided by McDonald (1982).

Although emphasis in much of the current life history theory is on aspects of reproductive biology, natural selection is a force in all phases of the life cycle. Comprehensive life history data (larval, postlarval, juvenile, and adult) are required to determine the origin or causes for life history variation within the decapod Crustacea, or for that matter, any other taxa. The possibilities for interpretation of comparative life history traits (Stearns 1976), once the necessary data are available, remain intriguing.

5.5 *Habitat trends*

Species with abbreviated developmental sequences have been recorded from a variety of habitat types (e.g., arboreal, terrestrial, montane, semiterrestrial, freshwater, estuarine, rocky intertidal shores, marine, deep sea, hypersaline, inquiline, and anchialine). Some general trends are apparent but do not form an encompassing theory of where one would expect to find abbreviated developmental patterns. The elimination or partial suppression of pelagic larval stages would seem advantageous in several of the above environs for a variety of reasons, such as lack of proper nutrition, physiological stress, dispersal and recruitment mechanisms, and lack of suitable nursery areas. Yet, other species which occupy similar habitats do not always possess similar developmental modes (nor should we expect evolution to have a particular endpoint).

In exclusively freshwater groups such as the crayfish (Parastacoidea and Astacoidea) and Potamoidean brachyuran crabs, there is complete elimination of larval stages, and development is direct (see section 3.1.1). These groups also represent degrees of terrestrial and semi-terrestrial existence. In other groups which span a range of hydrographic types from marine to estuarine-brackish to freshwater (e.g., Grapsidae, Palaemonidae, Atyidae), development in those groups occupying freshwater habitats is not limited to abbreviation of the direct mode, although in some cases (e.g., *Caridina mccullochi*, Benzie 1982; *C.brevirostris* and *C.denticulata ishigakiensis*, Shokita 1973b, 1976; *Macrobrachium hendersodayanum*, Jalihol & Sankolli 1975), the larval phase is so brief as to be almost direct. Other closely related freshwater species do undergo direct development (*Caridina singhalensis*, Benzie & de Silva 1983). Yet others retain a few larval stages (e.g., *Metopaulius depressus* and *Sesarma bidentatum*, Hartnoll 1964a; *S.jarvisi*, Abele & Means 1977; *Geosesarma perracae*, Soh 1969; *Macrobrachium asperulum*, Shokita 1977; *M.shokitai*, Shokita 1973a) although very brief and very advanced. Soh (1969) suggested that the complete elimination of larval stages would seem to be the outcome of this trend but felt that selective pressures towards this are slight, since semi-permanent sources of water are available. However, semi-permanent or permanent water sources are also available for many species with direct development where larval stages are curtailed. Other crabs, such as *Uca subcylindrica, Metopaulius depressus,* some *Sesarma,* and *Geosesarma* retain larval stages even though elimination of these would result in the hatching of young crabs more capable of benthic existence in areas where freshwater sources are limited.

The Palaemonidae, similar to the Atyidae, span a continuum of hydrographic types from freshwater to marine and a continuum of larval developmental patterns not always consistent with the idea that there is abbreviation in those species which inhabit freshwater habitats. For example, the freshwater species *Palaemonetes kadiakensis* develops through 5-8 larval stages while others, *P.paludosus* and *P.cummingi* have three larval stages (Hubschman & Broad 1974). Brackish and estuarine water inhabitants, *P.vulgaris* (Say) and *P.intermedius* Holthuis, have 7-11 and 6-8 larval stages, respectively. Some freshwater *Macrobrachium* have as many as nine larval stages (Kwon & Uno 1969).

Dobkin (1963) conjectured that the type of development may be related to the period of evolutionary time that the different species have inhabited freshwater. Those having been in freshwater for a greater length of time would have a condensed development, while those relatively new to this habitat would be expected to have longer larval life. He concluded that *P. kadiakensis* is a relatively new arrival to freshwater habitats. Dobkin's argument accepts as an unproven axiom, fundamental to its existence, that freshwater decapods will, ipso facto, evoke or have short development and begs the question of how and why freshwater exerts a retarding or abbreviating effect on some larvae and not others. Perhaps, the converse is true: Are those shrimp that have evolved shortened larval development with its associated characteristics better able to survive in freshwater habitats and can thus radiate to these environments? The possibility exists, also, that the required osmoregulatory capabilities are present in the 5-8 larval stages of *P. kadiakensis* but are absent in the similarly prolonged larval sequence of *P. vulgaris* and *P. intermedius*. These scenarios, while they seem logical, also fail to point out the advantages of such an evolutionary course.

Several suggestions have been given by many authors to 'account for' the abbreviation of pelagic larval stages in freshwater species. We have chosen instead to point out those aspects of abbreviated development that would enhance survivorship in these environments. In section 5.3.2 we discussed the adaptive significance of lecithotrophic larvae, particularly in habitats where food resources might be limited; in section 5.4.1, we dealt with the physiological advantages that may accrue to developmental stages that hatch in an advanced condition. The shortened duration of pelagic larval phase in species with abbreviated development would also be advantageous where microhabitats, which constitude nursery areas for many of the freshwater/terrestrial species such as the Potamoidea, may be in existence for only short periods. For *Pseudothelphusa richmondi*, the streams in which the crabs make their homes frequently dry up in the dry season (Boone 1929). The same is true for *Parathelphusa* spp. in India and eastern Asia (Fernando 1960, McCann 1937) where dry seasons alternate with monsoons. Similarly, fiddler crabs with abbreviated development which inhabit the banks of ephemeral freshwater creeks release larvae in puddles of rainwater which may last only a few days (Rabalais & Cameron 1983). The more probable microhabitat for larval development in *Sesarma jarvisi* would be ephemeral rainfall sources among rock rubble (Abele & Means 1977) where rapid larval development most likely occurs during periods of extensive rainfall. Ovigerous females of Potamoidea and *Geosesarma perracae* tend to be most numerous at the end of the rainy season (Fernando 1960, McCann 1937, Soh 1969), when nursery areas have short lives. Young crabs are less likely to be swept away from suitable parental populations by the floods of heavy monsoonal rain at this time, but behavioral modifications and reproductive ecology may minimize hazardous conditions for the adult crabs during the monsoon. Direct development might be more advantageous with regards to the ephemeral nature of nursery areas, but several species which produce young in these situations retain larval stages.

Abbreviation of larval development in freshwaters has been suggested by Lucas (1971) and others as a mechanism to reduce dispersal from optimal conditions occupied by the parental populations. He stated that in restricted freshwater habitats the larval stages are superfluous for dispersal and unsuitable where the postlarval habitat is running freshwater because of increased chances of being dispersed from the area. This logic is circuitous, like much in the literature. Limited dispersal by larvae is a consequence of shortened larval life, in conjunction with appropriate hydrographic regimes and behavioral responses of both larval and adult crabs. Where adaptive, one would expect selection for the appropriate developmental schemes and behavioral modifications. For instance, when conditions are appropriate for *Uca subcylindrica* to release larvae, following a substantial rainfall, the ephemeral creeks near which they live are often raging currents of water. The female crabs avoid these flowing streams, which are likely to sweep them away as well as the larvae, and instead hatch their eggs in standing pools of rainwater. The limited dispersal of larval forms in the case of this species is a result of behavioral adaptations of adult crabs as well as shortened larval life, neither of which may be a consequence of freshwater habitation. In fact, the presence of a larval and a megalopal stage which possess more natatory setae than most related forms, as well as a greater number of telsonal processes in the zoeal stages points to the active role of swimming larval and postlarval stages in dispersal and recruitment mechanisms in a species that is subject to flooding conditions (Rabalais & Cameron 1983).

Many of the freshwater forms described above are also terrestrial or semi-terrestrial in habit. Raja Bai Naidu (1951) considered abbreviated larval development as a more or less natural consequence in forms such as the ocypodid crab *Ocypode platytarsis* which, in her view (although unsubstantiated), was becoming progressively more terrestrial. Hartnoll (1965) also suggested that a semi-terrestrial group would have a greater reduction in the number of larval stages, but this is obviously not evident nor true. Such statements do not explain why other terrestrial or semi-terrestrial forms (Table 2) such as the mangrove tree crab *Aratus* (Warner 1968), the gecarcinid land crabs, *Gecarcinus* (Willems 1982, Kannupandi et al. 1980) and *Cardisoma* (Costlow & Bookhout 1968), terrestrial anomuran crabs, *Birgus* (Reese & Kinzie 1968) and *Coenobita* (Provenzano 1962), the many semi-terrestrial ocypodid crabs (with the notable exception of *Uca subcylindrica*), or the semi-terrestrial shrimp, *Merguia* (Gilchrist et al. 1983) have not moved toward abbreviating their larval development. In all of these cases, the adults return to permanent bodies of water and release larvae which develop through a prolonged developmental sequence. However, permanent water resources are always within reach of ovigerous females, although migration to the water requires energy expenditure (Wolcott & Wolcott 1982). In the semi-terrestrial crab, *Uca subcylindrica*, permanent waters are not always available, and often those that are are hypersaline with salinities consistently above 80 ‰ (Rabalais, unpublished data). Drawbacks for some terrestrial dwelling species, such as limited nursery areas, physiological stress, or lack of food resources, may be similar to freshwater inhabitants, and abbreviated development would be advantageous. In other terrestrial or semi-terrestrial species, these factors may not be important.

Wear suggested (1967) that crabs living in specialized habitats would benefit from abbreviated development, because that development would limit dispersal and tend to keep the young near the restricted ecological niche occupied by the parents. Dobkin (1965a) and Brooks & Herrick (1892) had previously noted that abbreviated development could aid inquiline *Synalpheus* species by retaining the larval stages near the host sponge. Some cave dwelling species are suspected of abbreviated larval development (Hartnoll 1964a, Fage & Monod 1936), based on both egg size and unusual habitat. The Jamaican pinnotherid crab, *Pinnotheres moseri,* is commensal with the sea squirt, *Ascidia nigra;* the single zoeal stage usually lasts no more than one day before metamorphosis to the megalopal stage (Goodbody 1960). The short larval time could prevent dispersal of the offspring away from the host animal. Bruce (1974) pointed to the apparent advantages of shortened larval histories in commensal shrimps, but also noted this type of development was distinctly and generally rare in commensal shrimps and in tropical marine shallow water shrimps.

In the New Zealand crab, *Pilumnus lumpinus,* development is rapid. Wear (1967) suggested that a short development time would keep the larvae near the supposedly restricted adult niche; however, other cold-temperate, cobbled-shore dwelling species (Table 2) have no such adaptation and seem to survive quite well (e.g., *Pilumnus hirtellus*, Salman 1982). The case for abbreviation of development and reduction of dispersal for other inhabitants of wave-washed shores (Hale 1925, 1931, Rathbun 1914, Wear 1967) is equally difficult to comprehend. Wear (1967) suggested that *Pilumnus lumpinus* with a single zoeal and megalopal stage was most recently evolved, compared to *P.novaezealandiae* without a zoeal stage, and suggested that, in this group, the evolutionary outcome of these trends would be direct development and complete elimination of larval stages. This is similar to the prediction of Dobkin (1963) for freshwater palaemonids. Wear made the important distinction that the shallow water rock platforms with deep crevices, which are the habitats of several species with abbreviated development, occur in *isolated* (our emphasis) areas of the coastline. In those species which also occupy similar habitats along other shores but develop through an extended larval period (e.g., some grapsids, Table 2), parental habitat types may not be as restrictive. Coastal current regimes and hydrographic conditions may also be important in dispersal and recruitment to optimal microhabitats.

A case in point is the family Pinnotheridae (Table 1) where members occupy microhabitats generally considered restrictive. Species range from free-living, or able to move freely to and from the tubes of polychaete worms with which they are associated (e.g., *Pinnixa sayana*, Williams 1965), to those commensal in the mantle cavity of bivalve molluscs and ascidians (see Table 1), or those considered obligate commensals or parasites (e.g., *Pinnotheres moseri*, Goodbody 1960; *Dissodactylus crinitichelis*, Pohle & Telford 1981). The number of zoeal stages for those species in which the complete larval development is known ranges from 0 to 5. Within a family where the majority of species pass through 3, 4 or 5 zoeal stages, the single zoeal stage of *Pinnotheres moseri*, in combination with its shorter duration, would obviously allow us to categorize it as abbreviated. Those with two zoeal stages are abbreviated in comparison to the majority of closely related taxa. If abbreviated development is truly an evolutionary response in limiting dispersal from a restricted habitat or from limited host organisms, then other pinnotherids may have fewer number of larval stages, depending on the degree of 'restrictiveness' of the habitat. A review of the habitats (although the 'restrictiveness' or relative abundance of hosts are not known) and the number of larval stages in a diverse sampling of pinnotherids (Table 1) does not support this contention. Rather, it appears (as stated earlier in our discussion of freshwater species with abbreviated development) that reduced dispersal is a correlate of abbreviated development, enhanced by behavioral modifications and hydrological conditions, rather than that abbreviated development is a response to a need for limited dispersal.

Lucas (1975) viewed the absence of the megalopal stage in estuarine *Halicarcinus* as an advantage in a reduced pelagic life and quicker transformation from the plankton to the benthos. Since there is restricted dispersal of the larvae away from the adult habitat of enclosed estuarine water he studied, the lack of a transitionary megalopa which can (in most species) locate a suitable habitat is superfluous. The reduced planktonic life is desirable, according to Lucas (1971), in decreasing the probability of encountering adverse conditions during larval life in this characteristically heterogeneous environment. In *Halicarcinus ovatus* and *H.rostratus* (Lucas 1975), this reduction would compensate for low fecundity by higher recruitment.

Saba et al. (1978) also proposed that the abbreviation of stages seemed to be related to the salinity of the sea water in which the adult lives. They described the abbreviated development of *Epixanthus dentatus* which lives in an estuary. Furthermore, most of the larvae, which died in pure sea water, survived in 80 % sea water. These 'time in the plankton' arguments were reviewed earlier (see section 5.4.1). Following similar lines of thought as those of Lucas (1971) and Saba et al. (1978), we might expect to see abbreviated development reduce dispersal from appropriate estuaries, since long planktonic life increases the probability (according to these authors) of larvae being lost from the estuary. However, investigations into larval behavior in the field and laboratory, combined with the complex interactions of export and retention of larval stages in the estuaries (see section 5.4.1), makes it obvious that estuarine forms have a variety of life patterns in estuarine environments. The larval development of many estuarine-dependent forms, in fact almost all, is prolonged; all these species are obviously successful.

The potential exists for other species with restrictive habitats, for example the coral inhabiting, gall forming crabs of the family Hapalocarcinidae (Scotto & Gore 1981) or the coral inhabiting xanthid crab, *Micropanope barbadensis,* to be removed by oceanic currents from an accessible and utilizable habitat already colonized by adults of the species (Makarov 1968, Gore et al. 1981). However, long planktonic development time may not necessarily be disadvantageous to these species. Gore et al. (1981) postulated a combination of longshore currents, upwelling systems, and developmental plasticity in relation to differential water temperatures to enhance the recruitment potential of late zoeal and megalopal stages to the appropriate coralline habitats. The general hypothesis that abbreviated development limits dispersal from restrictive habitats is not supported by the above scenario nor by existing knowledge of estuarine processes or oceanic transport of crab larval stages (Garth 1966). Presumably the mechanisms of inherited variation would aid in dispersal of larvae from successful parental genotypes as well as increasing recruitment to established populations (Sandifer & Smith 1979).

6 ABBREVIATED DEVELOPMENT AND EVOLUTION

Abbreviation of development is widespread among the decapod crustaceans and is evidently nonphylogenetic in nature. The advantages which accrue to species in which there is partial or complete elimination of larval stages are diverse and multiple, and 'may result from selection on several different combinations of traits in different taxa' (Strathmann & Strathmann 1982). The exact mechanisms or genetic control of development that cause stages (discrete ecdysial events) to be added or removed in a developmental series are not known. The acceleration in some aspects of growth and development, and a slowing, or normal, development in others creates variation in the number of stages and timing of acquisition of certain morphological characteristics. It is obvious that mechanisms exist which shorten developmental patterns. In families where the majority of the species consistently have longer developing sequences (e.g., Xanthidae with four or five, Pinnotheridae with four or five, Ocypodidae with five), there are isolated instances of abbreviation of development, often to the point of no free-swimming planktonic stage. These developmental patterns are strikingly similar and indicate a convergence of developmental patterns with similar mechanisms which shorten a developmental sequence.

The convergence in larval form in unrelated species, where certain characteristics may be highly modified as a result of abbreviation, may mask phylogenetic relationships. Rice (1981a) pointed to the 'potential danger in the uncritical use of larval information' in the interpretation of evolutionary histories. Similar thoughts were presented by Dobkin (1965b) as he referred to the larvae of two forms of *Glyphocrangon* (but applicable to many others):¯'. . . indeed the more we know about all the phases of the life-history of a species, the more easily can the relationships and true systematic position of this species be established'. Williamson (1982b) noted that it was rare to be able to draw evidence from larval morphology, adult morphology, and biochemical and paleontological characters when examining phylogenies, but that any case was strengthened with support from more than one category. These comprehensive sets of information are often difficult, but nonetheless important, to obtain.

We caution the use of 'advanced' and 'primitive' descriptors for morphological features, as they may imply a more or less derived or evolved state and concomitant systematic position. In species where numerous morphological variations are probably related to the abbreviated development sequence, the stages that are the most different and specialized (i.e., apomorphic) are the zoeal stages and to a lesser extent the megalopae in some species. First crab stages are more true to the expected plesiomorphic form (e.g., *Uca subcylindrica*, Rabalais & Cameron 1983). Such awareness may help clarify some of the current interpretations of evolutionary histories based on larval information. For example, phylogenetic relationships may be obscured in species which exhibit abbreviated development (see Rice 1981a). At present, analyses of 'advanced' or 'unusual' features, compared to those 'more primitive', may be overlooking the fact that there are many specialized features related to developmental patterns. Such features may well be inconsistent with the viewpoint that decapod larvae 'are all adapted for a relatively similar pelagic existence' (Williamson 1974, Rice 1980, 1981a). These views have been moderated more recently (Rice 1981a, Williamson 1982b) in phylogenetic studies where larval characteristics have been assessed equally with adult characteristics.

When characteristics of larval or early postlarval development promote the earlier attainment of a more adult form, any disadvantages (assuming such exist) inherent in prolonged planktonic existence are thereby avoided. The adaptive significance of shortened developmental patterns has been discussed in sections 5.3.2, 5.4.1, 5.4.3 and 5.5 and includes anti-predation, avoidance of suboptimal conditions for growth and development (such as inadequate nutrition or poor hydrographic conditions), reduced dispersal from specialized and restricted adult habitats, and reduced critical molt periods. Yet one wonders: if reduction of planktonic life is so adaptive to some species, why have not more evolved, especially those which retain one or two zoeal stages (see section 3.2.1)? Is the eventual outcome to be direct development, as suggested by Soh (1969)? (or vice versa)? Yet, a mutation does not arise in response to a need just because it would be advantageous at that time. The complex structure of an organism necessitates that the parts (including larval and adult) be contiguously coadapted. Some potentially crucial gene combinations may not appear, so that the extent of evolutionary change is consequently limited. The net adaptive value of the genotype determines its fate. Evolution is essentially random; natural selection only operates between adaptive and nonadaptive genotypes (which have arisen at random) in whatever conditions are current. Thus are evident the many parallel adaptations of abbreviated development. On the other hand, advantages accrued by longer-developing species that spend more time in a planktonic existence do not also accrue to those species which have partially or completely suppressed planktonic development. The number of species with longer-development planktotrophic larvae (the majority in decapod crustaceans) point to the overall success of this developmental pattern in the present environment.

The preceding review of abbreviated development in decapods, as well as treatments elsewhere in this volume of larval and postlarval growth and development, point out that the situation is not simple. The evidence is more than sufficient to illustrate that an array of developmental patterns exists from direct, with no larval form, to extended with a prolonged planktonic existence. This supports the continuum idea presented in earlier sections of this chapter. The ecological situations to which each developmental scheme is adaptive are equally diverse, and no 'best' developmental 'strategy' prevails over the rest. The uniqueness of each specific habitat demands differences in development as well as developmental plasticity intraspecifically, for the multiplicity of environmental variables that are likely to occur during growth and development. As with the variability in developmental patterns, more and more examples of variation in laboratory reared decapod crustaceans points to expected developmental plasticity *in vivo*. It would not be unexpected to find such variability in field populations. These data, however, are more difficult to acquire. Such phenotypic plasticity would itself be selected for as a singularly important adaptive feature, and larvae could complete development within a variety of environmental conditions.

Prolonged or reduced planktonic life do not present a set of discrete alternatives. The same is true for planktotrophy and lecithotrophy. There are degrees of all these and advantages and disadvantages to each. The dichotomy of r- versus K-selection has also been applied to considerations of variation in reproductive patterns among marine invertebrates [reviewed by Jablonski & Lutz (1983]. But the continuum of developmental patterns does not allow discrete alternatives for either of these categories which are, after all, only end points of a continuum and not strict alternatives. It has been suggested that in resource-limited populations that are often at their carrying capacity (K), the most fit genotypes are those that 1) most efficiently use resources, even if they grow slowly, 2) that have few offspring which are well prepared for competition, 3) that live long and reproduce repeatedly (Futuyma 1979). But these may not always hold true.

For example, resources (certainly food and available habitats) are limited in the semi-arid environments that are inhabited by *Uca subcylindrica,* a species with abbreviated development in which the adults produce few offspring of high survivability (section 5.4.1). Yet, this species (from what can be determined to date) does not live longer than its congeners (which face similar limited resources in intertidal habitats) and does not reproduce more often. In fact, these crabs probably have less opportunity to reproduce because of limited periods for hatching broods. Thus, seasonal and lifetime fecundity are less. However, based on survivorship to metamorphosis in laboratory-reared broods, the recruitment may be higher (section 5.4.5). Unfortunately, lifetime reproductive success of comparable species are not known.

Conversely, if the likelihood of an offspring's survival does not depend on how well provided it is with nutrients or defenses, the female will have the greatest number of surviving progeny if she has a great many small eggs (Futuyma 1979) and if there is a prolonged planktonic larval life. Crisp (1974) also showed the advantage of having planktotrophic larvae and a maximum number of offspring when there is density-independent mortality affecting isolated parts of the population unequally. Such reproductive patterns are usually those of an r-selected species in an environment having a variety and an abundance of resources.

Gould (1977) considered progenesis to be of some selective advantage in r-selected species. Instead of maintaining reproductive fitness by broadcasting hundreds to thousands of larvae, many of these species could have an accelerated maturational growth over that of somatic growth. One way is that instars could be lost either terminally or intermediately, so that maturity is more rapidly attained. The resulting effect would be shorter life span with a greater proportion of available resources committed to reproduction. Gould also considered K-selected species to trend toward neoteny, because these forms exist in relatively stable environments that have limited resources. These species, according to Gould's idea, should produce highly competitive offspring with a great deal of parental investment (e.g. many of the freshwater decapods should have few, large eggs producing highly advanced or completely suppressed larval stages). The resulting effect here would be late maturation, longer life, and lower reproductive effort. A strict application of Gould's views generates the hypothesis that r-selected decapod crustacean adults would more likely have a progenetic type of larval development, whereas K-selected adults would tend to have neotenous larvae. Either may be hard to demonstrate.

If we take oceanic caridean shrimps as an example of r-selected forms, it can be shown that many of these species do produce large numbers of larvae but may also have their larval developmental time extended beyond that when metamorphosis would normally occur. These neotenous forms (e.g., *Lucaya,* Gurney & Lebour (1941)) may continue to grow, and appear to develop anatomically (but perhaps not functionally) mature sexual appendages (cf. Coutière 1907). Still other carideans (e.g., Hippolytidae) seem to be near elimination of numerous stages, so that they appear to support Gould's suggestions.

Freshwater decapods often (but not always, cf. Shokita (1970)) produce advanced progenetic larvae, having nearly all adult attributes, but also retain some zoeal aspects (Shokita 1973b) and undergo rapid development in an environment normally considered supportive of K-selected forms. Furthermore, Herring (1974b) demonstrated that many bathypelagic forms produce large eggs with advanced larvae that pass through shortened development, again indicative of K-selection. These examples and many others that could be listed show that substantive evidence supporting Gould's hypothesis remains ambiguous as far as marine and freshwater decapods are concerned.

The considerable shortening of developmental time spent in the plankton, often towards a complete deletion of a pelagic phase, or in early postlarval periods, accelerates the attainment of more adult-like characteristics. The evidence now accumulated suggests that these shortened developmental patterns, while numerically less frequent, are not aberrant and that the evolutionary implications are far reaching. In some cases, condensation or elimination of stages in development proves to be advantageous only within a particular ecological situation; any disadvantages are outweighed by the net reproductive success of species with these types of developmental modes. Meanwhile, the majority of other decapod species continue to be successful with prolonged developmental sequences and extended planktonic existence. We do not pretend to judge which of the developmental patterns in the continuum is more successful. Instead, we prefer to acknowledge the array of developmental patterns that have resulted from natural selection acting on those patterns that have arisen in the past and which have the potential characteristics necessary to survive the unpredictable circumstances of the present and perhaps the future.

7 ACKNOWLEDGEMENTS

We thank D.L.Felder and J.W.Goy for helpful discussions during the preparation of this chapter and A.J.Provenzano,Jr, C.L.McLay and A.M.Wenner for critical reviews of the final manuscript. Much help was given by R.Ford and R.Grundy who located many obscure references and H.Garrett who typed the manuscript. During the writing of this chapter, support for NNR was provided by NSF Grant PCM 80-24358 to J.N.Cameron, University of Texas Marine Science Institute, and by the Louisiana Universities Marine Consortium. RHG acknowledges the aid and support provided by the Academy of Natural Sciences of Philadelphia.

University of Texas Marine Science Institute Contribution Number 578.

REFERENCES

Abele, L.G. 1970. Semi-terrestrial shrimp *(Merguia rhizophorae)*. *Nature* 226:661-662.

Abele, L.G. & D.B.Means 1977. *Sesarma jarvisi* and *Sesarma cookei:* Montane, terrestrial grapsid crabs in Jamaica (Decapoda). *Crustaceana* 32:91-93.

Al-Kholy, A.A. 1959a. Larval stages of three anoumuran Crustacea (from the Red Sea). *Publs. Mar. Biol. Sta. Al-Ghardaqa (Red Sea)* 10:83-89.

Al-Kholy, A.A. 1959b. Larval stages of four brachyuran Crustacea (from the Red Sea). *Publs. Mar. Biol. Sta. Al-Ghardaqa (Red Sea)* 10:239-246.

Andrews, E.A. 1907. The young of the crayfishes *Astacus* and *Cambarus*. *Smithson. Contr. Knowl.* 35 (1718):1-79.

Andryszak, B.L. & R.H.Gore 1981. The complete larval development in the laboratory of *Micropanope sculptipes* (Crustacea, Decapoda, Xanthidae) with a comparison of larval characteristics in western Atlantic xanthid genera. *Fishery Bull.* 79:487-506.

Anger, K. & R.R.Dawirs 1981. Influence of starvation on the larval development of *Hyas araneus* (Decapoda, Majdae). *Helgoländer Meeresunters* 34:287-311.

Atkins, D. 1955. The post-embryonic development of British *Pinnotheres* (Crustacea). *Proc. Zool. Soc. London* 124:687-715.

Baba, K. & Y.Fukuda 1972. Larval development of *Chasmagnathus convexus* de Haan (Crustacea, Brachyura) reared under laboratory conditions. *Mem. Fac. Educ. Kumamoto Univ. Nat. Sci.* 21:90-96.

Barnard, K.H. 1950. Descriptive catalogue of South African decapod Crustacea (crabs and shrimps). *Ann. S.Afr. Mus.* 38:1-837.

Bate, C.S. 1879. Report on the present state of our knowledge of the Crustacea. Part IV. On development. *Report Brit. Assoc. for 1878* 48:193-209.

Benzie, J.A.H. 1982. The complete larval development of *Caridina mccullochi* Roux 1926 (Decapoda Atyidae) reared in the laboratory. *J. Crust. Biol.* 2:493-513.

Benzie, J.A.H. & P.K.de Silva 1983. The abbreviated larval development of *Caridina singhalensis* Ortmann 1894 (Decapoda, Atyidae). *J. Crust. Biol.* 3:117-126.

Bookhout, C.G. & J.D.Costlow Jr 1974. Larval development of *Portunus spinicarpus* reared in the laboratory. *Bull. Mar. Sci.* 24:20-51.

Bookhout, C.G. & J.D.Costlow Jr 1977. Larval development of *Callinectes similis* reared in the laboratory. *Bull. Mar. Sci.* 27:704-728.

Boone, L. 1929. A collection of brachyuran Crustacea from the Bay of Panama and the fresh waters of the Canal Zone. *Bull. Amer. Mus. Nat. Hist.* 58:561-583.

Boone, L. 1935. Scientific results of the world cruise of the yacht 'Alva' 1931, William K.Vanderbilt, commanding. Crustacea: Anomura, Euphasiacea, Isopoda, Amphipoda and Echinodermata: Asteroidea and Echinoidea. *Bull. Vanderbilt Mar. Mus.* 6:1-583.

Bott, R. 1969. Die Süsswasserkrabben Süd-Amerikas und ihre Stammesgeschichte. *Abh. Senckenberg. Naturforsch. Ges.* 518:1-94.

Bourdillon-Casanova, L. 1960. Le meroplancton du Golfe de Marseille: Les larves de crustacés décapodes. *Recueil Trav. Sta. Mar. Endoume* Fasc.30, Bull. no.18:1-286.

Bousquette, G.D. 1980. The larval development of *Pinnixa longipes* (Lockington 1877) (Brachyura: Pinnotheridae), reared in the laboratory. *Biol. Bull.* 159:592-605.

Broad, A.C. 1957a. Larval development of *Palaemonetes pugio* Holthuis. *Biol. Bull.* 112:144-161.

Broad, A.C. 1957b. The relationship between diet and larval development of *Palaemonetes*. *Biol. Bull.* 112:162-170.

Broad, A.C. & J.H.Hubschman 1962. A comparison of larvae and larval development of species of east-ern US *Palaemonetes* with special reference to the development of *Palaemonetes intermedius* Hol-thuis. *Amer. Zool.* 2:394-395.

Broad, A.C. & J.H.Hubschman 1963. The larval development of *Palaemonetes kadiakensis* M.J.Rathbun in the laboratory. *Trans. Amer. Microscop. Soc.* 82:195-197.

Broekhuysen, G.J. 1955. The breeding and growth of *Hymenosoma orbiculare* Desm. (Crustacea, Brachyura). *Ann. S.Afr. Mus.* 41:313-343.

Brooks, W.K. & F.H.Herrick 1892. The embryology and metamorphosis of the Macroura. *Mem. Natn. Acad. Sci.* 5:321-576.

Bruce, A.J. 1972. Notes on some Indo-Pacific Pontoniinae, XVIII. A re-description of *Pontonia minuta* Baker 1907, and the occurrence of abbreviated development in the Pontoniinae (Decapoda Natantia, Palaemonidae). *Crustaceana* 23:65-75.

Bruce, A.J. 1974. Abbreviated larval development in the alpheid shrimp *Racilius compressus* Paulson. *J. East Africa Nat. Hist. Soc. & Natl. Mus.* 147:1-8.

Burns, J.W. 1972. The distribution and life history of South American freshwater crabs (*Aegla*) and their role in trout streams and lakes. *Trans. Amer. Fish. Soc.* 101:595-607.

Calman, W.T. 1909. Ruwenzori Expedition Reports. 5. Crustacea. *Trans. Zool. Soc. London* 19:51-56.

Calman, W.T. 1911. *The Life of Crustacea*. London: Methuen.

Campodonico, I. 1971. Desarrollo larval de la centrolla *Lithodes antarctica* Jacquinot, en condiciones de laboratorio. (Crustacea, Decapoda, Anomura: Lithodidae). *Ans. Inst. Patagonia, Punta Arenas (Chile)* 2:181-190.

Campodonico, I. & L.Guzman 1981. Larval development of *Paralomis granulosa* (Jacquinot) under laboratory conditions (Decapoda, Anomura, Lithodidae). *Crustaceana* 40:272-285.

Chace, F.A. & H.H.Hobbs 1969. The freshwater and terrestrial decapod crustaceans of the West Indies with special reference to Dominica. *Bull. US Natl. Mus.* 292:1-258.

Choudhury, P.C. 1970. Complete larval development of the palaemonid shrimp *Macrobrachium acan-thurus* (Wiegman 1836) reared in the laboratory (Decapoda, Palaemonidae). *Crustaceana* 18:113-132.

Choudhury, P.C. 1971a. Complete larval development of the palaemonid shrimp *Macrobrachium carcin-us* (L.), reared in the laboratory (Decapoda, Palaemonidae). *Crustaceana* 20:51-69.

Choudhury, P.C. 1971b. Response of larval *Macrobrachium carcinus* (L.) to variation in salinity and diet (Decapoda, Palaemonidae). *Crustaceana* 20:113-120.

Choudhury, P.C. 1971c. Laboratory rearing of larvae of the palaemonid shrimp *Macrobrachium acan-thurus* (Wiegman 1836) (Decapoda, Palaemonidae). *Crustaceana* 21:113-126.

Christiansen, M.E. & W.T.Yang 1976. Feeding experiments on the larvae of the fiddler crab *Uca pugi-lator* (Brachyura, Ocypodidae), reared in the laboratory. *Aquaculture* 8:91-98.

Christy, J.H. 1982. Adaptive significance of semilunar cycles of larval release in fiddler crabs (genus *Uca*): Test of an hypothesis. *Biol. Bull.* 163:251-263.

Copeland, D.E. 1968. Fine structure of salt and water uptake in the land crab, *Gecarcinus lateralis*. *Amer. Zool.* 8:417-432.

Copeland, D.E. & A.Fitzjarrell 1968. The salt absorbing cells in the gills of the blue crab (*Callinectes sapidus* Rathbun) with notes on modified mitochondria. *Z. Zellforsch.* 92:1-22.

Costlow, J.D.Jr 1965. Variability in larval stages of the blue crab, *Callinectes sapidus*. *Biol. Bull.* 128:58-66.

Costlow, J.D.Jr & C.G.Bookhout 1959. The larval development of *Callinectes sapidus* Rathbun reared in the laboratory. *Biol. Bull.* 116:373-396.

Costlow, J.D.Jr & C.G.Bookhout 1960. The complete larval development of *Sesarma cinereum* (Bosc) reared in the laboratory. *Biol. Bull.* 118:203-214.

Costlow, J.D.Jr & C.G.Bookhout 1962. The larval development of *Sesarma reticulatum* Say reared in the laboratory. *Crustaceana* 4:281-294.

Costlow, J.D.Jr & C.G.Bookhout 1966. Larval stages of the crab, *Pinnotheres maculatus*, under labora-tory conditions. *Chesapeake Sci.* 7:157-163.

Costlow, J.D.Jr & C.G.Bookhout 1968. The complete larval development of the land-crab, *Cardisoma guanhumi* Latreille in the laboratory (Brachyura, Gecarcinidae). *Crustaceana* suppl.2:259-270.

Costlow, J.D.Jr, C.G.Bookhout & R.J.Monroe 1966. Studies on the larval development of the crab, *Rhithropanopeus harrisii* (Gould). I. The effect of salinity and temperature on larval development. *Physiol. Zool.* 39:81-100.

Coutière, H. 1907. Sur la durée de la vie larvair des Eucyphotes. *C.R. Acad. Sci. Paris* 144:1170-1172.

Crane, J. 1941. Crabs of the genus *Uca* from the west coast of Central America. *Zoologica* 26:145-207.

Crisp, D.J. 1974. Energy relations of marine invertebrate larvae. *Thalassia Jugoslavica* 10:103-120.

Cronin, T.W. 1982. Estuarine retention of larvae of the crab *Rhithropanopeus harrisii. Estuar., Coast. Shelf Sci.* 15:207-220.

Dawirs, R.R. 1979. Effects of temperature and salinity on larval development of *Pagurus bernhardus* (Decapoda, Paguridae). *Mar. Ecol. Prog. Ser.* 1:323-329.

Dawirs, R.R. 1980. Elemental composition (C, H, N) in larval and crab-1 stages of *Pagurus bernhardus* (Decapoda, Paguridae) and *Carcinus maenas* (Decapoda, Portunidae). *Mar. Biol.* 57:17-23.

Dechancé, M. 1963. Développement direct chez un pagurid, *Paguristes abbreviatus* Dechancé; et remarques sur la développement des *Paguristes. Bull. Mus. National d'Histoire Naturelle* 35:488-495.

Dietrich, D.K. 1979. The ontogeny of osmoregulation in the fiddler crab, *Uca pugilator. Amer. Zool.* 19:972.

Dobkin, S. 1963. The larval development of *Palaemonetes paludosus* (Gibbes, 1850) (Decapoda, Palaemonidae), reared in the laboratory. *Crustaceana* 6:3-61.

Dobkin, S. 1965a. The first post-embryonic stage of *Synalpheus brooksi* Coutière. *Bull. Mar. Sci.* 15: 450-462.

Dobkin, S. 1965b. The early larval stages of *Glyphocrangon spinicauda* A.Milne Edwards. *Bull. Mar. Sci.* 15:872-884.

Dobkin, S. 1968. The larval development of a species of *Thor* (Decapoda: Caridea) from South Florida, USA. *Crustaceana,* suppl.2:1-18.

Dobkin, S. 1969. Abbreviated larval development in caridean shrimps and its significance in the artificial culture of these animals. *FAO Fisheries Reports* 57:935-945.

Elofsson, R. 1961. The larvae of *Pasiphea multidentata* (Esmark) and *Pasiphea trada* (Krøyer). *Sarsia* 4: 43-53.

Ewald, J.J. 1969. Observations on the biology of *Tozeuma carolinense* (Decapoda, Hippolytidae) from Florida, with special reference to larval development. *Bull. Mar. Sci.* 19:510-547.

Fage, L. & T.Monod 1936. Biospeologica. LXII. La faune marine du jameo de agua. Lac souterrain de l'Ile de Lanzarote (Canaries). *Arch. Zool. Exp. Gén.* 78:97-113.

Fagetti, E. & I.Campodonico 1971. The larval development of the crab *Cyclograpsus punctatus* H.Milne Edwards, under laboratory conditions (Decapoda, Brachyura, Grapsidae, Sesarminae). *Crustaceana* 21:183-195.

Feest, J. 1969. Morphophysiologische Untersuchungen zur Ontogenese und Fortpflazungsbiologie von *Uca annulipes* und *Uca triangularis* mit Vergleischbefunden an *Ilyoplax gangetica. Forma et Functio* 1:159-225.

Fernando, C.H. 1960. The Ceylonese freshwater crabs (Potamonidae). *Ceylon J. Sci. (Biol. Sci.)* 3:191-222.

Fielder, D.R. 1970. The larval development of *Macrobrachium australiense* Holthuis 1950 (Decapoda Palaemonidae), reared in the laboratory. *Crustaceana* 18:60-74.

Fielder, D.R. & J.G.Greenwood 1979. Larval development of the swimming crab *Thalamita danae* Stimpson 1858 (Decapoda, Portunidae), reared in the laboratory. *Proc. R.Soc. Queensland* 90:13-20.

Fielder, D.R., J.G.Greenwood & J.C.Ryall 1975. Larval development of the tiger prawn, *Penaeus esculentus* Haswell 1879 (Decapoda, Penaeidae), reared in the laboratory. *Austr. J. Mar Freshwater Res.* 26:155-175.

Fincham, A.A. 1977. Larval development of British prawns and shrimps (Crustacea: Decapoda: Natantia). 1. Laboratory methods and a review of *Palaemon (Palaeander) elegans* Rathke 1837. *Bull. Brit. Mus. (Nat. Hist.) Zool.* 32:1-28.

Fincham, A.A. 1979a. Larval development of British prawns and shrimps (Crustacea: Decapoda: Natantia). 3. *Palaemon (Palaemon) longirostris* H.Milne Edwards 1837 and the effect of antibiotics on morphogenesis. *Bull. Brit. Mus. (Nat. Hist.) Zool.* 37:17-46.

Fincham, A.A. 1979b. Larval development of British prawns and shrimps (Crustacea: Decapoda: Natantia). 2. *Palaemonetes (Palaemonetes) varians* (Leach 1814) and morphological variation. *Bull. Brit. Mus. (Nat. Hist.) Zool.* 35:163-182.

Forbes, A.T. 1973. An unusual abbreviated larval life in the estuarine burrowing prawn *Callianassa kraussi* (Crustacea: Decapoda: Thalassinidea). *Mar. Biol.* 22:361-365.

Forbes, A.T. 1977. Breeding and growth of the burrowing prawn *Callianassa kraussi* Stebbing (Crustacea: Decapoda, Thalassinidea). *Zoologica Africana* 12:149-161.

Forbes, A.T. & B.J.Hill 1969. The physiological ability of a marine crab *Hymenosoma orbiculare* Desm. to live in a subtropical freshwater lake. *Trans. Roy. Soc. S.Afr.* 38:271-283.

Foskett, J.K. 1977. Osmoregulation in the larvae and adults of the grapsid crab *Sesarma reticulatum* Say. *Biol. Bull.* 153:505-526.

Fukuda, Y. 1981. Larval development of *Trigonoplax unguiformis* (De Haan) (Crustacea, Brachura [sic]) reared in the laboratory. *Zool. Mag., Tokyo* 90:164-173.

Fukuda, Y. & K.Baba 1976. Complete larval development of the sesarminid crabs, *Chiromantes bidens, Holomentopus haematocheir, Parasesarma plicatum*, and *Sesarmops intermedius*, reared in the laboratory. *Mem. Fac. Educ., Kumamoto Univ. Nat. Sci.* 25:61-75.

Futuyma, D.J. 1979. *Evolutionary Biology*. Sunderland, Mass.: Sinauer Assoc.

Garth, J.S. 1966. On the oceanic transport of crab larval stages. In: Proc. of Symp. on Crustacea. Pt.1, *Symp. Ser. Mar. Biol. Ass. India* 2(1):443-448.

Gibbs, P.E. 1974. Notes on *Uca burgersi* Holthuis (Decapoda, Ocypodidae) from Barbuda, Leeward Islands. *Crustaceana* 27:84-91.

Gilchrist, S.L., L.E.Scotto & R.H.Gore 1983. The early zoeal stages of *Merguia rhizophorae* Rathbun, 1900 with notes on the larvae of the Hippolytidae. *Crustaceana* (in press).

Gilchrist, S.L., L.E.Scotto & R.H.Gore 1983. Early zoeal stages of the semiterrestrial shrimp *Merguia rhizophorae* cultured under laboratory conditions with a discussion of characters in the larval genus *Eretmocaris*. *Crustaceana* 45:238-259.

Gohar, H.A.F. & A.A.Al-Kholy 1957. The larvae of some brachyuran Crustacea (from the Red Sea). *Publs. Mar. Biol. Sta. Al-Ghardaqa (Red Sea)* 9:145-176.

Goodbody, I. 1960. Abbreviated development in a pinnotherid crab. *Nature* 185:704-705.

Gore, R.H. 1977. *Neopisosoma angustifrons* (Benedict 1901): The complete larval description under laboratory conditions, with notes on larvae of the related genus *Pachycheles* (Decapoda, Anomura, Porcellanidae). *Crustaceana* 33:284-300.

Gore, R.H. 1979. Larval development of *Galathea rostrata* under laboratory conditions, with a discussion of larval development in the Galatheidae (Crustacea, Anomura). *Fishery Bull.* 76:781-806.

Gore, R.H., C.L.Van Dover & K.A.Wilson 1981. Studies on decapod Crustacea from the Indian River region of Florida. XX. *Micropanope barbadensis* (Rathbun 1921): The complete larval development under laboratory conditions (Brachyura, Xanthidae). *J. Crust. Biol.* 1:28-50.

Gould, S.J. 1977. *Ontogeny and Phylogeny*. Cambridge: Belknap Press.

Graham, J.G. & G.F.Beaven 1942. Experimental sponge-crab plantings and crab larvae distribution in the region of Crisfield, Md. Board of Natural Resources, State of Maryland, Department of Research and Education, Publ.52, Chesapeake Biological Laboratory:1-18.

Gray, E.H. 1942. Ecological and life history aspects of the red-jointed fiddler crab, *Uca minax* (Le Conte), region of Solomons Island, Maryland. *Chesapeake Biol. Lab. Publ.* 51:3-20.

Greenwood, J.G. & D.R.Fielder 1979. The zoeal stages and megalopa of *Portunus rubromarginatus* (Lanchester) (Decapoda: Portunidae), reared in the laboratory. *J. Plankton Res.* 1:191-205.

Greenwood, J.G. & D.R.Fielder 1980. The zoeal stages and megalopa of *Charybdis callianassa* (Herbst) (Decapoda: Portunidae), reared in the laboratory. *Proc. R.Soc. Queensland* 91:61-76.

Guest, W.C. & P.P.Durocher 1979. Palaemonid shrimp, *Macrobrachium amazonicum:* Effects of salinity and temperature on survival. *Prog. Fish-Culturist* 41:14-18.

Gurney, R. 1937a. Notes on some decapod Crustacea from the Red Sea. I. The genus *Processa*. II. The larvae of *Upogebia savignyi*, Strahl. *Proc. Zool. Soc. London* (B)1937:85-101.

Gurney, R. 1937b. Notes on some decapod and stomatopod Crustacea from the Red Sea. III. The larvae of *Gonodactylus glabrous*, Brooks, and other Stomatopoda. IV: The larvae of *Callianassa*. V: The larvae of *Chlorotocella*, Balss. *Proc. Zool. Soc., London* (B)1937:319-336.

Gurney, R. 1938a. Larvae of decapod Crustacea. Part V. Nephropsidea and Thalassinidea. *Discovery Repts.* 17:293-344.

Gurney, R. 1938b. The larvae of the decapod Crustacea. Palaemonidae and Alpheidae. *Brit. Mus. Great Barrier Reef Exped. Rept.* 6:1-60.

Gurney, R. 1942. *Bibliography of the Larvae of Decapod Crustacea and Larvae of Decapod Crustacea*. London: Ray Society.

Gurney, R. 1949. The larval stages of the snapping shrimp, *Synalpheus goodei* Coutière. *Proc. Zool. Soc. London* 119:293-295.

Gurney, R. & M.V.Lebour 1941. On the larvae of certain Crustacea Macrura, mainly from Bermuda. *J. Linn. Soc. London, Zool.* 41:89-181.

Hagen, H.O.von 1970. Anpassungen an das spezielle Gezeitenzonen-Niveau bei Ocypodiden (Decapoda, Brachyura). *Forma et Functio* 2:361-413.

Hale, H.M. 1925. The development of two Australian sponge crabs. *Proc. Linn. Soc. New South Wales* 50:405-413.

Hale, H.M. 1927. *The Crustaceans of South Australia, Part 1.* Adelaide: British Science Guild.

Hale, H.M. 1931. The post-embryonic development of an Australian xanthid crab (*Pilumnus vestitus* Haswell). *Rec. South Aust. Mus.* 4:321-331.

Harges, P.L. & R.B.Forward jr 1982. Salinity perception by larvae of the crab *Rhithropanopeus harrisii* (Gould). *Mar. Behav. Physiol.* 8:311-331.

Harrison, K.E., P.L.Lutz & L.Farmer 1981. The ontogeny of osmoregulatory ability of *Macrobrachium rosenbergii. Amer. Zool.* 21:1015.

Hart, J.F.L. 1935. The larval development of British Columbia Brachyura. I. Xanthidae, Pinnotheridae (in part) and Grapsidae. *Canadian J. Res.* 12:411-432.

Hart, J.F.L. 1965. Life history and larval development of *Cryptolithodes typicus* Brandt (Decapoda, Anomura) from British Columbia. *Crustaceana* 8:255-276.

Hartnoll, R.G. 1964a. The freshwater grapsid crabs of Jamaica. *Proc. Linn. Soc. London* 175:145-169.

Hartnoll, R.G. 1964b. The zoeal stages of the spider crab *Microphrys bicornutus* (Latr.). *Ann. Mag. Nat. Hist.* (13)7:241-246.

Hartnoll, R.G. 1965. Notes on the marine grapsid crabs of Jamaica. *Proc. Linn. Soc. London* 176:113-147.

Hartnoll, R.G. 1971. *Sesarma cookei* n.sp., a grapsid crab from Jamaica (Decapoda, Brachyura). *Crustaceana* 20:257-262.

Haynes, E. 1976. Description of zoeae of coonstripe shrimp, *Pandalus hypsinotus*, reared in the laboratory. *Fish. Bull.* 74:323-342.

Haynes, E. 1980. Larval morphology of *Pandalus tridens* and a summary of the principal morphological characteristics of North Pacific pandalid shrimp larvae. *Fish. Bull.* 77:625-640.

Haynes, E. 1981. Early zoeal stages of *Lebbeus polaris, Eualus suckleyi, E.fabricii, Spirontocaris arcuata, S.ochotensis,* and *Heptacarpus camtschaticus* (Crustacea, Decapoda, Caridea, Hippolytidae) and morphological characterization of zoeae of *Spirontocaris* and related genera. *Fish. Bull.* 79:421-440.

Haynes, E. 1982. Description of larvae of the golden king crab, *Lithodes aequispina*, reared in the laboratory. *Fish. Bull.* 80:305-313.

Herring, P.J. 1974a. Size, density and lipid content of some decapod eggs. *Deep-Sea Res.* 21:91-94.

Herring, P.J. 1974b. Observations on the embryonic development of some deep-living decapod crustaceans with particular reference to species of *Acanthephyra. Mar. Biol.* 25:25-33.

Herrnkind, W.F. 1968. The breeding of *Uca pugilator* (Bosc) and mass rearing of the larvae with comments on the behavior of the larval and early crab stages (Brachyura, Ocypodidae). *Crustaceana* suppl.2:214-224.

Hines, A.H. 1982. Allometric constraints and variables of reproductive effort in brachyuran crabs. *Mar. Biol.* 69:309-320.

Hoffman, E.G. 1968. Description of laboratory-reared larvae of *Paralithodes platypus* (Decapoda, Anomura, Lithodidae). *J. Fish.Res. Bd. Canada* 25:439-455.

Hong, S.Y. 1974. The larval development of *Pinnaxodes major* Ortmann (Decapoda, Brachyura, Pinnotheridae) under the [sic] laboratory conditions. *Publ. Mar. Lab. Busan Fish. Coll.* 7:87-99.

Hong, S.Y. 1976. Zoeal stages of *Orithyia sinica* (Linnaeus) (Decapoda, Calappidae) reared in the laboratory. *Publs. Inst. Mar. Sci. Nat. Fish. Univ. Busan* 9:17-23.

Hubschman, J.H. & A.C.Broad 1974. The larval development of *Palaemonetes intermedius* Holthius 1942 (Decapoda, Palaemonidae) reared in the laboratory. *Crustaceana* 26:89-103.

Hubschman, J.H. & J.A.Rose 1969. *Palaemonetes kadiakensis* Rathbun: Post embryonic growth in the laboratory (Decapoda, Palaemonidae). *Crustaceana* 16:81-87.

Huni, A.A. 1979. The effect of salinity and temperature on the larval development of the porcellanid crab, *Petrolisthes galathinus* (Bosc, 1802), reared in the laboratory. *Libyan J. Sci.* 9B:41-58

Hunte, W. 1979. The complete larval development of the freshwater shrimp *Micratya poeyi* (Guérin-Méneville) reared in the laboratory (Decapoda, Atyidae). *Crustaceana* suppl.5:153-166.

Hyman, O.W. 1920. The development of *Gelasimus* after hatching. *J. Morphol.* 33:487-501.

Hyman, O.W. 1924. Studies on larvae of crabs of the family Grapsidae. *Proc. US Natn. Mus.* 65:1-8.

Hyman, O.W. 1925. Studies on the larvae of the family Xanthidae. *Proc. US Natn. Mus.* 67:1-22.

Ingle, R.W. 1979. The larval development of the spider crab *Rochinia carpenteri* (Thomson) [Oxyrhyncha: Majidae] with a review of majid subfamilial larval features. *Bull. Brit. Mus. (Nat. Hist.) Zool.* 37:47-66.

Irvine, J. & H.G.Coffin 1960. Laboratory cultures and early stages of *Fabia subquadrata* (Dana), (Crustacea, Decapoda). *Walla Walla Coll. Publ.* 28:1-22.

Jablonski, D. & R.A.Lutz 1983. Larval ecology of marine benthic invertebrates: Paleobiological implications. *Biol. Rev.* 58:21-89.

Jacoby, C.A. 1982. Behavioral responses of the larvae of *Cancer magister* Dana (1852) to light, pressure, and gravity. *Mar. Behav. Physiol.* 8:267-283.

Jalihal, D.R. & K.N.Sankolli 1975. On the abbreviated metamorphosis of the freshwater prawn *Macrobrachium hendersodayanum* (Tiwari) in the laboratory. *J. Karnatak Univ. (Sci.)* 20:283-291.

Johns, D.M. 1981. Physiological studies on *Cancer irroratus* larvae. I. Effect of temperature and salinity on survival, development rate and size. *Mar. Ecol. Prog. Ser.* 5:75-83.

Johnson, D.S. 1965. Land crabs. *J. Malay. Brch. R. Asiat. Soc.* 38:43-66.

Jones, M.B. 1978. Aspects of the biology of the big-handed crab, *Heterozius rotundifrons* (Decapoda: Brachyura), from Kaikoura, New Zealand. *New Zealand J. Zool.* 5:783-794.

Jones, M.B. 1980. Reproductive ecology of the estuarine burrowing mud crab *Helica crassa* (Grapsidae). *Est. Coast. Mar. Sci.* 11:433-443.

Jones, M.B. & M.J.Simons 1982. Responses of embryonic stages of the estuarine mud crab, *Macrophthalmus hirtipes* (Jacquinot) to salinity. *Intl. J. Invert. Reprod.* 4:273-279.

Kakati, V.S. & K.N.Sankolli 1975. Larval development of the pea crab *Pinnotheres gracilis* Burger, under laboratory conditions (Decapoda, Brachyura). *Bull. Dept. Mar. Sci. Univ. Cochin* 7:965 979.

Kalber, F.A. & J.D.Costlow Jr 1966. The ontogeny of osmoregulation and its neurosecretory control in the decapod crustacean, *Rhithropanopeus harrisii* (Gould). *Amer. Zool.* 6:221-229.

Kannupandi, T., J.A.Khan M.Thomas, S.Sundaramoorthy & R.Naturajan 1980. Larvae of the land crab *Cardisoma carnifex* (Herbst) (Brachyura: Gecarcinidae) reared in the laboratory. *Indian J. Mar. Sci.* 9:271-277.

Kemp, S.W. 1910a. The Decapoda Natantia of the coast of Ireland. *Sci. Invest. Fish. Br. Ire.* 1908, 1: 1-190.

Kemp, S.W. 1910b. The Decapoda collected by the 'Huxley' from the north side of the Bay of Biscay in August, 1906. *J. Mar. Biol. Ass. UK* 8:407-420.

Kircher, A.B. 1970. The zoeal stages and glaucothoe of *Hypoconcha arcuata* Stimpson (Decapoda: Dromiidae) reared in the laboratory. *Bull. Mar. Sci.* 20:769-792.

Knight, M.D. 1970. The larval development of *Lepidopa myops* Stimpson, (Decapoda, Albuneidae) reared in the laboratory, and the zoeal stages of another species of the genus from California and the Pacific coast of Baja California, Mexico. *Crustaceana* 19:125-156.

Knowlton, R.E. 1965. Effects of some environmental factors on the larval development of *Palaemonetes vulgaris* (Say). *J. Elisha Mitchell Sci. Soc.* 81(2):87.

Knowlton, R.E. 1970. Effects of environmental factors on the larval development of *Alpheus heterochaelis* Say and *Palaemonetes vulgaris* (Say) (Crustacea Decapoda Caridea), with ecological notes on larval and adult Alpheidae and Palaemonidae. PhD thesis, Univ. North Carolina. (Diss. Abstr.31: 5076-13).

Knowlton, R.E. 1973. Larval development of the snapping shrimp *Alpheus heterochaelis* Say, reared in the laboratory. *J. Nat. Hist.* 7:273-306.

Knowlton, R.E. 1974. Larval developmental processes and controlling factors in decapod Crustacea, with emphasis on Caridea. *Thalassia Jugoslavica* 10:138-158.

Knudsen, J.W. 1960. Reproduction, life history, and larval ecology of the California Xanthidae, the pebble crabs. *Pacific Sci.* 14:3-17.

Koba, K. 1936. Preliminary notes on the development of *Geotelphusa dehaani* (White). *Proc. Imp. Acad. Japan* 12:105-107.

Konishi, K. 1981a. A description of laboratory-reared larvae of the pinnotherid crab *Sakaina japonica* Serène (Decapoda, Brachyura). *J. Fac. Sci. Hokkaido Univ., Ser.VI, Zool.* 22:165-176.

Konishi, K. 1981b. A description of laboratory-reared larvae of the commensal crab *Pinnaxodes mutuensis* Sakai (Decapoda, Brachyura) from Hokkaido, Japan. *Annotnes Zool. Japan* 54:213-229.

Kurata, J. 1969. Larvae of Decapoda Brachyura of Arasaki, Sagami Bay – IV. Majidae. *Bull. Tokai Reg. Fish. Res. Lab.* 57:81-127.

Kurata, J. 1970. *Studies on the life histories of decapod Crustacea of Georgia: Part III. Larvae of deca-pod Crustacea of Georgia. Final Rep.* Sapelo Island, Georgia: Univ. of Georgia Mar. Sci. Inst.

Kurata, H., R.W.Heard & J.W.Martin 1981. Larval development under laboratory conditions of the xanthid mud crab *Eurytium limosum* (Say, 1818) (Brachyura: Xanthidae) from Georgia. *Gulf Res. Repts.* 7:19-25.

Kurata, H. & T.Midorikawa 1975. The larval stages of the swimming crabs, *Portunus pelagicus* and *P. sanguinolentus* reared in the laboratory. *Bull. Nansei Reg. Fish. Res. Lab.* 8:29-38.

Kurata, H. & H.Omi 1969. The larval stages of a swimming crab, *Charybdis acuta. Bull. Tokai Reg. Fish. Res. Lab.* 57:129-136.

Kwon, C.S. & Y.Uno 1969. The larval development of *Macrobrachium nipponense* (De Haan) reared in the laboratory. *La Mer (Bull. Soc. Franco Japonaise d'Oceanographie)* 7:278-294.

Lambert, R. & C.E.Epifanio 1982. A comparison of dispersal strategies in two genera of brachyuran crab in a secondary estuary. *Estuaries* 5:182-188.

Lanchester, W.F. 1901. *Potamon (Parathelphusa) improvisum,* sp.n. *Proc. Zool. Soc. London* 1901: 546-547.

Lang, W.H. & A.M.Young 1980. Larval development of *Hypoconcha sabulosa* (Decapoda: Dromiidae). *Fishery Bull.* 77:851-863.

Lebour, M.V. 1922. The food of plankton organisms. *J. Mar. Biol. Ass., UK* 12:644-677.

Lebour, M.V. 1928a. Studies of the Plymouth Brachyura. II. The larval stages of *Ebalaia* and *Pinnotheres. J. Mar. Biol. Ass., UK* 15:109-123.

Lebour, M.V. 1928b. The larval stages of the Plymouth Brachyura. *Proc. Zool. Soc. London* 2:473-560.

Lebour, M.V. 1930. The larvae of the Plymouth Galatheidae. I. *Munida banffica, Galathea strigosa* and *Galathea dispersa. J. Mar. Biol. Ass., UK* 17:175-187.

Lebour, M.V. 1934. The life history of *Dromia vulgaris. Proc. Zool. Soc. London* 1934:241-249.

Lee, C.L. & D.R.Fielder 1981. The effect of salinity and temperature on the larval development of the freshwater prawn, *Macrobrachium austaliense* Holthuis, 1950 from south eastern Queensland, Australia. *Aquaculture* 26:167-172.

Lucas, J.S. 1971. The larval stages of some Australian species of *Halicarcinus* (Crustacea, Brachyura, Hymenosomatidae). I. Morphology. *Bull. Mar. Sci.* 21:471-490.

Lucas, J.S. 1972. The larval stages of some Australian species of *Halicarcinus* (Crustacea, Brachyura, Hymenosomatidae). II. Physiology. *Bull. Mar. Sci.* 22:824-840.

Lucas, J.S. 1975. The larval stages of some Australian species of *Halicarcinus* (Crustacea, Brachyura, Hymenosomatidae). III. Dispersal. *Bull. Mar. Sci.* 25:94-100.

Lucas, J.S. & E.P.Hodgkin 1970. Growth and reproduction of *Halicarcinus australis* (Haswell) (Crustacea, Brachyura) in the Swan estuary, western Australia. II. Larval stages. *Austr. J. Mar. Freshwat. Res.* 21:163-173.

MacDonald, J.D., R.B.Pike & D.I.Williamson 1957. Larvae of the British species of *Diogenes, Pagurus, Anapagurus* and *Lithodes* (Crustacea, Decapoda). *Proc. Zool. Soc. London* 128:209-257.

Makarov, R.R. 1968. Ob ukorochyenii lichinochnogo razvitiya u decyatinogikh rakoobraznykh. *Zool. Zh.* 47:348-359.

Makarov, R.R. 1973. Lichinochnoye razvitiye antarkticheskoy krevetki *Notocrangon antarcticus* (Decapoda, Crangonidae). *Zool. Zh.* 52:1149-1152.

Martin, J.W., D.L.Felder & F.M.Truesdale (1984). A comparative study of morphology and ontogeny in juvenile stages of four western Atlantic xanthoid crabs (Crustacea: Decapoda: Brachyura). *Phil. Trans. R.Soc. London.* (B)303:537-604.

Mayer, P. 1881. Die Metamorphose von *Palaemonetes varians. Mitt. Zool. Sta. Neapel* 2:197-221.

McCann, C. 1937. Notes on the common land crab *Paratelphusa (Barytelphusa) guerini* (M. Eds.) of Salsette Island. *J. Bombay Nat. Hist. Soc.* 39:531-543.

McDonald, J. 1982. Divergent life history patterns in co-occurring intertidal crabs *Panopeus herbstii* and *Eurypanopeus depressus* (Crustacea: Brachyura: Xanthidae). *Mar. Ecol. Prog. Ser.* 8:173-180.

Milne Edwards, A. & E.L.Bouvier 1897. Descriptions des Crustacés de la famille des Galathéidés recueillis pendant les Expéditions du 'Blake' et du 'Hassler'. *Mem. Mus. Comp. Zool. Harvard* 19:1-141.

Mizue, K. & Y.Iwamoto 1961. On the development and growth of *Neocaridina denticulata* de Haan. *Bull. Fac. Fish., Nagasaki Univ.* 10:15-24.

Montague, C.L. 1980. A natural history of temperate western Atlantic fiddler crabs (genus *Uca*) with reference to their impact on the salt marsh. *Contrib. Mar. Sci.* 23:25-55.

Montgomery, S.K. 1922. Direct development in a dromiid crab. *Proc. Zool. Soc. London* 1922:193-196.

Moreira, C. 1912. Embryologie du *Cardisoma guanhumi* Latr. *Mém. Soc. Zool. France* 25:155-161.

Muraoka, K. 1976. The post-larval development of *Uca lactea* (de Haan) and *Macrophthalmus (Mareotis) japonicus* (de Haan) (Crustacea, Brachyura, Ocypodidae). *Zool. Mag., Tokyo* 85:40-51.

Muraoka, K. 1977a. The larval stages of *Halicarcinus orientalis* Sakai and *Rhynchoplax messor* Stimpson reared in the laboratory (Crustacea, Brachyura, Hymenosomatidae). *Zool. Mag., Tokyo* 86:94-99.

Muraoka, K. 1977b. Larval development of *Pinnotheres boniensis* Stimpson reared in the laboratory (Crustacea, Brachyura, Pinnotheresidae [sic]). *Proc. Jap. Soc. Syst. Zool.* 13:72-80.

Nayak, V.N. 1981. Larval development of the hermit crab *Diogenes planimanus* Henderson (Decapoda, Anomura, Diogenidae) in the laboratory. *Indian J. Mar. Sci.* 10:136-141.

Nayak, V.N. & V.S.Kakati 1977. Metamorphosis of the hermit crab *Diogenes diogenes* (Herbst) (Decapoda, Anomura) in the laboratory. *Indian J. Mar. Sci.* 6:31-34.

Ngoc-Ho, N. 1977. The larval development of *Upogebia darwini* (Crustacea, Thalassinidea) reared in the laboratory, with a description of the adult. *J. Zool., London* 181:439-464.

Novak, A. & M.Salmon 1974. *Uca panacea*, a new species of fiddler crab from the Gulf coast of the United States. *Proc. Biol. Soc. Washington* 87:313-326.

Ong, K.S. 1966. The early developmental stages of *Scylla serrata* Forskal (Crustacea Portunidae), reared in the laboratory. *Indo-Pac. Fish. Counc. Proc. 11th Sess.* 2:135-146.

Ong, K. & J.D.Costlow Jr 1970. The effect of salinity and temperature on the larval development of the stone crab, *Menippe mercenaria* (Say), reared in the laboratory. *Chesapeake Sci.* 11:16-29

Pace, F., R.R.Harris & V.Jaccarini 1976. The embryonic development of the Mediterranean freshwater crab, *Potamon edulis (= P.fluviatile)* (Crustacea, Decapoda, Potamonidae). *J. Zool., Lond.* 180:93-106.

Pandian, T.J. 1967. Changes in chemical composition and caloric content of developing eggs of the shrimp *Crangon crangon*. *Helgoländer Meeresunters* 16:216-224.

Pandian, T.J. 1970. Ecophysiological studies on the developing eggs and embryos of the European lobster *Homarus gammarus*. *Mar. Biol.* 7:249-254.

Pandian, T.J. & K.-H.Schumann 1967. Chemical composition and caloric content of egg and zoea of the hermit crab *Eupagurus bernhardus*. *Helgoländer Meeresunters.* 16:225-230.

Perkins, H.C. 1973. The larval stages of the deep-sea red crab *Geryon quinquedens*. *Fishery Bull.* 71:69-82.

Pesta, O. 1930. Zur Kenntnis der Land- und Süsswasserkrabben von Sumatra und Java. *Arch. Hydrobiol. Suppl.* 8:92-108.

Pike, R.B. & R.G.Wear 1969. Newly hatched larvae of the genera *Gastroptychus* and *Uroptychus* (Crustacea, Decapoda, Galatheidea) from New Zealand waters. *Trans. Roy. Soc. New Zealand, Bio. Sci.* 11:189-195.

Pike, R.B. & D.I.Williamson 1961. The larvae of *Spirontocaris* and related genera (Decapoda, Hippolytidae). *Crustaceana* 2:187-208.

Pike, R.B. & D.I.Williamson 1964. The larvae of some species of Pandalidae (Decapoda). *Crustaceana* 6:265-284.

Pohle, G. & M.Telford 1981. The larval development of *Dissodactylus crinitichelis* Moreira 1901 (Brachyura: Pinnotheridae) in laboratory culture. *Bull. Mar. Sci.* 31:753-773.

Pohle, G. & M.Telford 1983. The larval development of *Dissodactylus primitivus* Bouvier, 1917 (Brachyura: Pinnotheridae) reared in the laboratory. *Bull. Mar. Sci.* 33:257-273.

Powell, C.B. 1979. Suppression of larval development in the African freshwater shrimp *Desmocaris trispinosa* (Decapoda, Palaemonidae). *Crustaceana* suppl.5:185-194.

Powers, L.W. 1975. Fiddler crabs in a nontidal environment. *Contrib. Mar. Sci.* 19:67-78.

Provenzano, A.J. Jr 1962. The larval development of the tropical land hermit, *Coenobita clypeatus* (Herbst) in the laboratory. *Crustaceana* 4:207-228.

Provenzano A.J. Jr 1967. Recent advances in the laboratory culture of decapod larvae. Proceedings Symp. on Crustacea, Part 2, *J. Mar. Biol. Assoc. India* 1967:940-945.

Provenzano, A.J. Jr 1968. *Lithopagurus yucatanicus*, a new genus and species of hermit crab with a distinctive larva. *Bull. Mar. Sci.* 18:627-644.

Provenzano, A.J. Jr & W.N.Brownell 1977. Larval and early post-larval stages of the West Indian spider crab, *Mithrax spinosissimus* (Lamarck) (Decapoda: Majidae). *Proc. Biol. Soc. Washington* 90:735-752.

Provenzano, A.J.Jr, K.B.Schmitz & M.A.Boston 1978. Survival, duration of larval stages, and size of postlarvae of grass shrimp, *Palaemonetes pugio*, reared from Kepone® contaminated and uncontaminated populations in Chesapeake Bay. *Estuaries* 1:239-244.

Rabalais, N.N. & J.N.Cameron 1981. Larval development of *Uca subcylindrica*. *Amer. Zool.* 21:990.

Rabalais, N.N. & J.N.Cameron 1982. The effects of temperature, salinity and nutrition on the larval development of *Uca subcylindrica*. *Amer. Zool.* 22:409.

Rabalais, N.N. & J.N.Cameron 1983. Abbreviated development in *Uca subcylindrica* (Stimpson 1859) (Crustacea, Decapoda, Ocypodidae) reared in the laboratory. *J. Crust. Biol.* 3:519-541.

Raja Bai Naidu, K.G. 1951. Some stages in the development and bionomics of *Ocypoda platytarsis*. *Proc. Indian Acad. Sci.* 33:32-40.

Rajabai, K.G. 1960. Studies on the larval development of Brachyura. 1. The early and post larval development of *Dotilla blanfordi* Alcock. *Ann. Mag. Nat. Hist.* (13)2:129-135.

Rathbun, M.J. 1914. Stalk-eyed crustaceans collected at the Monte Bello Islands. *Proc. Zool. Soc. London* 1914:653-664.

Rathbun, M.J. 1918. The grapsoid crabs of America. *Bull. US Natl. Mus.* 97:1-445.

Reese, E.S. & R.A.Kinzie 1968. The larval development of the coconut or robber crab *Birgus latro* (L.) in the laboratory (Anomura, Paguridae). *Crustaceana* suppl.2:117-144.

Rice, A.L. 1968. Growth 'rules' and the larvae of decapod crustaceans. *J. Nat. Hist.* 2:525-530.

Rice, A.L. 1980. Crab zoeal morphology and its bearing on the classification of the Brachyura. *Trans. Zool. Soc. London* 35:271-424.

Rice, A.L. 1981a. Crab zoeae and brachyuran classification: A reappraisal. *Bull. Brit. Mus. (Nat. Hist.) Zool.* 40:287-296.

Rice, A.L. 1981b. The megalopa stage in brachyuran crabs. The Podotremata Guinot. *J. Nat. Hist.* 15:1003-1011.

Rice, A.L., R.W.Ingle & E.Allen 1970. The larval development of the sponge crab, *Dromia personata* (L.) (Crustacea, Decapoda, Dromiidea) reared in the laboratory. *Vie et Milieu* (A)21:223-240.

Rice, A.L. & A.J.Provenzano Jr 1965. The zoeal stages and the glaucothoe of *Paguristes sericeus* A. Milne Edwards. *Crustaceana* 8:239-254.

Rice, A.L. & A.J.Provenzano Jr 1966. The larval development of the West Indian sponge crab *Dromidia antillensis* (Decapoda: Dromiidae). *J. Zool., London* 149:297-319.

Ringold, P.L. 1979. Costs to reproduction in an egg carrying mobile crustacean. *Amer. Zool.* 19:928.

Roberts, M.H.Jr 1969. Larval development of *Bathynectes superba* (Costa) reared in the laboratory. *Biol. Bull.* 137:338-351.

Roberts, M.H.Jr 1971. Larval development of *Pagurus longicarpus* Say reared in the laboratory. II. Effects of reduced salinity on larval development. *Biol. Bull.* 140:104-116.

Roberts, M.H.Jr 1975. Larval development of *Pinnotheres chamae* reared in the laboratory. *Chesapeake Sci.* 16:242-252.

Rodrigues, S.de A. 1976. Sobre a reprodução embriologia e desenvolvimento larval de *Callichirus major* Say, 1818 (Crustacea, Decapoda Thalassinidea). *Bolm. Zool., Univ. S.Paulo* 1:85-104.

Rothlisberg, P.C. 1979. Combined effects of temperature and salinity on the survival and growth of the larvae of *Pandalus jordani* (Decapoda: Pandalidae). *Mar. Biol.* 54:125-134.

Rothlisberg, P.C. 1980. A complete larval description of *Pandalus jordani* Rathbun (Decapoda, Pandalidae) and its relation to other members of the genus *Pandalus*. *Crustaceana* 38:19-48.

Saba, M., M.Takeda & Y.Nakasone 1978. Larval development of *Epixanthus dentatus* (White) (Brachyura, Xanthidae). *Bull. Natn. Sci. Mus. (Tokyo)* 4:151-161.

Salman, S.D. 1982. Larval development of the crab *Pilumnus hirtellus* (L.) reared in the laboratory (Decapoda Brachyura, Xanthidae). *Crustaceana* 42:113-126.

Samuelsen, T.J. 1972. Larvae of *Munidopsis tridentata* (Esmark) (Decapoda, Anomura) reared in the laboratory. *Sarsia* 48:91-98.

Sandifer, P.A. 1972. Morphology and ecology of Chesapeake Bay decapod crustacean larvae. PhD dissertation, Univ. of Virginia.

Sandifer, P.A. 1973a. Mud shrimp (*Callianassa*) larvae (Crustacea, Decapoda, Callianassidae) from Virginia plankton. *Chesapeake Sci.* 14:149-159.

Sandifer, P.A. 1973b. Larvae of the burrowing shrimp, *Upogebia affinis*, (Crustacea, Decapoda, Upogebiidae) from Virginia plankton. *Chesapeake Sci.* 14:98-104.

Sandifer, P.A. 1973c. Effects of temperature and salinity on larval development of grass shrimp, *Palaemonetes vulgaris* (Decapoda, Caridea). *Fishery Bull.* 71:115-123.

Sandifer, P.A. 1973d. Distribution and abundance of decapod crustacean larvae in the York River estuary and adjacent lower Chesapeake Bay, Virginia, 1968-1969. *Chesapeake Sci.* 14:235-257.

Sandifer, P.A. 1975. The role of pelagic larvae in recruitment to populations of adult decapod Crustaceans in the York River estuary and adjacent lower Chesapeake Bay, Virginia. *Estuar. Coastal. Mar. Sci.* 3:269-279.

Sandifer, P.A. & T.I.J.Smith 1979. Possible significance of variation in the larval development of palaemonid shrimp. *J. Exp. Mar. Biol. Ecol.* 39:55-64.

Sandoz, M. & S.H.Hopkins 1947. Early life history of the oyster crab, *Pinnotheres ostreum* (Say). *Biol. Bull.* 93:250-258.

Sankolli, K.N. & S.Shenoy 1968. Larval development of a dromiid crab, *Conchoecetes artificiosus* (Fabr.) (Decapoda, Crustacea) in the laboratory. In: Proceedings Symp. on Crustacea, Part 3. *J. Mar. Biol. Assc. India* 1967, 9:96-110.

Sankolli, K.N. & S.Shenoy 1976. Laboratory behavior of a crangonid shrimp *Pontocaris pennata* Bate and its first three larval stages. *J. Mar. Biol. Assc. India* 18:62-70.

Sars, G.O. 1889. Bidrag til Kundskabenom Decapodernes Forvandlinger. II. *Lithodes – Eupagurus – Spiropagurus – Galathodes – Munida – Porcellana – Nephrops. Arch. Math. Naturvidensk.* 3:133-201.

Sastry, A.N. & J.F.McCarthy 1973. Diversity in metabolic adaptation of pelagic larval stages of two sympatric species of brachyuran crabs. *Netherlands J. Sea Res.* 7:434-446.

Scelzo, M.A. & V.Lichtschein de Bastida 1978. Desarrallo larval y metamorphosis del cangrejo *Cyrtograpsus altimanus* Rathbun, 1914 (Brachyura, Grapsidae) en laboratorio, con observaciones sobre la ecologia de la especie. *Physis* (A)38:103-126.

Schmitt, W.L. 1942. The species of *Aegla*, endemic South American freshwater crustaceans. *Proc US Natl. Mus.* 91:431-520.

Scotto, L.E. & R.H.Gore 1981. Studies on decapod Crustacea from the Indian River region of Florida. XXIII. The laboratory culture of the coral gall-forming crab *Troglocarcinus corallicola* Verrill, 1908 (Brachyura: Hapalocarcinidae) and its familial position. *J. Crust. Biol.* 1:486-505.

Sekiguchi, H. 1978. Larvae of a pinnotherid crab, *Pinnixa rathbuni* Sakai. *Proc. Jap. Soc. Syst. Zool.* 15:36-46.

Shen, C.J. 1939. The larval development of some Peiping Caridea – the *Caridina* (Atyidae), the *Palaemonetes* and the *Palaemon* (Palaemonidae). *40th Anniv. Pap. Nat. Univ. Peking* 1:169-201.

Shenoy, S. & K.N.Sankolli 1967. Studies on the larval development in Anomura (Crustacea, Decapoda). In: Proceedings Symp. on Crustacea, Part 3. *J. Mar. Biol. Assc., India* 2:805-814.

Shinkarenko, L. 1979. Development of the larval stages of the blue swimming crab *Portunus pelagicus* L. (Portunidae: Decapoda: Crustacea). *Austr. J. Mar. Freshwat. Res.* 30:485-503.

Shokita, S. 1970. Studies on the multiplication of the freshwater prawn *Macrobrachium formosense* Bate. I. The larval development reared in the laboratory. *Biol. Mag. Okinawa* 6:1-12.

Shokita, S. 1973a. Abbreviated larval development of fresh-water atyid shrimp *Caridina brevirostris* Stimpson from Iriomote Island of the Ryukyus (Decapoda, Atyidae). *Bull. Sci. Engng. Div., Univ Ryukyus (Math. Nat. Sci.)* 16:222-231.

Shokita, S. 1973b. Abbreviated larval development of the fresh-water prawn, *Macrobrachium shokitai* Fujino et Baba (Decapoda, Palaemonidae) from Iriomote Island of the Ryukyus. *Annotnes Zool. Jap.* 46:111-126.

Shokita, S. 1976. Early life-history of the land-locked atyid shrimp, *Caridina denticulata ishigakiensis* Fujino et Shokita, from the Ryukyu Islands. *Res. Crustacea Carcinol. Soc. Japan* 7:1-10.

Shokita, S. 1977. Abbreviated metamorphosis of land-locked fresh-water prawn, *Macrobrachium asperulum* (Von Martens) from Taiwan. *Annotnes Zool. Jap.* 50:110-122.

Simons, M.J. & M.B.Jones 1981. Population and reproductive biology of the mud crab, *Macrophthalmus hirtipes* (Jacquinot, 1853) (Ocypodidae), from marine and estuarine habitats. *J. Nat. Hist.* 15:981-994.

Siqueira Bueno, S.L.de 1980. Notas preliminares sobre o desenvolvimento larval de *Macrobrachium iheringi* (Ortmann 1897) (Crustacea, Decapoda, Palaemonidae) obtido em laboratório. *Ciência e Cultura* 32:486-488.

Smart, G.C. jr 1962. The life history of the crayfish *Cambarus longulus longulus*. *Amer. Midland Naturalist* 68:83-94.

Snodgrass, R.E. 1956. Crustacean metamorphoses. *Smithson. Misc. Coll.* 131:1-78.

Soh, C.L. 1969. Abbreviated development in a non-marine crab, *Sesarma (Geosesarma) perracae* (Brachyura; Grapsidae), from Singapore. *J. Zool., London* 158:357-370.

Sollaud, E. 1923. Le dévelopment larvaire des 'Palaemoninae'. *Bull. Biol. Fr. Belg.* 57:509-603.

Stearns, S.C. 1976. Life history tactics: A review of the ideas. *Quart. Rev. Biol.* 51:3-47.

Strathmann, R.R. & M.F.Strathmann 1982. The relationship between adult size and brooding in marine invertebrates. *Amer. Nat.* 119:91-101.

Sulkin, S.D. 1978. Nutritional requirements during larval development of the portunid crab, *Callinectes sapidus* Rathbun. *J. Exp. Mar. Biol. Ecol.* 34:29-41.

Sulkin, S.D. & K.Norman 1976. A comparison of two diets in the laboratory culture of the zoeal stages of the brachyuran crabs *Rhithropanopeus harrisii* and *Neopanope* sp. *Helgoländer Meeresunters.* 28: 183-190.

Sulkin, S.D. & W.F.van Heukelem 1980. Ecological and evolutionary significance of nutritional flexibility in planktonotropic larvae of the deep sea red crab *Geryon quinquedens* and the stone crab *Menippe mercenaria. Mar. Ecol. Prog. Ser.* 2:91-95.

Sulkin, S.D., W.van Heukelem, P.Kelly & L.van Heukelem 1980. The behavioral basis of larval recruitment in the crab *Callinectes sapidus* Rathbun: A laboratory investigation of ontogenetic changes in geotaxis and barokinesis. *Biol. Bull.* 159:402-417.

Talbot, P., W.Clark & A.Lawrence 1972. Light and electron microscopic studies of osmoregulatory tissue in the developing brown shrimp, *Penaeus aztecus. Tissue and Cell* 4:271-286.

Terada, M. 1979. On the zoea larvae of five crabs of the family Ocypodidae. *Zool. Mag., Tokyo* 88:57-72.

Terada, M. 1981. Zoeal development of five crabs (Brachyura, Majidae, Majinae) in the laboratory. *Zool. Mag., Tokyo* 90:283-289.

Terada, M. 1982. Zoeal development of the chlorodinid crab, *Pilodius nigrocrinitus* Stimpson. *Zool. Mag, Tokyo* 91:23-28.

Thurman, C.L. 1979. Fiddler crabs of the Gulf of Mexico. PhD dissertation, Univ. of Minnestoa.

Uno, Y. & C.S.Kwon 1969. Larval development of *Macrobrachium rosenbergi* (de Man) reared in the laboratory. *J. Tokyo Univ. Fisheries* 55:179-190.

Van Dover, C.L., J.R.Factor & R.H.Gore 1982. Developmental patterns of larval scaphognathites: An aid to the classification of anomuran and brachyuran Crustacea. *J. Crust. Biol.* 2:48-53.

Warner, G.F. 1968. The larval development of the mangrove tree crab, *Aratus pisonii* (H.Milne Edwards), reared in the laboratory (Brachyura, Grapsidae). *Crustaceana* suppl.2:249-258.

Wear, R.G. 1965. Larvae of *Petrocheles spinosus* Miers, 1876 (Crustacea, Decapoda, Anomura) with keys to New Zealand porcellanid larvae. *Trans. R.Soc. New Zealand Zool.* 5:147-168.

Wear, R.G. 1967. Life-history studies on New Zealand Brachyura. 1. Embryonic and post-embryonic development of *Pilumnus novaezealandiae* Filhol, 1886, and of *P.lumpinus* Bennett, 1964 (Xanthidae, Pilumninae). *New Zealand J. Mar. Freshwat. Res.* 1:482-535.

Wear, R.G. 1968. Life-history studies on New Zealand Brachyura. 2. Family Xanthidae. Larvae of *Heterozius rotundifrons* A.Milne Edwards, 1867, *Ozius truncatus* H.Milne Edwards, 1834, and *Heteropanope (Pilumnopeus) serratifrons* (Kinahan, 1856). *New Zealand J. Mar. Freshwat. Res.* 2: 293-332.

Wear, R.G. 1970a. Some larval stages of *Petalomera wilsoni* (Fulton & Grant, 1902) (Decapoda, Dromiidae). *Crustaceana* 18:1-12.

Wear, R.G. 1970b. Notes and bibliography on the larvae of xanthid crabs. *Pacific Sci.* 24:84-89.

Wear, R.B. 1970c. Life history studies on New Zealand Brachyura. 4. Zoea larvae hatched from crabs of the family Grapsidae. *New Zealand J. Mar. Freshwat. Res.* 4:3-35.

Wear, R.G. 1974. Incubation in British decapod Crustacea, and the effects of temperature on the rate and success of embryonic development. *J. Mar. Biol. Assc., UK* 54:745-762.

Wear, R.G. 1976. Studies on the larval development of *Metanephrops challengeri* (Balss, 1914) (Decapoda, Nephropidae). *Crustaceana* 30:113-122.

Wear, R.G. 1977. A large megalopa attributed to *Petalomera wilsoni* (Fulton & Grant, 1902) (Decapoda, Dromiidae). *Bull. Mar. Sci.* 27:572-577.

Wear, R.G. & E.J.Batham 1975. Larvae of the deep sea crab *Cymonomus bathamae* Dell, 1971 (Decapoda, Dorippidae) with observations on larval affinities of the Tymolinae. *Crustaceana* 28:113-120.

Webb G.E. 1919. The development of the species of *Upogebia* from Plymouth Sound. *J. Mar. Biol. Assc., UK* 12:81-135.

Willems, K.A. 1982. Larval development of the land crab *Gecarcinus lateralis lateralis* (Fréminville, 1835) (Brachyura: Gecarcinidae) reared in the laboratory. *J. Crust. Biol.* 2:180-201.

Williams, A.B. 1965. Marine decapod crustaceans of the Carolinas. *Fish. Bull.* 65:1-298.

Williams, A.B. 1980. A new crab family from the vicinity of submarine thermal vents on the Galapagos Rift (Crustacea: Decapoda: Brachyura). *Proc. Biol. Soc. Washington* 93:443-472.

Williamson, D.I. 1960. Larval stages of *Pasiphaea sivado* and some other Pasipheidae (Decapoda). *Crustaceana* 1:331-341.

Williamson, D.I. 1974. Larval characters and the origin of crabs (Crustacea, Decapoda, Brachyura). *Thalassia Jugoslavica* 10:401-414.

Williamson, D.I. 1982a. Larval morphology and diversity. In: L.G.Abele (ed.), *Biology of the Crustacea. Vol.2:*43-110. New York: Academic Press.

Williamson, D.I. 1982b. The larval characters of *Dorhynchus thomsoni* Thomson (Crustacea, Brachyura, Majoidea) and their evolution. *J. Nat. Hist.* 16:727-744.

Wilson, K.A. 1980. Studies on decapod Crustacea from the Indian River region of Florida. XV. The larval development under laboratory conditions of *Euchirograpsus americanus* A.Milne Edwards, 1880 (Crustacea, Decapoda: Grapsidae) with notes on grapsid subfamilial larval characteristics. *Bull. Mar. Sci.* 30:756-775.

Wilson, K.A. & R.H.Gore 1980. Studies on decapod Crustacea from the Indian River region of Florida. XVII. Larval stages of *Plagusia depressa* (Fabricius, 1775) cultured under laboratory conditions (Brachyura: Grapsidae). *Bull. Mar. Sci.* 30:776-789.

Wolcott, T.G. & D.L.Wolcott 1982. Larval loss and spawning behavior in the land crab *Gecarcinus lateralis* (Fréminville). *J. Crust. Biol.* 2:477-485.

Yang, W.T. & P.A.McLaughlin 1979. Development of the epipodite in the second maxilliped and gills in *Libinia erinacea* (Decapoda, Brachyura, Oxyrhyncha). *Crustaceana* suppl.5:47-54.

Yatsuzuka, K. 1962. Studies on the artificial rearing of the larval Brachyura, especially of the larval blue-crab, *Neptunus pelagicus* Linnaeus. *Rept. Usa Mar. Biol. Sta.* 9:1-88.

Yatsuzuka, K. & N.Iwasaki 1979. On the larval development of *Pinnotheres* aff. *sinensis* Shen. *Rept. Usa Mar. Biol. Inst. Kochi Univ.* 1:79-96.

JOHN R.McCONAUGHA
Department of Oceanography, Old Dominion University, Norfolk, Virginia, USA

NUTRITION AND LARVAL GROWTH

ABSTRACT

Decapod larval nutrition has recently generated considerable interest, especially as it relates to ecological parameters. Available evidence indicates that most larvae require animal food, especially long chain fatty acids, to complete development. There is some indication that phytoplankton may be used as a supplement, especially during periods of low zooplankton abundance. All species studied are well adapted to withstand short periods of starvation. While starvation inhibits molting in late stage larvae, morphogenesis appears to proceed under specific circumstances. Often larval or postlarval intermediates or the omission of the last larval stage results. Additional research on a wider group of larvae is needed before the ecological impact of decapod larvae can be fully appreciated.

1 INTRODUCTION

The primitive pattern of crustacean post-embryonic development is characterized by a series of planktotrophic larval stages. While there has been a trend towards extended embryonic development, especially in estuarine and freshwater habitats, the vast majority of crustaceans have retained some form of planktonic phase (see Gore, this volume). Retention of a planktonic larval phase may facilitate colonization of new habitats due to increased dispersal (Thorson 1950, Mileikovsky 1971, Chia 1974). Alternatively, a long planktotrophic larval period may permit larval migration into a habitat more favorable to larval survival, with a subsequent reinvasion of the adult habitat (Strathman 1982). While planktonic larvae may have certain advantages, successful reproduction in these species is dependent on the release of larvae during periods when both abiotic and biotic factors are optimal for larval survival. Thorson (1950) suggested that temperature, predation and food limitations were the major factors affecting larval survival. Considerable attention has been directed towards understanding the role of abiotic factors, especially temperature and salinity (e.g., Costlow 1967, Costlow & Bookhout 1967, 1969). Although Thorson (1950) 'expected that most pelagic larvae living under natural conditions would starve', 'nutrition' as an ecological factor in crustacean larval development has received much less attention.

Although lecithotrophic and direct development do occur (Dobkin 1963, Rice & Provenzano 1965, Ware 1967, Brownell et al. 1977, Rabalais & Gore, this volume), most crustacean larvae hatch with little or no yolk reserve and are dependent on the availability of a suitable food source for growth and metamorphosis. Suitable prey items must meet three general criteria: 1) an appropriate size for easy capture and consumption, 2) an adequate concentration and 3) any essential dietary nutrients. The appropriate prey size and concentration for larval development will vary among species and within a species, dependent on larval stage. Our understanding of the essential dietary requirements is still rudimentary. However, mounting evidence suggests that basic essential dietary nutrients foster decapod larval growth and development.

This review of our current understanding of larval food types, nutritional requirements and bio-energetics as they relate to crustacean larval ecology is not comprehensive but is intended only to summarize current knowledge and identify productive pathways for future research. Most larval studies have dealt with decapods; this review will thus concentrate on that group. Readers interested in holoplanktonic nutrition are referred to other recent reviews (Conover 1978, Paffenhoffer & Harris 1979).

2 FEEDING STUDIES

Our early knowledge of decapod larval nutrition was based largely on attempts to maintain recently hatched larvae or material captured in the plankton and held in the laboratory to clarify taxonomy. With the notable exception of Lebour (1928), who utilized the eggs and larvae of other invertebrates as food, most of the early attempts to maintain larvae for more than a day or two ended in failure (Gurney 1942). These early studies did elucidate the planktotrophic and carnivorous nature of most decapod larvae.

2.1 *Starvation*

Starved larvae rarely molted (Fig.1). In those instances where molting did occur, a second molt was rare and metamorphosis was usually inhibited.

Both lecithotrophic and direct development have been reported in Decapoda. Wear (1967) described direct development in two xanthid species from New Zealand, *Pilumnus novaezealandiae* and *P.lumpinus*. *P.novaezealandiae* hatched as megalopae retained by the female. *P.luminus* hatched as a late non-motile zoeal stage. The zoea quickly molted (15-30 min.) to a megalopa stage, which had limited locomotion.

Lecithotropic development has been reported in several species of Anomura (Pike & Williams 1961, Rice & Provenzano 1965) and Caridea (Dobkin 1963, 1968). In all cases where lecithotrophic development has been reported, larvae could capture food. Dobkin (1963, 1968) reported no difference in survival or rate of development between starved and fed *Palaemonetes paludosus* and *Thor* sp.larvae under 'optimal' laboratory conditions. In a study of the hermit crab *Paguristes sericeus,* lecithotrophic development only occurred under optimal abiotic conditions (Rice & Provenzano 1965). Under less than optimal temperatures larvae required food to complete development. Starved zoeae metamorphosed at 25°C but not at 20 or 15°C, and survival to metamorphosis was comparable to that of fed larvae at 25°C. None of the starved animals molted to the crab stage, while 85 % of the fed glaucothoe developed through the third crab stage. Starvation delayed metamorphosis for a period of six to eight days, compared with four to six days in the fed control.

Starvation for even short periods during larval development may adversely affect both larval and juvenile survival and growth (Kon 1979, Anger & Nair 1979, Anger & Dawirs

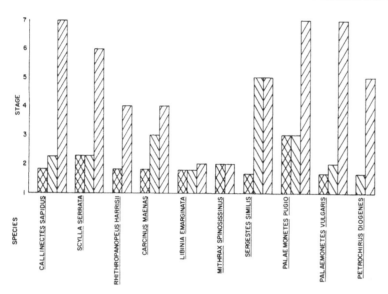

Figure 1. Development of decapod larvae under three feeding regimes. First bar for each species (cross-hatched), starvation; second bar, phytoplankton; third bar, animal food. Data obtained from: *Callinectes* (Sandoz & Rogers 1944, Rust & Carlson 1960, Sulkin 1975), *Scylla* (Brick 1974), *Rhithropanopeus* (Welch & Sulkin 1974), *Carcinus* (Williams 1968), *Libinia* (Bigford 1977), *Mithrax* (Brownell et al. 1977), *Sergestes* (Omori 1979), *Palaemonetes* (Broad 1957), and *Petrochirus* (Rice & Provenzano 1965).

1981, Paul & Paul 1980). As is the case with most starved animals, crustacean larvae reach a critical point at which survival and growth is no longer possible even if food is made available. This has been termed the point-of-no-return (PNR) (Blaxter & Hempel 1963, Anger & Dawirs 1980, Gore, this volume). Survival time of starved zoea was approximately equal to or slightly greater than normal developmental time for that stage under optimal abiotic conditions (Kon 1979, Anger et al. 1981a, McConaugha 1982). The PNR for early stage larvae was generally reached after approximately 70 % of the normal developmental time had elapsed (Kon 1970, Anger & Dawirs 1981, Anger et al. 1981a). Late stage *Rhithropanopeus harrisii* larvae did not reach PNR until after at least 150 % of the normal developmental time had elapsed (McConaugha 1982). Since an accumulation of organic reserves during successive larval stages has been demonstrated for a number of brachyuran species (Mootz & Epifanio 1974, Levine & Sulkin 1979, Anger & Nair 1979, Johns 1982, Anger et al. 1983, McConaugha unpublished observations) the extension of the PNR beyond the normal development time for late stage larvae can probably be related to utilization of these reserves.

Long term starvation has often been associated with catabolism of proteins. Starved stage I *Hyas* larvae used 52 % of their initial protein content within eight days of starvation, while fed larvae increased protein by 52 % (Anger & Nair 1979). Lipid content also fell during starvation (15.4 %), as opposed to an increase in fed larvae (53 %). The PNR probably represents the point at which protein degradation has surpassed the organism's ability to repair structural damage. Providing food after this point only extended survival. Histological observations of the hepatopancreas of starved larvae support this hypothesis

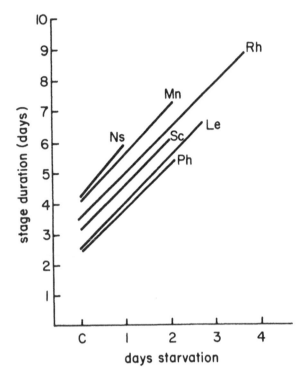

Figure 2. Effect of starvation on larval stage duration. First zoeae: Mm – *Mnippe mercenaria*, Sc – *Sesarma cinareum*, Ph – *Panopeus herbstii*, Le – *Libinia emarginata*, Ns – *Neopanope sayi* (Anger et al. 1981b). Third zoeae: Rh – *Rhithropanopeus harrisii* (McConaugha 1982).

(Storch & Anger 1983, McConaugha, unpublished observations). Ultrastructural changes occurred shortly after the onset of starvation and were not completely reversed by refeeding (Storch & Anger 1983).

Starvation for periods of less than the PNR resulted in an extension of the particular stage (Kon 1979, Paul & Paul 1980, Anger & Dawirs 1981a, McConaugha 1982). This delay takes the form of:

$$D_t = D_N + at$$

where D_t is total time of development, D_N is the normal stage duration, a is a constant and t is the time of starvation. When larvae were starved for short periods of time (1-3 days), a = 1. Starvation for periods greater than three days produced a value of >1, which suggests a recovery period, presumably while energy reserves are re-established (Fig.2). There is now evidence in at least two species of brachyurans to suggest that larvae starved in one stage shorten the next molt cycle (Anger & Dawirs 1981, McConaugha 1982). Although this truncation of the next molt cycle reduces the total developmental time, it does not completely compensate for a starvation induced delay. This phenomenon will be discussed further in section 5.3.

Starvation can also alter crustacean larval behavior, particularly phototaxis. Studies on nauplii of two species of barnacles (Singarajah et al. 1967), zoeae of the anomuran *Emerita analoga* (Burton 1979), and the xanthid *Rhithropanopeus harrisii* (Cronin & Forward 1980) indicated that starved larvae became more positively phototactic. The altered phototaxis moved larvae higher in the water column where food is presumably more abundant. However, the reduced swimming speeds of starved larvae (Cornin & Forward 1980) would reduce their ability to capture prey. Paul & Paul (1980) found that the ability of stage I larvae of the king crab, *Paralithodes camtschatica* to capture prey was directly related to the duration of starvation. Adverse abiotic factors further reduced larval feeding success.

It appears that decapod larvae are well adapted both physiologically and behaviorally for short periods of starvation. This is undoubtedly a response to the often sparse and patchy nature of plankton distributions.

2.2 *Phytoplankton*

Attempts to rear brachyuran and anomuran larvae on phytoplankton in the laboratory have had only limited success. While the presence of phytoplankton in the culture media can extend the period of survival over that observed for starved larvae, it does not generally support larval growth or extensive molting (Sandoz & Rogers 1944, Kurata 1959, Brick 1974, Roberts 1974, Sulkin 1975, Bigford 1977) (Fig.1). Failure to maintain larval growth may be due to mechanical difficulties or nutritional limitations or both. In most studies, either diatoms, dinoflagellates or both have been offered to the larvae. While larvae reportedly can consume algae, its small size (6-100 μm) may limit ingestion rates. Intact diatoms have been reportedly egested by some larvae (Roberts 1974). This suggests that the larval digestive systems may lack enzymes capable of extracting nutrients from these armored forms. Conversely, algae may simply lack some essential nutrients.

In most instances where molting occurs among algae fed animals, there was a significant delay when compared to larvae fed animal foods (Rust & Carlson 1960, Williams 1968, Brick 1974). A notable exception to this general rule was reported for *Cancer anthonyi* larvae reared on a mixture of dinoflagellates and diatoms (Yazdandoust 1983). In that series of experiments stage I larvae which were fed algae molted faster and had a higher survival rate than did larvae fed *Artemia* nauplii. A mixture of algal species supported development through the five zoeal stages of that species (Yazdandoust, personal communication).

The role of phytoplankton in the diet of brachyuran larvae is further complicated by reports of enhanced growth and survival of larvae fed a mixture of algae and animal foods (Williams 1968, Brick 1974). The exact mechanism that leads to enhancement is not known. Brick (1974) suggested that algae may 'condition' the media by removing metabolites and releasing oxygen.

A similar situation regarding the nutritional role of phytoplankton is found in caridean larval development. Broad (1957) found an algal diet alone equivalent to starvation for *Palaemonetes* larvae. Sandifer (1972) reported increased survival and fewer molts to post-larvae when *Thor floridanus* larvae were reared on an equal mixture of *Monochrysis lutheri* and *Phaeodctylum tricornutum* compared to *Artemia* nauplii. The beneficial effects of an algal diet were most pronounced during the first five zoeal stages. After the fifth stage, *Artemia* fed larvae developed faster than algal fed larvae. A mixture of *Artemia* and algae was found superior to either algal or *Artemia* diets for larvae of both *T.floridanus* (Sandifer 1972) and *Hippolyte inermis* (Regnault 1969). *Macrobrachium rosenbergii* larval survival was enhanced by the use of 'green water' in conjunction with *Artemia* (Fiyanrusa 1966, Wickins 1972, Manzi, Maddox & Sandifer 1977). It is not clear whether this enhancement was due to direct uptake of algae or an enhancement of the nutritional quality of the nauplii (Wickins 1972, Joseph 1977). Larvae, particularly large stage larvae, were reported to have large numbers of algal cells in their guts (Manzi, Maddox & Sandifer 1977). While ingestion does not constitute assimilation, it suggests a direct nutritional role.

The use of phytoplankton in penaeid culture appears to be essential, at least during the early larval stages. *Skeletonema costatum* (50 000 cells/ml) maintained protozoeal stages of a number of species of *Penaeus* (Mock, Fontaine & Revera 1980). *Sergestes similis* proto-zoeal stages survived well on an equal mixture of the flagellate *Dunaliella* and the diatoms *Thalassiosira* and *Ditylum* (Omori 1979). Algal diets alone did not support development to postlarvae.

Despite claims by early authors that decapod larvae, especially brachyurans, could not derive nutrients from phytoplankton, it is becoming increasingly clear that decapod larvae can gain some form of nutrients from algae. However, in no species yet studied has phyto-plankton been sufficient to support development through metamorphosis. An ability to utilize phytoplankton even for a limited period of time may have considerable ecological importance, given the often dilute and patchy distribution of zooplankton in the world's oceans.

2.3 *Artemia*

With the realization that most decapod larvae are carnivorous or omnivorous, the search for convenient food supplies was increased. Lebour's (1928) success at rearing crab larvae on other invertebrate larval forms was a start in the right direction. However, most of those foods were difficult to obtain and were dependent on the timing of natural spawning, which did not always correspond with spawning in decapods. The discovery that freshly hatched *Artemia* nauplii were an excellent food source for larval decapods marked a new era in the study of crustacean development. The ease of obtaining *Artemia* cysts makes it a highly desirable food for laboratory studies. Despite the fact that *Artemia* nauplii would never be encountered by marine decapod larvae under natural conditions, it has proven to be an adequate diet for larvae of hundreds of species of decapods in most major sub-divisions.

2.3.1 *Variability among strains*

Increased usage of *Artemia* nauplii as food for decapod larvae led to the realization that certain strains of *Artemia* supported larval growth better than other strains. This variability was attributed to an increase in the level of pollutants (Bookhout & Costlow 1970) or to biochemical variation among strains (Wickins 1972). Bookhout & Costlow (1970) attributed the low survival and high numbers of abnormal *Rhithropanopeus* larvae reared on Great Salt Lake *Artemia* to the high DDT content. While there is no question that pesticides can adversely affect decapod larval growth and survival (Bookhout et al. 1972, 1976, 1980), the available data (Table 1) indicate that variations in survival of decapod larvae cannot be totally attributed to differences in pesticide pollution among *Artemia* strains. Italian *Artemia* nauplii, which contain levels of DDT greater than that of the Utah strain (Olney et al. 1980), resulted in excellent survival to megalopae when fed to *Rhithropanopeus* larvae (Johns et al. 1980). Johns et al. (1981), using Brazilian *Artemia* nauplii artificially conta-minated with various levels of the pesticides chlorodane and dieldrin, demonstrated that the adverse effects of current Utah *Artemia* stocks cannot be attributed to pesticide con-tamination.

Table 1. Total lipid, energy content, and chlorinated hydrocarbons in different strains of *Artemia*.

	Total lipid mg/g dry weight	Energy content joules/g dry weight	Total chlorinated hydrocarbon Ng/g wet weight
Australia	185 ± 9	2.5 ± 0.16	18.3
Brazil	202 ± 8	2.4 ± 0.04	24.7
Canada	142 ± 34	–	32.6
China	201 ± 0.3	–	578.8
France	152 ± 29	–	213.9
Italy	224 ± 14	2.2 ± 0.06	928.5
San Francisco	174 ± 4	–	50.9
Utah	224 ± 14	2.3 ± 0.08	55.1
San Pablo Bay	160 ± 3	2.4 ± 0.11	242.6

Wickins (1972) found that Great Salt Lake (GSL) nauplii fed on *Isochrysis galbana* for 24 hours proved to be a satisfactory food for larvae of the prawn *Palaemon serratus*. He suggested that GSL nauplii were deficient in some essential dietary component. Fujita et al. (1980) concluded that the nutritional effectiveness of *Artemia* strains are linked with their fatty acid contents. *Artemia* high in $20:5\omega3$ polyunsaturated fatty acids (PUFA) support better growth and survival of marine organisms than those low in $20:5\omega3$ and high in $18:3\omega3$ PUFA. *Artemia* Strains can be arbitrarily divided between freshwater type (Utah, San Pablo Bay) characterized by low $20:5\omega3$ content and marine strains characterized by high $20:5\omega3$ levels (Table 2) (Fujita et al. 1980). Larvae reared on freshwater strains consistently resulted in extremely low survival (Bookhout & Costlow 1970, Provenzano & Goy 1976, Goy & Costlow 1980, Johns et al. 1980).

2.4 *Other animal foods*

While the ease in use of *Artemia* nauplii facilitated laboratory rearing, that practice suppressed investigations of other types of natural foods perhaps suitable for larval development. Only if *Artemia* nauplii failed to support development were other foods examined. A number of investigators experimented with a variety of small motile animals as food for decapod larvae. Those that proved successful included various polychaete larvae (Lebour 1928, Kurata 1959, Reese & Kinzie 1968, Roberts 1974, Sulkin 1975, Paul et al. 1979) echinoderm eggs and larvae (Lebour 1928, Costlow & Bookhout 1952, Warner 1968, Sulkin & Epifanio 1975), copepod nauplii and copepodites (Roberts 1974, Bigford 1977, Paul et al. 1979), barnacle nauplii (Reed 1969) and the rotifer *Brachionus plicatilis* (Brick 1974, Sulkin & Epifanio 1975, Sulkin & Norman 1976, Omori 1979, Bigford 1977). Mixed wild zooplankton was used in several studies (e.g., Broad 1957, Brick 1974). Larval survival was reduced possibly because of zooplankton mortality and subsequent rapid fouling of the culture water. Molluscan veliger larvae also proved to be of little value for decapod larval culture (Reed 1969, Roberts 1974), presumably because of the larval shell.

Many species reared on 'natural' diets have also been reared on *Artemia* nauplii diets, often under identical conditions (Table 3). If one assumes that an excess of *Artemia* nauplii represents an adequate feeding condition, then one can compare those studies with others using 'natural' prey items.

Table 2. Fatty acid composition of ten strains of *Artemia* plus *Acartia* and *Brachionus*. Data derived from Schauer et al. 1980, Seidel et al. 1982, Fujita et al. 1980.

Fatty acid methyl esters	Australia	Brazil	Italy	Utah	San Pablo	San Francisco	Canada	China	France	RAC	Acartia clausi	Brachionus plicatilis
14:0	1.34	1.57	1.53	0.93	0.43	1.57	0.83	1.80	1.73	1.79	–	1.3
14:1	2.23	0.81	3.30	1.45	2.26	0.74	1.67	2.24	3.03	2.92	–	–
15:0	0.34	0.67	0.11	0.11	0.25	0.58	–	–	–	–	–	–
15:1	0.15	0.24	0.54	0.37	0.46	0.13	–	–	–	–	–	–
16:0	13.45	15.42	15.23	11.78	7.79	12.13	9.99	11.40	11.90	12.70	20.5	15.4
16:1	9.97	10.79	10.38	5.64	5.24	19.52	9.03	19.06	11.34	16.78	4.9	2.8
16:2ω7	–	–	2.94	–	1.51	–	–	–	–	–	–	–
16:3ω4	3.87	3.88	3.28	2.90	2.44	2.32	1.47	2.54	2.20	4.33	–	–
18:0	3.07	2.79	3.17	4.07	3.08	2.90	5.12	3.99	4.21	4.07	6.7	4.1
18:1ω9	28.23	35.86	29.05	28.58	29.15	31.20	28.24	26.81	24.73	30.37	3.2	–
18:2ω6	5.78	9.59	6.79	4.60	4.60	3.69	7.95	4.68	6.14	9.62	1.7	5.0
18:3ω3	14.77	4.87	6.35	31.46	35.59	5.16	19.87	7.38	20.90	2.55	2.5	4.0
18:4ω3	4.37	0.96	1.01	3.10	4.88	1.28	1.60	1.26	2.04	–	2.7	trace
20:1ω9	0.37	0.52	0.42	0.37	0.35	0.35	–	–	–	–	0.3	–
20:2ω6/ω9	0.12	0.06	0.20	0.09	0.24	–	0.44	0.15	1.13	0.20	–	0.7
20:3ω6	0.79	2.76	1.47	0.48	0.05	2.23	–	–	–	–	–	1.3
20:3ω3	–	–	–	–	1.48	2.69	4.21	3.34	2.45	5.82	0.8	–
20:5ω3	10.50	8.98	13.63	3.55	1.68	12.44	9.52	15.35	8.01	8.45	18.7	1.2
22:6ω3	0.26	0.06	–	–	–	–	–	–	–	–	20.7	1.0

Table 3. Comparison of food value of natural prey versus *Artemia*.
* Improves survival for early stages only; NA—data not available.

Species	Prey	Stage	Sustain metamor-phosis	Comparison with *Artemia*	
				Survival	Duration
Rithropanopeus harrisii	*Brachionus plicatilis*	Juvenile, adult	Yes	–	–
Neopanope sp.	*B.plicatilis*	Juvenile	Yes	–	–
Hyas araneus	*Polydora ciliata*	Trochophore	Yes	+	+
Libinia emarginata	*B.plicatilis*	Juvenile, adult	Yes	–	–
	Eurytemora affinis	Copepodite	No	–	–
	Mixture:				
	Artemia	Nauplii	Yes	–	NA
	B.plicatilis	Juvenile			
	E.affinis	Copepodite			
	Dunaliella				
Pagurus longicarpus	*Arenicola*	Trochophore	Yes	+	–
	Crassostrea	Veliger	No	–	NA
Paralithodes camtschatia	*Chone*	Trochophore	Yes	+	NA
	Polydora	Trochophore	Yes	–	NA
Scylla serrata	*B.plicatilis*	Juvenile, adult	No	+/–*	NA
	Wild zooplankton	73-200 μm	No	–	NA
	Wild zooplankton	200-300 μm	No	–	NA
Callinectes sapidus	*B.plicatilis*	Juvenile, adult	No	+/–*	–
	Lytechnius variegatus	Eggs	No	–	NA
	Hydroides dianthus	Trochophore	Yes	+	–
	Arbacia punctulata	Eggs	No	+/–*	NA
Palaemonetes pugio	Fresh killed zoo-plankton	–	Yes	–	–

Of all the natural prey types offered, only polychaete trochophores seem to have consistently supported development to metamorphosis. In general, polychaete trochophores appeared to be an excellent dietary source for decapod larvae, as indicated by the increased survival to metamorphosis compared with *Artemia* nauplii. Larval duration was slightly longer for larvae fed polychaete trochophores than for individuals fed *Artemia* (Roberts 1974, Sulkin 1975, Paul et al. 1979).

The acceptance of *Brachionus plicatilis* and *Arbacia punctulata* embryos by early stage *Callinectes spidus* larvae is apparently based on the predator-prey size ratio, since the absence of small prey during the early zoeal stages results in high mortality. The small size of some portunid stage I larvae prohibits them from adequately feeding on the larger *Artemia* nauplii (200-250 μm) (Bookhout & Costlow 1974). The smaller size and slower swimming speed of *B.plicatilis* (45-180 μm) and *A.punctulata* (110 μm) apparently allow small zoea to capture and manipulate these prey more readily.

Sulkin (1975, 1978, Sulkin & van Heukelem 1980) suggested that the inability of *B.plicatilis* to support development to metamorphosis in *C.sapidus* and *Menippe mercenaria* is based on a nutritional limitation that is somehow related to the evolutionary position of the species. He suggested that species with high fecundity and long planktotrophic larval phases have increased nutritional requirements. While there is no doubt that rotifers and *Artemia* nauplii differ nutritionally (see section 4, Table 2) a more plausible explanation is that of size. *Brachionus plicatilis* sustained development from hatching through metamorphosis in two species of xanthids, *Rhithropanopeus harrisii* and *Neopanopeus* sp. (Sulkin & Norman 1976) and *Uca pugilator* (Christiansen & Yang 1976). Larvae fed rotifers throughout development had a slower rate of development and higher mortalities than *Artemia* fed larvae. Survival of stage I *Uca* larvae which were fed rotifers was greater than that of *Artemia* fed larvae. Brick (1974) found that zoeae of the portunid *Scylla serrata* failed to develop to metamorphosis on a rotifer diet. In his opinion the large size difference between stage I *S. serrata* (650 μm) and the rotifer (200 μm) may have reduced the feeding efficiency. The results of Hue et al. (1972) support that concept, since larvae of a smaller portunid, *Portunus trituberculatus,* developed equally well on diets of rotifers or *Artemia*. More indepth analyses of the predator-prey size relationship is needed to clarify these observations.

3 PREY DENSITY

Prey density greatly affected larval survival and development in a number of decapod species (Brick 1974, Welch & Sulkin 1974, Bigford 1977, Omori 1979, Paul et al. 1979, Carlberg & Van Olst 1980). Low food density generally resulted in lower survival or increased duration to metamorphosis or both. Brachyuran and anomuran species which tend to have a definitive number of stages (isochronal development) had an increased larval development due to an increase in individual instar duration. *Rhithropanopeus harrisii* larvae developed significantly slower when fed *Artemia* at low densities of 2-5/ml (Welch & Sulkin 1974), with the delay in development restricted to the first two zoeal stages. That result may represent a critical point in development, with energy divided between molting and metamorphosis (see section 5.2). There was no significant difference in mortality among *R. harrisii* larvae fed at various food densities (Fig.3). Brick (1974) reported the opposite results for the portunid *Scylla serrata,* with larval survival increasing with an

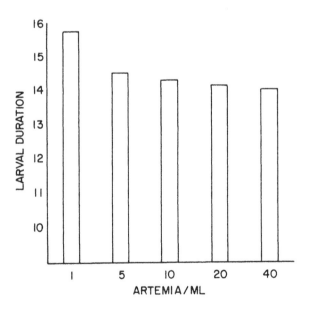

Figure 3. Developmental duration of *Rhithropanopeus harrisii* larvae fed five concentrations of *Artemia nauplii* (data from Welch & Sulkin 1974).

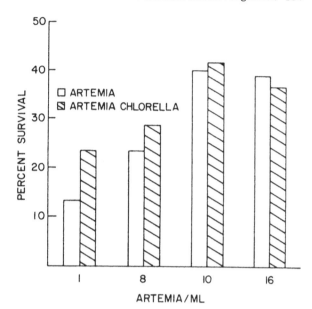

Figure 4. Survival of *Scylla serrata* larvae fed four concentrations of *Artemia*, or *Artemia* and alga *Chlorella* (data taken from Brick 1974).

increase in food concentration but with no significant difference in duration with variations in food density (Fig.4). Mootz (1973) reported that larvae of the stone crab, *Menippe mercenaria,* displayed both reduced survival and increased larval duration when fed low food concentrations. On the basis of limited data, the physiological response to low food levels appears to be species-specific. Data from studies on a larger number of species will be required before any generalizations can be made.

Carideans which have a wide variation in the number of larval stages (heterochronal development) increased both the intermolt period and the number of larval stages prior to metamorphosis under low food density (Broad 1957, Reeve 1969, Knowlton 1974). Both Reeve (1969) and Knowlton (1974) reported reduced survival under low food density. *Palaemonetes vulgaris* fed 'low' food levels failed to reach a post-larval stage even though they passed through 14 molts before dying (Knowlton 1974).

Most species studied had an increase in survival and/or rate of development with increasing food densities, up to an optimal level. Increases beyond this optimal level failed to significantly increase either survival or development rate. Optimal food densities obtained in laboratory studies were consistently higher than those which larvae would normally encounter under natural environmental conditions (Kon 1979, Paul et al. 1979, Anger & Nair 1979). This raised some question as to whether the number of stages and rate of development found in laboratory studies truly reflects conditions and stage morphology in the field, especially for caridean species. More studies that closely parallel field feeding conditions will be needed to evaluate this problem.

3.1 *Density dependent feeding rates*

Most decapod larvae are assumed to be particulate feeders that rely on chance encounters with prey. If correct, a direct relationship should exist between the number of prey cap-

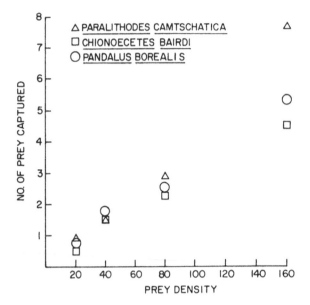

Figure 5. Relationship between prey density and prey captured by larvae of three species of decapods (data from Paul et al. 1979).

tured and prey density. This relationship has been demonstrated for a number of species, including the king crab, *Paralithodes camtschatica,* the snow crab, *Chionoecetes bairdi,* the pink shrimp, *Pandalus borealus* (Paul et al. 1979) and the spider crab, *Hyas araneus* (Anger & Nair 1979) (Fig.5). At low food concentrations, ranging from 40-320 prey/l, consumption increased over the entire range for all four species tested. A prey density of 160/1 was needed before all larvae captured at least one prey/day. Omori (1979) found a similar relationship in feeding studies of *Sergestes similis.* While prey concentrations were generally higher than those normally found in the plankton, there was no evidence that maximum ingestion had been reached. Spider crab larvae fed high concentrations of polychaete larvae approached a feeding plateau at about 2 000 polychaete larvae/l. Feeding rate decreased rapidly with increasing predator density (10-100 predators/l), presumably because of high predation rates at the beginning and reduced rates over time (Anger & Nair 1979). The authors suggested that predator concentrations of 10-100/l were unrealistic levels and suggested that predator concentrations of 1-10/l be used to determine natural predation rates. Recent reports of larval concentrations in the range of 1 000-2 000 per cubic meter (Provenzano et al. 1983, McConaugha et al. 1983) indicated that the experimental levels used may reflect natural conditions. Larvae in these concentrations could devour all available food, even in a dense patch of zooplankton, resulting in low feeding rates and reduced growth and survival.

4 NUTRITIONAL REQUIREMENTS

Despite recent advances in adult crustacean nutrition (New 1976, 1980, Conklin 1980) specific information concerning essential nutritional requirements for larval decapod growth has been limited to a few studies on lipid biosynthesis (Whitney 1969, Jones et al. 1976,

1979). Additional information has been extrapolated from feeding studies with food of known biochemical composition (i.e., percent total protein, lipid and carbohydrate) (Sulkin 1975, Johns 1980a). These studies, in conjunction with work on the biochemical changes associated with larval development (Costlow & Sastry 1966, Regnault 1969, Frank et al. 1975, Sulkin et al. 1975, Tucker 1978, Dawirs 1980, 1981, Anger et al. 1983) permitted us to tentatively identify some basic nutritional needs.

4.1 *Sterol synthesis*

As in the case of most arthropods, both adult and larval decapod crustaceans appear to lack the metabolic pathways to synthesize cholesterol from acetate (Whitney 1969, 1970, Teshima 1972). Free sterols, mainly as cholesterol, have been reported to be one of the major constituents of neutral lipids in some prawns (Teshima et al. 1977) and play an important role in normal growth and metabolic maintenance (Kanazawa et al. 1979a,b, Guany & Kanazawa 1973, Castell et al. 1975). Cholesterol is a precursor of the molting hormone, ecdysone (Kanazawa & Teshima 1971, Spaziani & Kater 1973). Ecdysone apparently regulates molting in both adult and larval decapods (McConaugha 1980, McConaugha & Costlow 1981, Chang & Bruce 1981, Freeman & Costlow 1983) and has been implicated in the regulatory processes of other essential functions such as regeneration. Based on all available evidence chloesterol must be considered to be an essential dietary requirement for larval crustaceans.

4.2 *Fatty acids*

There is ample evidence that lipids in general, and long chain PUFA in particular, are essential dietary requirements for adult crustaceans. Phosphatidyl choline has been identified as an essential dietary component for prawns (Kamazawa et al. 1979a) and lobsters (D'Abramo et al. 1981). The addition of $20:5\omega3$ and $22:6\omega3$ fatty acids to diets of the prawn, *Penaeus japonicus,* increased survival (Kanazawa et al. 1977) and weight gain (Kanazawa et al. 1979a,b) compared to diets high in $18:2\omega6$ and $18:3\omega3$ fatty acids. Radioactive tracer studies indicated that prawns can elongate $18:3\omega3$ to $20:5\omega3$ and $22:6\omega3$ (Kanazawa & Teshima 1977, Bottino 1980); however, the conversion rates were relatively low. Jones et al. (1979), using microcapsules containing ^{14}C-palmitic acid, showed that *P.japonicus* larvae can synthesize $20:5\omega3$ and $22:6\omega3$ from $18:3\omega3$. The rate of conversion appeared to be too slow to meet larval requirements. Thus both zoea and juvenile prawns appear to have similar rates of synthesis and both require some dietary long chain PUFA to achieve maximum growth and survival.

Circumstantial evidence suggests that brachyuran larvae also have dietary requirements for long-chain PUFA before completing development and metomorphosis. In a series of studies, *R.harrisii* larvae were fed *Artemia* nauplii high in $20:5\omega3$ or low in $20:5\omega3$ (Bookhout & Costlow 1970, Johns et al. 1980). No significant differences in survival were reported for the early stages of development on the two types of diets. Larvae fed low levels of $20:5\omega3$ showed a dramatic drop in survival during the last zoeal stage. Those larvae fed *Artemia* low in $20:5\omega3$ which survived to metamorphosis displayed a number of abnormalities including partial molting, unusual carapace spination, zoeal maxillipeds and irregularly formed thoracic appendages. Survival and larval duration were not statistically

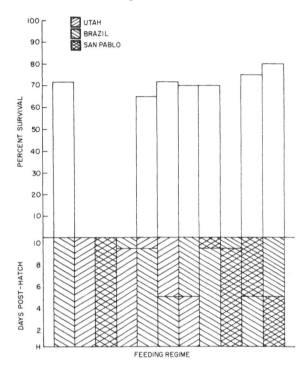

Figure 6. Survival of *Rhithropa-nopeus harrisii* larvae fed combinations of three strains of *Artemia* (data taken from Johns et al. 1981, Bookhout & Costlow 1970).

different for larvae fed on various strains of *Artemia* high in 20:5ω3. Similar results were reported for larvae of another brachyuran *Cancer irroratus* (Johns et al. 1980). In addition to poor survival and irregularities at metamorphosis, *Cancer* larvae showed a significant increase in larval duration from hatch to megalopa.

While these differences might be attributed to other chemical differences among *Artemia* strains, such as total lipids, energy content, pesticide or heavy metal contamination, a survey of the literature suggests that these factors are secondary or inconsequential. Seidal et al. (1982), who reared *R.harrisii* larvae on five strains of *Artemia* high in 20:5ω3 fatty acid (8%) but with varying levels of pesticide, heavy metals and total lipids, found no statistical difference in survival among strains. The French strain, which was low in total lipid and high in chlorinated hydrocarbons, produced the highest survival and the largest larvae (Table 1).

A series of cross-feeding experiments, in which *R.harrisii* larvae were fed a 'freshwater' type (low-PUFA) of *Artemia* for half the normal zoeal development and then switched to a 'marine' type (high PUFA), resulted in high survival and normal development (Johns et al. 1981). Whether 'freshwater' *Artemia* were used during the first half or the last half of larval development was immaterial (Fig.6). This suggests that the essential factor found in 'marine' type *Artemia,* presumably PUFA, is not required during the complete development period and can be stored during the early developmental stages for use during metamorphosis. Sulkin (1975, 1978), relying on circumstantial evidence, suggested that *C.sapidus* larvae similarly require lipids. Larvae fed the low lipid, diet (9%) of *Brachionus plicatilis* did not complete metamorphosis, and only 50% of the larvae molted to a supernumerary stage. Fertilized eggs of the polychaete *Hydroides dianthus* (20 % lipid/w) supported development through metamorphosis, but 35 % of the larvae molted to supernumerary stage. Of larvae fed rotifers for 14 days, and then switched to *Artemia* (30 % lipid/w), 29 % reached metamorphosis with no reported supernumerary stages. Larvae fed *Artemia* nauplii for as short a time as the last third of development had some individuals complete metamorphosis. Larvae receiving *Artemia* for less than 10 % of normal development time did not survive.

The number of supernumerary stages increased somewhat with prolonged periods on the rotifer diet. Supernumerary larvae also more often successfully completed metamorphosis. Feeding *C.sapidus* larvae on polychaete larvae for 15 days prior to switching to rotifers led to high survival and successful metamorphosis. In all groups of larvae fed on polychaetes some larvae metamorphosed one instar early. The number of larvae completing early metamorphosis was directly proportional to the duration of the polychaete-*Artemia* feeding regime (Sulkin 1978).

Thus a food (e.g., polychaete larvae) that in itself does not support normal growth and development, accelerates metamorphosis even when the larvae are fed a suboptimal rotifer diet. The inability of polychaete larvae to support complete development may be associated with their small size (70 μm), which might reduce feeding efficiency. The effectiveness of a short application of polychaete larvae supports the concept of storage of essential nutrients for use during metamorphosis. These essential factors, presumably PUFA, appear to be required for metamorphosis but not for completion of the molt cycle, since diets low in PUFA result in supernumerary larval stages. These studies, along with other recent work (Knowlton 1974, McConaugha 1982), suggest that the manner of energy reserve allocation between growth, molting and metamorphosis may be the important factor in regulating the rate of development and instar number.

Hernandorena (1979, 1980, 1981) presented evidence that dietary alteration of the purine/pyrimidine ratio can alter the postembryonic developmental program of *Artemia salina*. Morphogenesis of supernumerary gonopodes was induced by a reduced AMP/CMP ratio. This response appears to be associated with changes in RNA and protein synthesis (Hernandorena 1981). Since *Artemia* cannot synthesize *de novo* the purine ring (Clegg et al. 1967) it is considered a nutritional requirement. Information on the nutritional requirements for AMP and CMP in other crustacean larvae and their role in morphogenesis is lacking.

5 BIOENERGETICS

Animals require energy sources for normal metabolic maintenance and growth requirements. In the case of developmental stages the amount of energy available for growth (net growth efficiency) from the total energy intake is of greatest concern. Energy for growth depends upon the availability of food, assimilation efficiency of the animal, and metabolic demands, all of which can be altered by environmental variables. Only in the last ten years, when studies of crustacean larval development began to move beyond taxonomic studies towards an understanding of the role of decapod larvae in the marine environment, have investigators examined the question of energy utilization in decapod development.

The limited information on larval energy budgets is summarized in Table 4. In all studies the larvae were fed on *Artemia*, making the data comparable. All brachyuran species had

Table 4. Energetic efficiency of representative decapod larvae. a – Johns 1982, b – Mootz & Epifanio 1974, c – Levine & Sulkin 1979, d – Capuzzo & Lancaster 1979.

Species	Assimilation efficiency	Gross growth efficiency	Net growth efficiency
Cancer irroratus[a]	54%	27%	51%
Menippe mercenaria[b]	58%	30%	52%
Rhithropanopeus harrisii[c]	45%	30%	65%
Homarus americanus[d]	81%	28%	50%

Table 5. Changes in net growth efficiencies during larval development. See Table 4 for sources.

Species	Stage				
	1	2	3	4	5
Cancer irroratus	27.1	28.1	76.4	78.9	61.8
Menippe mercinaria	28.4	38.3	46.2	46.8	59.9
Rhithropanopeus harrisii	32.0	55.4	65.7	71.4	–

similar assimilation efficiencies over the complete larval development, but that for *Homarus* larvae was substantially higher. All assimilation efficiencies reported for larvae were within the range reported for holoplanktonic crustaceans (Conover 1978). An examination of changes in assimilation efficiency at various stages in development revealed a general increase in later stages for brachyuran larvae, probably related to the increased enzyme complexity found in later stage zoeae (Frank et al. 1975, Morgan et al. 1978). Assimilation remained constant throughout the larval stages of *Homarus.*

Assimilation efficiencies can vary greatly depending on the types of foods available (Conover 1978). While no data exist for larval crustaceans, Nelsen et al. (1977) found that assimilation in juvenile *Macrobrachium rosenbergii* varied according to the trophic level at which the animal functioned, herbivore, carnivore, or omnivore.

Gross and net growth efficiencies ranged from 27-30% and 50-65%, respectively, and were similar for all species. An examination of growth rates for individual stages indicated an increase in later stages and maximum growth in the last larval stage (Table 5). This is consistent with measurements of changes in lipid, protein and DNA/RNA in larvae of various species (Frank et al. 1975, Sulkin et al. 1975, Capuzzo & Lancaster 1979, Anger & Nair 1979). These studies indicated a rapid increase in total protein and lipid during the last two larval stages (Table 6). Sulkin et al. (1975) found DNA content increased in the last two zoeal stages of *Rhithropanopeus harrisii,* suggesting rapid cell proliferation presumably associated with morphogenesis.

5.1 *Growth efficiencies*

The growth efficiencies reported in these studies were derived from animals at basal metabolism and do not include increased metabolic demands associated with feeding or specific dynamic action (SDA). SDA is the increased calorigenic effect of catabolism (Krebs 1964).

Table 6. Changes in protein and lipid levels during larval development. a.– Recalculated from author's data. b – Based on dry weight. c – Based on wet weight. d – Postlarval stage.

Species	Stage	Larval weight	Protein	Lipid	Source
Homarus americanus[a, b]	1	1.0 mg	0.843 mg	61 μg	Cappuzo & Lancaster 1979
	2	1.9 mg	1.54 mg	118 μg	
	3	2.9 mg	2.37 mg	197 μg	
	4	7.2 mg	5.89 mg	396 μg	
	5[d]	8.9 mg	7.19 mg	365 μg	
Rhithropanopeus harrisii[a, c]	1	50 μg	2 μg	2 μg	Frank et al 1975
	2	100 μg	4 μg	4 μg	
	3	200 μg	13 μg	25 μg	
	4	335 μg	22 μg	35 μg	
	5[d]	395 μg	25 μg	31 μg	
Hyas araneus[b]	1	159 μg	72.6 μg	29.3 μg	Anger et al. 1983
	2	304 μg	167 μg	50.7 μg	
	3[d]	518 μg	205.4 μg	64.2 μg	

Table 7. Oxygen:Nitrogen ratio. Calculated from energetic data of: a – Capuzzo & Lancaster (1979), b – Johns (1982), c – Mootz & Epifanio (1974), d – Levine & Sulkin (1979).

Species	Stage				
	1	2	3	4	5
Homarus americanus[a]	26.7	26.3	26.6	22.1	–
Cancer irroratus[b]	28.6	31.2	12.9	12.2	19.1
Menippe mercenaria[c]	18.0	14.4	17.7	16.0	–
Rhithropaneus harrisii[d]	8.9	8.5	5.4	16.5	–

This calorigenic effect is believed to be associated with the amount of amino acid deamination and proportional to ingestion, given a constant food composition. Values for SDA in *Homarus* larvae ranged from 23-35 % for late stage larvae and 59-64 % for the first three larval stages (Capuzzo & Lancaster 1979).

Since SDA generally depends on the amount of dietary protein used for metabolic energy, knowledge of the substrate used for energy production is valuable for determining energy budgets. Substrate utilization (protein, carbohydrate, lipid) can be estimated from the oxygen consumption/nitrogen excretion ratio (O:N). This ratio was directly measured in two species of decapod larvae, *Homarus* (Capuzzo & Lancaster 1979) and *Cancer irroratus* (Johns 1981a,b). However, using a value of 7.37 x 10^{-3} joule/μg NH_4–N as the heat of formation of ammonia (Johns 1982), the O:N ratio could be estimated from energetics data for *Rhithropanopeus harrisii* (Levine & Sulkin 1979) and *Menippe mercenaria* (Mootz & Epifanio 1974) (Table 7). An O:N ratio of 7 indicates protein is the sole substrate used for energy production. The O:N ratios found for decapod larvae indicate protein to be the primary energy substrate with some utilization of lipid.

Increasing the levels of carbohydrate while reducing the level of protein in juvenile crustacean feeds results in increased growth and reduced SDA (e.g., Capuzzo & Lancaster 1979,

Table 8. Variations in number of larval stages.

Species	Stress type	Type of variation
Menippe mercenaria	Pollutant	Supernumerary stage
Rhithropanopeus harrisii	Endocrine/Metabolic imbalance	Supernumerary stage
	Starvation	Supernumerary or reduction
Callinectes sapidus	Food type	Supernumerary or reduction
Sesarma reticulatum	Endocrine/Metabolic imbalance	Supernumerary stage
Palaemonetes pugio	Food level	Supernumerary stage
Cyclograpsus integer	Temperature	Supernumerary stage

Clifford & Brick 1979, Capuzzo 1980). Since currently available larval energetics are based on *Artemia* fed animals the O:N ratios may be reflective of the high protein diet. Caution should be exercised in extrapolating these laboratory data to field conditions.

5.2 *Environmental stress*

Stressful environmental conditions, both natural and anthropogenic, can alter decapod larval energetics. Johns (1982) found that adverse temperatures and salinities could reduce growth of *Cancer irroratus* larvae. Net growth efficiencies ranged from 20-51 % depending on the culture temperature or salinity. Efficiencies of 32-34 % were found in those larvae which had low overall survival to the crab stage. Groups with survival rates greater than 50 % had net growth efficiencies of 48 to 51 %. A similar reduction in growth efficiency was reported for *C.irroratus* larvae exposed to 0.1 ppm water accommodated fraction of number 2 fuel oil (Johns & Pechenik 1980). In both cases, the reduction in growth was associated with an increased cost of metabolic maintenance. Larval lobsters exposed to naturally or chemically dispersed oil had increased rates of protein catabolism measured as a reduction in the O:N ratio (Capuzzo & Lancaster 1981, 1982). Larvae had a simultaneous increase in protein content and a decrease in lipid levels. These studies suggest that although increased amounts of protein were being metabolized for immediate energy needs, less protein was utilized for *de novo* lipid synthesis (see also Fingerman, Volume 3).

Alterations in metabolic costs may also explain the presence of supernumerary larval stages in a number of decapod species reared under stress conditions (Table 8). The mechanisms regulating larval growth and development are not well known. Marine invertebrate larval development and metamorphosis depend on efficient utilization of energy reserves, especially lipids (Crisp 1974, Holland 1978). Increased metabolic cost which alter the energy balance may limit growth, especially in the later zoeal stages. Assuming that a critical size is required to initiate metamorphosis (see 5.3.3), failure to reach a critical mass may result in a supernumerary stage.

5.3 *Molting, nutrition and morphogenesis*

Based on evidence accumulated over the last 20 years, it is clear that morphogenesis and molting are independent processes normally synchronized during larval development (Costlow 1968, Knowlton 1974, McConaugha 1982). It is also evident that larval nutrition can directly affect the rates of molting and morphogenesis. The concept of a 'Critical Point' in development has been applied to both processes. As previously discussed, certain nutritional and energetic requirements must be met before either process can proceed to completion.

The mechanism regulating morphogenesis in crustaceans resembles those of the insects in that both allometric and hormonal factors appear to be involved. The lack of detailed studies on crustaceans compared with the vast literature on insects prohibits an indepth comparison. However, the basic concepts underlying both systems will be reviewed.

5.3.1 *Larval molt cycle and hormonal regulation*

Larval crustaceans undergo a diecdysis type of molt cycle. This consists of a post-ecdysial period, followed by a short intermolt period that leads into proecdysis (Fig.7). The duration of the molt cycle appears to be species specific, but generally each phase can be measured in terms of hours or days as opposed to days, weeks or months in adult organisms. *Rhithropanopeus harrisii* larvae passed through these phases in a matter of hours at 25°C. Both endogenous and exogenous factors such as regeneration of limbs (McConaugha & Costlow 1981) and temperature (Costlow 1968, Costlow & Bookhout 1971), can alter the duration of any or all phases of the cycle.

Larval crustacean molting appears to be under the same endocrine controls as those found in adult crustaceans. Morphological studies of the larval Y-organ indicates that cyclic secretion is temporally related to the physical changes associated with ecdysis (McConaugha 1980). Ecdysone has been tentatively identified in larval lobsters by radioimmunoassay (Chang & Bruce 1981). Incubation studies indicate that β-ecdysone has an active role in regulating larval molting (McConaugha 1979, McConaugha & Costlow 1981, Freeman & Costlow 1983). Although early work (Costlow 1966a,b) suggested that larval eyestalks lack a molt inhibiting factor (MIH), recent work (Freeman & Costlow 1980) provided evidence for an active MIH factor located in larval eyestalks.

5.3.2 *Larval ecdysis and nutrition*

Knowlton (1974) suggested that food energy was partitioned into survival, molting and growth, with priority placed on survival. Experimental evidence now supports that hypothesis. Anger & Dawirs (1981) introduced the concept of 'point of reserve saturation' (PRS), the minimal feeding time before ecdysis to the next stage. Those authors found that stage I *Hyas araneus* larvae had a PRS_{50} of approximately three days or 30 % of the developmental time for stage I. Starvation after that point did not inhibit ecdysis to the

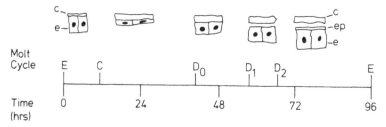

Figure 7. Temporal changes in epidermis and cuticle during larval molt cycle of *Rhithropanopeus harrisii*. c – cuticle, ep – epicuticle, E – epidermis. E, C, D_0, D_1, D_2, stages of molt cycle according to Drach (1942).

Table 9. Percentage of stage III animals completing a molt to stage IV following starvation initiated at different times in the molt cycle.

Time after molt (hours)	N	Number of molting to stage IV	Percent molting to stage IV
0	30	0	0%
5	25	0	0%
10	45	1	2%
13	38	15	40%
14	41	14	34%
16	37	14	38%
19	36	25	70%
24	43	43	100%

next stage. The earlier larvae were starved, the longer the duration of stage I and stage II even though food was not limiting during the second stage. Similar results have been found for stage I larvae of four other species of brachyuran larvae (Anger et al. 1981b). In all species, mortality rates through metamorphosis were consistently higher than in fed controls. The PRS_{50} for *R.harrisii* stage III larvae was 16-19 hours after ecdysis with a PRS_{100} at 24 hours post-ecdysis (Table 9) (McConaugha 1982). The acquisition of PRS apparently occurs abruptly between 10 and 12 hours after molting. While larvae starved prior to the acquisition of PRS failed to molt, these animals proceeded through premolt, stage D (Drach 1939), prior to suspension of the molt cycle. When food deprivation was maintained until stage D the molt cycle was reinitiated when food became available (McConaugha 1982).

5.3.3 *Nutrition and morphogenesis*

Costlow (1968) was the first to suggest that molting and morphogenesis were independent processes that were normally synchronized, a hypothesis based on a series of eyestalk ablation experiments (Costlow 1966a,b). Bilateral extirpation prior to a critical point in larval development resulted in a supernumerary larval stage. Knowlton (1974), using elevated temperature or variation in food level to alter molting frequency, also found a supernumerary stage in the larval development of *Palaemonetes pugio*. When starvation inhibited molting in stage III *R.harrisii* larvae, morphogenesis proceeded in the absence of molting. Morphogenesis, once initiated, was a time dependent process (Fig.8), as indicated by the gradation in morphology of larvae starved for 72 to 144 hours, culminating in the loss of the last zoeal stage. Thus, studies in which the molt cycle is accelerated or inhibited support the concept that crustacean morphogenesis is independent of larval molting. This raises questions regarding allocation of energy between molting, growth and morphogenesis.

A survey of the literature suggests that decapod larval development can be divided into two phases, an early slow growth phase and a later metamorphic phase. The latter phase can be characterized by rapid growth and preparation for the morphological changes associated with metamorphosis. In the brachyurans this phase is normally restricted to the penultimate and ultimate larval stages.

Growth during the early larval stages consists of a gradual increase in lipid and protein content over two or more stages. Based on DNA/RNA ratios, increase in size is due to an increase in cell volume with little increase in cell number (Sulkin et al. 1975). Although the evidence is equivocal, it appears that growth has an energetic priority over molting during this phase. However, it is not a simple relationship. Failure to initiate growth and reach the minimum value (PRS) during the first zoeal stage results in death of the animal. Once the larvae surpasses the minimum growth level (PRS) the animal enters the molt cycle, an obligatory step for long term growth. In the event that food deprivation occurred after this commitment, the animal completes ecdysis but prolongs the subsequent molt cycle and to compensate for lost feeding time. Starvation immediately after eclosion may result in molt suspension and require a recovery period equal to or greater than the normal feeding period (Fig.2). The animal can compensate by reducing the duration of the following molt cycle.

Brachyuran larvae appear to have an abrupt transition from the slow growth phase to rapid preparation for metamorphosis. The critical period occurs during the first 24 hours of the penultimate larval stage and is characterized by rapid accumulation of lipid and protein. Disruption of the normal hormonal pattern (Costlow 1966a,b) or energetics (McConaugha 1982) prior to this critical point can alter the normal developmental sequence. After this critical point larvae are committed to metamorphosis.

In the insect system, metamorphosis is dependent on attainment of a specific weight (Allegret 1964, Nijhout 1975, Blankley & Goodner 1978). Allegret found that in the wax moth, *Galleria mellonella,* larvae are committed to metamorphosis during the penultimate larval stage and metamorphosis will proceed, provided the energetic needs of the larvae are met. Starvation during the last larval stage produced nonviable larval-pupal intermediates, attributed to high levels of juvenile hormone (Nijhout 1975). Experimental evidence from at least one species of brachyuran suggests that a similar mechanism may regulate crustacean metamorphosis. Feeding for 24 hours postmolt is sufficient to commit penultimate *R. harrisii* larvae to metamorphosis, provided food is available through development of the ultimate larval stage. Starvation for a period of three to five days during the last stage resulted in a small percentage of larvae molting to a supernumerary stage. Feeding for 48 hours postmolt committed larvae to metamorphosis, regardless of a short period of food deprivation during the last stage (Table 10). The development of partially metamorphosed animals, when the molt cycle is inhibited is consistent with the hypothesis that molting and metamorphosis are independent processes; however, a time dependent factor may allow for cell differentiation and mitosis.

Once the critical mass has been obtained in the insect system, there is a nervous-hormonal feedback system that results in the release of ecdysones in the absence of juvenile hormone (Nijhout 1975). Our knowledge of juvenile factors in crustaceans is extremely limited. Gomez et al. (1973) found that exposure to a juvenile hormone mimic (ZR-512) for one to three hours stimulated premature metamorphosis of cyprid *Balanus galeatus*

Table 10. Effect of starvation after critical point of stage III *Rhithropanopeus harrisii.*

Stage III Time of starvation	Fed % megalopa	Stage IV Starved % megalopa	% zoea V
24	100	64	36
48	100	100	–

larvae into unattached adults. Similar results were reported for three other JH mimics (Cheung & Nigrelli 1973, Tighe-Ford 1977) and insect JH (Ramenofsky et al. 1974). These cyprid/adult intermediates are similar to the pupal/adult intermediates formed by holometabolous insects in response to JH analogues. Exposure of sixth stage nauplii of *Elminus modestus* to farnesyl methyl ether, another JH analogue, resulted in delayed or reduced metamorphosis and the retenion of some naupliar characteristics to form nauplii/cyprid intermediates (Tighe-Ford 1977).

Studies in which decapod larvae were subjected to JH mimics include chronic exposures (Christiansen et al. 1977a,b), which resulted in a reduced survival and an increased duration of larval life with increasing concentration. However, there were rare instances of supernumerary larval stages following exposure to the juvenile mimic ZR-512. Freeman & Costlow (1979) reported the presence of a water-soluable juvenile factor in stage III. *R.harrisii* larvae, which inhibited metamorphosis of the dorsal spine. The highest titer of this factor occurred in stage III larvae. It is interesting to speculate about the role of this factor in regulation of brachyuran metamorphosis (McConaugha 1982). However, further work on characterization of this factor is needed.

6 SUMMARY

Recent interest in the ecological role of meroplankton in zooplankton communities has sparked interest in larval nutrition and energetics. Meroplanktonic larvae of benthic invertebrates often dominate nearshore and estuarine zooplankton communities in temperal and boreal seas during spring and summer. Crustacean larval forms often contribute a major portion of these larval forms. During this period, they may have a major influence on the ecology and productivity of the pelagic system. Additional interest in larval nutrition has been generated from intense aquaculture of shellfish.

Available information suggests that most crustacean larvae can withstand short periods of starvation or sub-optimal feeding. Although most decapod larvae are carnivorous for most or all of their larval phase, there is now evidence to suggest that phytoplankton may provide limited nutrition. This may play a critical role under natural conditions, since 'optimal' prey densities reported from laboratory studies are consistently higher than natural prey concentrations. Maximum concentrations of decapod larvae reported in the literature is one to two thousand *Callinectes sapidus* larvae per cubic meter (McConaugha et al. 1983). Under these conditions most prey would be consumed rapidly, presumably resulting in high mortality and the presence of zoeal/megalopal intermediates.

Based on available information, crustacean aquaculturists should be concerned with maximizing the bioenergetics of larval culture. Particular concern should be directed towards maximizing the amounts of PUFA, reducing environmental stress which can ad-

---

Figure 8. Varying degree of morphogenesis associated with starvation delayed molting in stage III larvae of *Rhithropanopeus harrisii* following molt to stage IV. (A) Thoracic appendages of larva starved 72 hr. Note formation of dactyl spines. (B) Thoracic appendages of larva starved for 120 hr. Initial phase of segmentation and myomere formation. Presence of several sensory hairs. (C) Functional cheliped of larva starved for 144 hr. Note larva number of sensory hairs. (D) Functional pleopods of larva starved for 144 hr. All morphological structures characteristic of megalopa features A_b − Abdomen, C − Cheliped, DS − Dactyl spine, M − Myomere, SH − Sensory hair, T − Thoracic appendage.

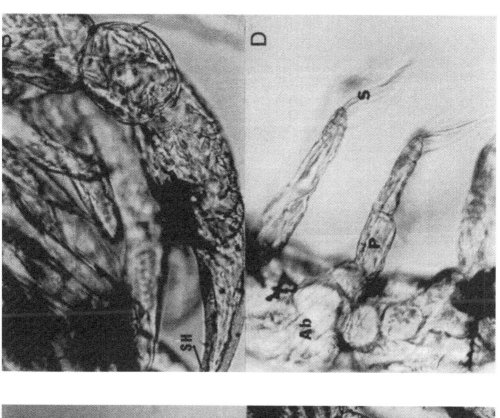

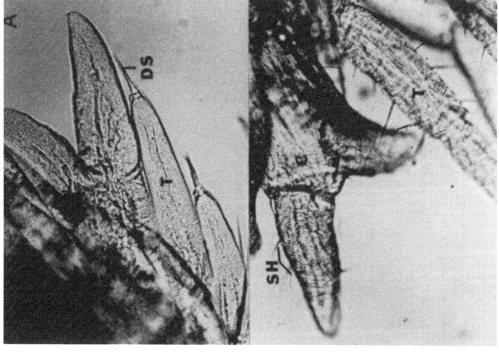

versely alter growth efficiencies, and examine predator-prey size ratio and density to maintain larval feeding rates at optimal levels.

REFERENCES

Allegret, P. 1964. Interrelationship of larval development, metamorphosis and age in a pyralid lepidopteran, *Galleria mellonella* (L.) under the influence of dietitic factors. *Exp. Geront.* 1:49-66.
Anger, K. & K.K.C.Nair 1979. Laboratory experiments on the larval development of *Hyas araneus* (Decapoda, Mijidae). *Helgolander wiss. Meeres.* 32:36-54.
Anger, K. & R.R.Dawirs 1981. Influence of starvation on the larval development of *Hyas araneus* (Decapoda, Majidae). *Helgolander wiss. Meeres.* 34:287-311.
Anger, K., R.R.Dawirs, V.Angler, J.W.Goy & J.D.Costlow 1981a. Starvation resistance in first stage zoeae of brachyuran crabs in relation to temperature. *J. Crust. Biol.* 1:518-525.
Anger, K., R.R.Dawirs, V.Angler & J.D.Costlow 1981b. Effects of early starvation periods on zoeal development of brachyuran crabs. *Biol. Bull.* 161:199-212.
Anger, K., N.Loasch, C.Puschel & F.Schorn 1983. Changes in biomass and chemical composition of spider crab (*Hyas araneus*) larvae reared in the laboratory. *Mar. Ecol. Prog. Ser.* 12:91-101.
Bigford, T.E. 1977. Effect of several diets on survival development time and growth of laboratory reared spider crab, *Libinia emarginata*, larvae. *Fish. Bull. US* 76:59-64.
Blankley, N. & S.R.Goodner 1978. Size-dependent timing of metamorphosis in milkweed bugs (Oncopeltus) and its life history implications. *Biol. Bull.* 155:499-510.
Blaxter, J.H. & G.Hempel 1963. The influence of egg size on herring larvae (*Clupea harengus*). *J. Con. Perm. Int. Explor. Mer.* 28:211-240.
Bookhout, C.G. & J.D.Costlow 1970. Nutritional effects of Artemia from different locations on larval development of crabs. *Helolander wiss. Meeres.* 20:435-442.
Bookhout, C.G. & J.D.Costlow 1974. Crab development and effects of pollutants. *Thalassia Jugosl.* 10:77-87.
Bookhout, C.G., A.J.Wilson jr, T.W.Duke & J.I.Love 1972. Effects of mirex on the larval development of two crabs. *Water, Air, Soil Pollution* 1:165-180.
Bookhout, C.G., J.D.Costlow & R.Monroe 1976. Effects of methoxychlor on larval development of mud crab and blue crab. *Water, Air, Soil Pollution* 5:349-365.
Bookhout, C.G., J.D.Costlow & R.Monroe 1980. Kepone effects on larval development of mud crab and blue crab. *Water, Air, Soil Pollution* 13:57-77.
Brick, R.W. 1974. Effects of water quality, antibiotics, phytoplankton and food on survival and development of larvae of *Scylla serrata* (Crustacea: Portunidae). *Aquaculture* 3:231-244.
Broad, A.C. 1957. The relationship between diet and larval development of *Palaemonetes*. *Biol. Bull.* 112:162-170.
Brownell, W.N., A.J.Provenzano & M.Martinez 1977. Culture of the West Indian spider crab, *Mithrax spinosissimus* at Los Roques, Venezuela. *Proc. Wld. Mariculture Soc.* 8:157-163.
Burton, R.S. 1979. Depth regulation at the first stage zoea larvae of the sand crab *Emerita analoga* (Stimpson) (Decapoda, Hippidae). *J. Exp. Mar. Biol. Ecol.* 37:255-270.
Capuzzo, J.M. 1980. The effect of low-protein feeds on bioenergetics of juvenile lobsters (*Homarus americanus*). In: R.C.Bayer & A'Dgostino (eds.), *1980 Lobster Nutrition Workshop Proceedings:*11-20. Maine Sea Grant Technical Report 58.
Capuzzo, J.M. & B.A.Lancaster 1979. Some physiological and biochemical considerations of larval development in the American lobster, *Homarus americanus* Milne Edwards. *J. Exp. Mar. Biol. Ecol.* 40:53-62.
Capuzzo, J.M. & B.A.Lancaster 1981. Physiological effects of South Louisiana crude oil on larvae of the American lobster (*Homarus americanus*). In: F.J.Vernberg, A.Calabrese, F.P.Thurberg & W.B. Vernberg (eds.), *Biological Monitoring of Marine Pollutants:*405-423. New York: Academic Press.
Capuzzo, J.M. & B.A.Lancaster 1982. Physiological effects of petroleum hydrocarbons on larval lobsters *(Homarus americanus):* Hydrocarbon accumulation and interference with lipid metabolism. In: *Physiological Mechanisms of Marine Pollutant Toxicity:*477-499. New York: Academic Press.
Carlberg, J.M. & J.C.VanOlst 1980. Brine shrimp (*Artemia salina*) consumption by the larval stages of

the American lobster (*Homarus americanus*) in relation to food density and water temperature. *Proc. Wld. Mariculture Soc.* 11:379-389.

Castell, J.D.E., E.G.Mason & J.F.Covey 1975. Cholesterol requirements of American lobster (*Homarus americanus*). *J. Fish. Res. Bd. Can.* 32:1431-1435.

Chang, E.S. & M.J.Bruce 1981. Ecdysteroid titers of larval lobsters. *Comp. Biochem. Physiol.* 70A:239-241.

Cheung, P.J. & R.F.Nigrelli 1973. The development of barnacles from cyprids in pre-heated sea waters with and without farnesol. *Amer. Zool.* 13:1339-1340.

Chia, F.S. 1974. Classification and adaptive significance of developmental patterns in marine invertebrates. *Thalassia Jugosl.* 10:121-130.

Christiansen, M.E. & W.T.Yang 1976. Feeding experiments on the larvae of the fiddler crab *Uca pugilator* (Brachyura, Ocypodidae), reared in the laboratory. *Aquaculture* 8:91-98.

Christiansen, M.E., J.D.Costlow & R.J.Monroe 1977a. Effects of the juvenile hormone mimic ZR-515 (Altosid) on larval development of the mud crab, *Rhithropanopeus harrisii* in various salinities and cyclic temperatures. *Mar. Biol.* 39:269-279.

Christiansen, M.E., J.D.Costlow & R.J.Monroe 1977b. Effects of the juvenile hormone mimic ZR-512 (Altozar) on larval development of the mud crab, *Rhithropanopeus harrisii* at various cyclic temperatures. *Mar. Biol.* 39:281-288.

Clegg, J.S., A.H.Warner & F.J.Finamore 1967. Evidence for the function of P^1, P^4-diguanosine 5 -tetraphosphate in the development of *Artemia salina*. *J. Biol. Chem.* 242:1938-1967.

Clifford, H.C. & R.W.Brick 1979. Protein utilization in the freshwater shrimp, *Macrobrachium rosenbergii*. *Proc. Wld. Mariculture Soc.* 8:841-852.

Conklin, D.E. 1980. Nutrition. In: J.S.Cobb & B.F.Phillips (eds.), *The Biology and Management of Lobsters:*277-300. New York: Academic Press.

Conover, R.J. 1978. Transformation of organic matter. In: O.Kinne (ed.), *Marine Ecology Vol.IV.:*221-456. New York: John Wiley & Sons.

Costlow, J.D. 1966a. The effect of eyestalk extirpation on larval development of the mud crab, *Rhithropanopeus harrisii* (Gould). *Gen. Comp. Endocrinol.* 7:255-274.

Costlow, J.D. 1966b. The effect of eyestalk extirpation on the larval development of the crab, *Sesarma reticulatum* Say. In: H.Barnes (ed.), *Some Contemporary Studies in Marine Science:*209-224. London: George Allen & Unwin.

Costlow, J.D. 1967. The effect of salinity and temperature on survival and metamorphosis of megalopa of the blue crab, *Callinectes sapidus*. *Helgolander wiss. Meeres.* 15:84-97.

Costlow, J.D. 1968. Metamorphosis in crustaceans. In: W.Etkin & L.Gilbert (eds.), *Metamorphosis a problem in Developmental Biology:*3-41. New York: Academic Press.

Costlow, J.D. & C.G.Bookhout 1962. The effects of environmental factors on larval development of crabs. In: *Biological Problems in Water Pollution 3rd Seminar:*77-86. New York: Academic Press.

Costlow, J.D. & C.G.Bookhout 1969. Temperature and meroplankton. *Chesapeake Sci.* 10:252-257.

Costlow, J.D. & C.G.Bookhout 1971. The effect of cyclic temperatures on larval development in the mud crab *Rhithropanopeus harrisii*. In: D.J.Crisp (ed.), *Fourth European Marine Biology Symposium:* 211-220. Cambridge: Cambridge University Press.

Costlow, J.D. & A.Sastry 1966. Free amino acids in developing stages of two crabs, *Callinectes sapidus* Rathbun and *Rhithropanopeus harrisii* (Gould). *Acta Embryol. Morphol. Exper.* 9:44-55.

Crisp, D.J. 1974. Energy relations of marine invertebrate larvae. *Thalassia Jugosl.* 10:77-87.

Cronin, T. & R.B.Forward jr 1980. The effects of starvation on phototaxis and swimming of larvae of the crab, *Rhithropanopeus harrisii*. *Biol. Bull.* 158:283-294.

D'Abramo, L.R., C.E.Bordner, D.E.Conklin & N.A.Baum 1981. Essentiality of dietary phosphatidylcholine for the survival of juvenile lobsters. *J. Nutrition* 111:425-431.

Dawirs, R.R. 1980. Elemental composition (C, H, N) in larval and crab-1 stages of *Pagurus bernhardus* (Decapoda, Paguridae) and *Carcinus maenas* (Decapoda, Portunidae). *Mar. Biol.* 57:17-23.

Dawirs, R.R. 1981. Elemental composition (C, H, N) and energy in the development of *Pagurus bernhardus* (Decapoda, Paguridae) megalopa. *Mar. Biol.* 64:117-123.

Dobkin, S. 1963. The larval development of *Palaemonetes paludosus* (Gibbes 1850) (Decapoda, Palaemonidae) reared in the laboratory. *Crustaceana* 6:41-61.

Dobkin, S. 1968. The larval development of a species of *Thor* (Caridea Hippolytidae) from south Florida, USA. *Crustaceana* suppl.2:1-18.

Drach, P. 1939. Mue et cycle d'intermue chez crustaces decapods. *Ann. Inst. Oceanog.* 19:103-391.

Frank, J.R., S.D.Sulkin & R.P.Morgan 1975. Biochemical changes during larval development of the xanthid crab, *Rhithropanopeus harrisii*. I. Protein, total lipid, alkaline phosphatase, and glutamic oxaloacetic transaminase. *Mar. Biol.* 32:105-111.

Freeman, J.A. & J.D.Costlow 1979. Endocrine regulation of metamorphosis in crab larvae. *Am. Zool.* 19:979.

Freeman, J.A. & J.D.Costlow 1980. The molt cycle and its hormonal control in *Rhithropanopeus harrisii* larvae. *Devel. Biol.* 74:479-485.

Freeman, J.A. & J.D.Costlow 1983. The cyprid molt cycle and its hormonal control in the barnacle *Balanus amphitrite*. *J. Crust. Biol.* 3:173-182.

Fujiamura, T. 1966. Notes on the development of a practical mass culturing technique of the giant prawn *Macrobrachium rosenbergii*. *Indo-Pac. Fish. Counc. Proc.* 12:1-4.

Fujita, S., T.Watanabe & C.Kitajima 1980. Nutritional quality of Artemia from different localities as a living feed for marine fish from the viewpoint of essential fatty acids. In: L.G.Persoone, P.Sorgeloos, O.Roels & E.Jaspers (eds.), *The Brine Shrimp Artemia, Vol.3:*277-290. Wettesen, Belgium: Universal Press.

Gomez, E.D., D.J.Faulkner, W.A.Newman & C.Ireland 1973. Juvenile hormone mimics: Effect on cirriped crustacean metamorphosis. *Sci.* 179:813-814.

Goy, J.W. & J.D.Costlow 1980. Nutritional effects of *Artemia* from different geographical strains on larval development of decapod crustaceans. *Am. Zool.* 20:888.

Guany, J.-C.B. & A.Kanazawa 1973. Distribution and fate of exogenous cholesterol during the molting cycle of the prawn, *Penaeus japonicus*, Bate. *Comp. Biochem. Physiol.* 46A:5-10.

Gurney, R. 1942. *Larvae of Decapod Crustacea*. London: Ray Soc.

Hernandorena, A. 1979. Relationship between purine and pyrimidine dietary requirements and *Artemia salina* morphogenesis. *Comp. Biochem. Physiol.* 62B:7-12.

Hernandorena, A. 1980. Programmation of postembryonic development in *Artemia* by dietary supplies of purine and pyrimidine. In: G.Persoone, P.Sorgeloos, O.Roels & E.Jaspers (eds.), *The Brine Shrimp Artemia, Vol.2:*209-218. Wetteren, Belgium: Universa Press.

Hernandorena, A. 1981. Activite enzymatique et espression d'une information genetique chez *Artemia* (Crustace, Brachiopode). *C.R. Acad. Sci., Paris* 292:705-708.

Holland, D.L. Lipid reserves and energy metabolism in the larvae of benthic marine invertebrates. In: D.C.Malins & J.R.Saigent (eds.), *Biochemical and Biophysical Perspectives in Marine Biology, Vol.4:* 85-123. New York: Academic Press.

Hue, J.S., K.S.Bang & Y.K.Rho 1972. Studies on growth and artificial rearing of the larval blue crab, *Portunus trituberculatus*. *Bull.Korean Fish.Res.Dev. Agency* 9:55-70.

Johns, D.M. 1981a. Physiological studies on *Cancer irroratus* larvae. I. Effects of temperature and salinity on survival, development rate and size. *Mar. Ecol. Prog. Ser.* 5:75-83.

Johns, D.M. 1981b. Physiological studies on *Cancer irroratus* larvae. II. Effects of temperature and salinity on physiological performance. *Mar. Ecol. Prog. Ser.* 6:309-315.

Johns, D.M. 1982. Physiological studies on *Cancer irroratus* larvae. III. Effects of temperature and salinity on the partitioning of energy resources during development. *Mar. Ecol. Prog. Ser.* 8:75 85.

Johns, D.M., M.E.Peters & A.D.Beck 1980. International study on *Artemia*. VI. Nutritional value of geographical and temporal strains of *Artemia:* effects on survival and growth of two species of brachyuran larvae. In: G.Personne, P.Sorgeloos, O.Roels & E.Jaspers (eds.), *The Brine Shrimp Artemia. Vol.1:* 291-304. Wetteren, Belgium: Universa Press.

Johns, D.M, W.J.Berry & S.McLean 1981. International study on *Artemia*. XXI. Investigations into why some strains of *Artemia* are better food sources than others. Further nutritional work with larvae of the mud crab *Rhithropanopeus harrisii*. *J. Wld. Maricul. Soc.* 12:303-314.

Johns, D.M. & J.A.Pechenik 1980. Influence of the water accommodated fraction of No.2 fuel oil on energetics of *Cancer irroratus* larvae. *Mar. Biol.* 55:247-254.

Jones, D.A., T.H.Moeller, R.J.Campbell, J.G.Munford & P.A.Gabbot 1976. Studies on the design and acceptability of micro-encapsulated diets for marine particulate feeders. I. Crustacea. In: G.Persoone & E.Jaspers (eds.), *Proc. 10th Eur. Symp. Mar. Biol. Ostend, Belgium:*229-239. Wetteren, Belgium: Universa Press.

Jones, D.A., A.Kanazawa & K.Ono 1979. Studies on the nutritional requirements of the larval stages of *Penaeus japonicus* using microencapsulated diets. *Mar. Biol.* 54:261-267.

Joseph, J.D. 1977. Assessment of the nutritional role of algae in the culture of larval prawns (*Macrobrachium rosenbergii*). *Proc. Wld. Mariculture Soc.* 8:853-864.

Kanazawa, A. & S.Teshima 1971. In vivo conversion of cholesterol to steroid hormones in the spiny lobster, *Panulirus japonica. Bull. Jap. Soc. Sci. Fish.* 37:891-898.

Kanazawa, A. & S.Teshima 1977. Biosynthesis of fatty acids from acetate in the prawn, *Penaeus japonicus. Mem. Fac. Fish., Kagoshima Univ.* 26:49-53.

Kanazawa, A., N.Tanaka, S.Teshima & K.Kashiwada 1971. Nutritional requirements of prawn. II. Requirements for sterols. *Bull. Jap. Soc. Sci. Fish.* 37:211-215.

Kanazawa, A., S.Teshima & S.Tokiwa 1977. Nutritional requirements of prawn – VII. Effect of dietary lipids on growth. *Bull. Jap. Soc. Sci. Fish.* 43:849-856.

Kanazawa, A., S.Teshima, S.Tokiwa, M.Kayama & M.Hirata 1979a. Essential fatty acids in the diet of prawn-II. Effect of docosahexaenoic acid on growth. *Bull. Jap. Soc. Sci. Fish.* 45:1151-1153.

Kanazawa, A., S.Teshima & K.Ono 1979b. Relationship between essential fatty acid requirements of aquatic animals and the capacity for bioconversion of linolenic acid to highly unsaturated fatty acids. *Comp. Biochem. Physiol.* 63B:295-298.

Knowlton, R.E. 1974. Larval developmental processes and controlling factors in decapod crustacea with emphasis on Caridea. *Thalassia Jugosl.* 10:138-158.

Kon,T. 1979. Ecological studies on larvae of the crabs belonging to the genus *Chionoecetes* I. The influence of starvation on the survival and growth of the Zuwai crab. *Bull. Jap. Soc. Sci. Fish.* 45:7-9.

Krebs, H.A. 1964. The metabolic fate of amino acids. In: H.N.Munro & B.Allison (eds.), *Mamalian Protein Metabolism:* 125-176. New York: Academic Press.

Kurata, H. 1959. Studies on the larva and post-larva of *Paralithodes camtschatica* I. Rearing of the larvae with special reference to the food of the zoea. *Bull. Hokkaido Reg. Fish. Res. Lab.* 20:76-83.

Lebour, M.V. 1928. The larval stages of the Plymouth Brachyura. *Proc. Zool. Soc. London* 1928:473-560.

Levine, D.M. & S.D.Sulkin 1979. Partitioning and utilization of energy during larval development of the xanthid crab, *Rhithropanopeus harrisii* (Gould). *J. Exp. Mar. Biol. Ecol.* 40:247-257.

Maddox, M.B. & J.J.Manzi 1976. The effects of algal supplements on static system culture of *Macrobrachium rosenbergii* (de Man) larvae. *Proc. Wld. Mariculture Soc.* 7:677-678.

Manzi, J.J., M.B.Maddox & P.A.Sandifer 1977. Algal supplement enhancement of *Macrobrachium rosenbergii* (de Man) larviculture. *Proc. Wld. Mariculture Soc.* 8:207-223.

McConaugha, J.R. 1979. The effects of 20 hydroxecdysone on survival and development of first and third stage *Cancer anthonyi* larvae. *Gen. Comp. Endocrinol.* 37:421-427.

McConaugha, J.R. 1980. Identification of the Y-organ in the larval stages of *Cancer anthonyi. J.Morph.* 664:83-88.

McConaugha, J.R. 1982. Regulation of crustacean morphogenesis in larvae of the mud crab, *Rhithropanopeus harrisii. J. Exp. Zool.* 223:155-163.

McConaugha, J.R. & J.D.Costlow 1981. Ecdysone regulation of larval crustacean molting. *Comp. Biochem. Physiol.* 68A:91-93.

McConaugha, J.R., D.F.Johnson, A.J.Provenzano & R.Maris 1983. Seasonal distribution of larvae of *Callinectes sapidus* (Crustacea, Decapoda) in the waters adjacent to the Chesapeake Bay. *J. Crust. Biol.* 3:582-591.

Mileikovsky, S.A. 1971. The 'pelagic larvation' and its role in the biology of the world ocean, with special reference to pelagic larvae of marine bottom invertebrates. *Mar. Biol.* 16:13-21.

Mock, C.R., C.T.Fontaine & D.B.Revera 1980. Improvements in rearing larval penaeid shrimp by the Galveston Laboratory method. In: G.Persoone, P.Sorgeloos, O.Roels & E.Jaspers (eds.), *The Brine Shrimp Artemia Vol.3:* 331-342. Wetteren, Belgium: Universa Press.

Morgan, R.P. II, E.Kramarsky & S.D.Sulkin. Biochemical changes during the zoeal development of the xanthid crab, *Rhithropanopeus harrisii* III. Isozyme changes during ontogeny. *Mar. Biol.* 48:223-226.

Mootz, C.A. 1973. Energetics and feeding rates of the larvae of *Menippe mercenaria* Say. Masters thesis, University of Delaware.

Mootz, C.A. & C.E.Epifanio 1974. An energy budget for *Menippe mercenaria* larvae fed *Artemia nauplii. Biol. Bull.* 146:44-55.

Nelsen, S.G., A.W.Knight & H.W.Li 1977. The metabolic cost of food utilization and ammonia production by juvenile *Macrobrachium rosenbergii. Comp. Biochem. Physiol.* 57A:67-72.

New, M.B. 1976. A review of dietary studies with shrimps and prawns. *Aquaculture* 9:101 144.

New, M.B. 1980. A bibliography of shrimp and prawn nutrition. *Aquaculture* 21:101-128.

Nijhout, H.F. 1975. Dynamics of juvenile hormone action in larvae of the tobacco hornworm *Manduca sexta. Biol. Bull.* 149:568-579.

Omori, M. 1979. Growth, feeding and mortality of larval and early postlarval stages of the oceanic shrimp *Sergestes similis* Hansen. *Limnol. Oceanogr.* 24:273-288.

Olney, C.E., P.S.Schauer, S.McLean, Y.Li & K.L.Simpson 1980. International study on *Artemia* VIII. Comparison of the chlorinated hydrocarbons and heavy metals in five different strains of newly hatched *Artemia* and a laboratory reared marine fish. In: G.Persoone, P.Sorgeloos, O.Roels and E.Jaspers (eds.), *The Brine Shrimp Artemia Vol.3:*343-352. Wetteren, Belgium: Universa Press.

Paffenhoffer, G. & R.P.Harris 1979. Laboratory culture of marine holozooplankton and its contribution to studies of marine planktonic food webs. *Adv. Mar. Biol.* 16:211-308.

Paul, A.J. & J.M.Paul 1980. The effect of early starvation on later feeding success of king crab zoeae. *J. Exp. Mar. Biol. Ecol.* 44:247-251.

Paul, A.J., J.M.Paul, P.A.Shoemaker & H.M.Feder 1979. Prey concentrations and feeding response in laboratory reared stage one zoeae of king crab snow crab and pink shrimp. *Trans. Am. Fish Soc.* 108:440-443.

Pike, R.B. & D.I.Williamson 1961. The larvae of *Spirontocaris* and related genera (Decapoda, Hippolytidae). *Crustaceana* 2:187-208.

Provenzano, A.J. & J.Goy 1976. Evaluation of a sulphate lake strain of *Artemia* as a food for larvae of the grass shrimp, *Palaemonetes pugio. Aquaculture* 9:343-350.

Provenzano, A.J., J.R.McConaugha, K.B.Philips, D.F.Johnson & J.Clark 1983. Vertical distribution of first stage larvae of the blue crab, *Callinectes sapidus* at the mouth of Chesapeake Bay. *Estuarine Coastal and Shelf Sci.* 16:489-499.

Ramenofsky, M., D.J.LFaulkner & C.Ireland 1974. Effect of juvenile hormone on cirriped metamorphosis. *Biochem. Biophys. Res. Commun.* 60:172-177.

Reed, P.H. 1969. Culture methods and effects of temperature and salinity on survival and growth of Dungeness crab (*Cancer magister*) larvae in the laboratory. *J. Fish. Res. Bd. Can.* 26:389-397.

Reeve, M.R. 1969. Growth, metamorphosis and energy conversion in the larvae of the prawn *Palaemon serratus. J. Mar. Biol. Ass. UK* 49:77-96.

Reese, E.S. & R.A.Kinzie 1968. The larval development of the coconut or robber crab *Birgus latro* (L.) in the laboratory (Anomura, Paguridea). *Crustaceana* suppl.2:117-144.

Regnault, M. 1969. Etude experimentale de la nutrition d'*Hippolyte inermis* Leach (Decapoda, *Natantia*) au cours de son developpement larvaire, au laboratoire. *Int. Revue ges. Hydrobiol.* 54:749-764.

Rice, A.L. & A.J.Provenzano 1965. The zoeal stages and the glaucothoe of *Paguristes sericeus* A.Milne Edwards (Anormura, Diogenidae). *Crustaceana* 8:239-254.

Roberts, M.H.Jr 1974. Larval development of *Pagurus longicarpus* Say reared in the laboratory V. Effect of diet on survival and molting. *Biol. Bull.* 146:67-77.

Rust, J.D. & F.Carlson 1960. Some observations on rearing blue crab larvae. *Chesapeake Sci.* 1:196-197.

Sandifer, P.A. 1972. Effects of diet on larval development of *Thor floridanus* (Decapoda, Caridea) in the laboratory. *Va. J. Sci.* 23:5-8.

Sandoz, M. & R.Rogers 1944. The effect of environmental factors on hatching, molting and survival of zoea larvae of the blue crab, *Callinectes sapidus* Rathbun. *Ecol.* 25:216-228.

Schauer, P.S., D.M.Johns, C.E.Olney & K.L.Simpson 1980. International Study on *Artemia* IX. Lipid level, energy content and fatty acid composition of the cysts and newly hatched nauplii from five geographical strains of *Artemia.* In: G.Persoone, P.Sorgeloos, O.Roels & E.Jaspers (eds.), *The Brine Shrimp Artemia Vol.3:*365-373. Wetteren, Belgium: Universa Press.

Seidel, C.R., D.M.Johns, P.S.Schauer & C.E.Olney 1982. International study on *Artemia* XXVI. Food value of nauplii from reference *Artemia* cysts and four geographical collections of *Artemia* for mud and crab larvae. *Mar. Ecol. Prog. Ser.* 8:309-312.

Singarajah, K.V., J.Moyse & E.W.Knight-Jones 1967. The effect of feeding upon the phototactic behavior of cirriped nauplii. *J. Exp. Mar. Biol. Ecol.* 1:143-153.

Spaziani, E. & S.B.Kater 1973. Uptake and turnover of cholesterol-^{14}C in Y-organ of the crab, *Hemigrapsus* as a function of molt cycle. *Gen. Comp. Endocrinol.* 20:534-549.

Strathman, R.R. 1982. Selection for retention or export of larvae in estuaries. In: V.S.Kennedy (ed.), *Estuarine Comparisons:*521-551. New York: Academic Press.

Sulkin, S.D. 1975. The significance of diet on the growth and development of larvae of the blue crab, *Callinectes sapidus* Rathbun, under laboratory conditions. *J. Exp. Mar. Biol. Ecol.* 20:119-135.

Sulkin, S.D. 1978. Nutritional requirements during larval development of the portunid crab *Callinectes sapidus* Rathbun. *J. Exp. Mar. Biol. Ecol.* 34:29-41.

Sulkin, S.D. & C.E.Epifanio 1975. Comparison of rotifers and other diets for rearing early larvae of the blue crab *Callinectes sapidus* Rathbun. *Estuarine Coastal Mar. Sci.* 3:109-113.

Sulkin, S.D. & K.Norman 1976. A comparison of two diets in the laboratory culture of the zoeal stages of the brachyuran crabs *Rhithropanopeus harrisii* and *Neopanope* sp. *Helgolander wiss. Meeres.* 28: 183-190.

Sulkin, S.D., R.P.Morgan II & L.L.Minasian 1975. Biochemical changes during the larval development of the xanthid crab, *Rhithropanopeus harrisii* (Gould). II. Nucleic acids. *Mar. Biol.* 32:113-118.

Sulkin, S.D. & W.F.van Heukelem 1980. Ecological and evolutionary significance of nutritional flexibility in planktotrophic larvae of the deep sea red crab, *Geryon quinquedens* and the stone crab, *Menippe mercenaria*. *Mar. Ecol. Prog. Ser.* 2:91-95.

Teshima, S., A.Kanazawa & H.Okamoto 1977. Variation in lipid classes during the molting cycle of the prawn *Penaeus japonicus*. *Mar. Biol.* 39:1219-136.

Teshima, S. 1972. Studies on the sterol metabolism in marine crustaceans. *Mem. Fac. Fish. Kagoshima Uni.* 21:69-147.

Thorson, G. 1950. Reproductive and larval ecology of marine bottom invertebrates. *Biol. Rev.* 25:1-45.

Tighe-Ford, D.J. 1977. Effects of juvenile hormone analogues on larval metamorphosis in the barnacle, *Elminius modestus* Darwin (Crustacea, Cirripedia). *J. Exp. Mar. Biol. Ecol.* 26:163-176.

Tucker, R.K. 1978. Free amino acids in developing larvae of the stone crab, *Menippe mercenaria*. *Comp. Biochem. Physiol.* 60A:169-172.

Warner, G.G. 1968. The larval development of the mangrove tree crab, *Aratus pisonie* (H.Milne Edwards) reared in the laboratory (Brachyura, Grapsidae). *Crustaceana* suppl.2:249-258.

Wear, R.G. 1967. Life history studies of New Zealand Brachyura I. Embryonic and post-embryonic development of *Pilumnus novaezelandiae'* Filhow, 1886, and of *P.lumpinus* Bennett 1964 (Xanthidae, Pilumninae). *NZ J. Mar. Freshwater Res.* 1:482-535.

Welch, J. & S.D.Sulkin 1974. The effect of diet concentration on the survival and growth of the zoeal stages of *Rhithropanopeus harrisii* (Gould). *J. Elisha Mitch. Sci. Soc.* 90:69-72.

Whitney, J.O. 1969. Absence of sterol synthesis in larvae of the mud crab *Rhithropanopeus harrisii* and of the spider crab, *Libinia emarginata*. *Mar. Biol.* 3:134-135.

Whitney, J.O. 1970. Absence of sterol biosynthesis in the blue crab, *Callinectes sapidus* Rathbun and in the barnacle *Balanus nubilus* Darwin. *J. Exp. Mar. Biol. Ecol.* 4:229-237.

Wickins, J.F. 1972. The food value of brine shrimp *Artemia salina* L. to larvae of the prawn, *Palaemon serratus* Pennant. *J. Exp. Mar. Biol. Ecol.* 10:151-170.

Williams, B.G. 1968. Laboratory rearing of the larval stages of *Carcinus maenas* (Crustacea, Decapoda). *J. Nat. Hist.* 2:121-126.

BRENDA SANDERS* / ROY B.LAUGHLIN, JR.** / JOHN D.COSTLOW, JR.*
* Duke University, Marine Laboratory, Beaufort, North Carolina, USA
** Naval Biosciences Laboratory, University of California, Naval Supply Center, Oakland, USA

GROWTH REGULATION IN LARVAE OF THE MUD CRAB RHITROPANOPEUS HARRISII

ABSTRACT

Growth and molting are intimately linked in crustaceans, making comparisons of growth between crustaceans and other organisms difficult. However, in this chapter we review an aspect of growth physiology which appears common to all organisms, the regulation of biomass accumulation. With a cybernetic model of growth we test the hypothesis that biomass accumulation is internally regulated by a negative feedback mechanism in *Rhithropanopeus harrisii* larvae. We review previously published data, present new data consistent with that hypothesis, and discuss both the usefulness and limitations of a model for growth regulation.

1 INTRODUCTION

Growth may be defined as an increase in length, volume, wet weight or dry weight (Hartnoll 1982). Scientists studying growth in crustaceans commonly use an increase in length as a measure of growth, and since crustaceans grow by molting, two growth components are often distinguished: 1) the molt increment is a measure of the increase in size occurring at a molt, and 2) the molt interval is the duration between two successive molts (Hartnoll 1982). In this paper, we address a very different aspect of growth, an increase in dry weight or biomass with time. We strive to understand how biomass accumulations may be regulated within the context of, but independent from, the molt cycle.

Larvae of the brachyuran crab, *Rhithropanopeus harrisii* have been used as a model system for work on the regulation of growth. These crabs are relatively easy to culture in the laboratory (Costlow & Bookhout 1971) and have been used extensively to understand how a variety of physical and chemical perturbations affect growth, survival and duration of larval development (Costlow et al. 1966, Christiansen & Costlow 1975, Bookhout et al. 1976, Laughlin et al. 1977, 1978, Laughlin & Neff 1979). Large numbers of these hardy animals can be reared for comparisons of growth rates. Except under very unusual conditions there are four zoeal and a megalopa stage (Costlow & Bookhout 1971). Zoeal development, which averages 12 days at 25°C, is a clearly defined interval with high, easily measured growth rates.

Growth rate is the product of the rate of availability of resources and the efficiency of conversion of these resources to biomass (Calow & Townsend 1981). It is often assumed that organisms are adapted to maximum growth rates. If organisms do grow at the fastest rate possible for available nutrients, any environmental stress would divert energy away from growth, i.e., increased biomass, and redirect it toward responses which would counteract the effects of stress (Newell 1976). This redirection of energy should result in a linear reduction in growth with increasing levels of stress.

Linear reductions in growth have been observed in *R.harrisii* larvae exposed to high levels of toxicants (Laughlin et al. 1978, Laughlin et al. 1981, Sanders et al. 1983). However, at very low levels, non-linear growth responses (increases in dry weight relative to controls) were observed when larvae were exposed to both petroleum hydrocarbons (Laughlin et al. 1981) and copper (Sanders et al. 1983). Furthermore, growth rates for larvae exposed to a 5°C diurnal temperature cycle oscillated around growth rates for the controls (Sanders & Costlow 1981). These oscillations decreased in amplitude with time, a response characteristic of self regulating systems (Hubbel 1971, Calow 1973).

Similar growth responses have been reported for other organisms when they were stressed, including vertebrates, invertebrates and bacteria (Smyth 1967, Hubbel 1971, Calow 1973, Stebbing 1979, 1981a,b, Ramkrishna 1983). These responses were similar regardless of the physical or chemical characteristics of the toxicant; a linear reduction in growth was observed at high levels, but at very low levels growth was greater than it was for the controls. This increased growth in response to stress indicates that we cannot assume that growth occurs at the greatest rates possible for available nutrients.

1.1 *Regulation of growth*

If the regulation of growth is a homeostatic process, non-linear growth responses could be attributed to transient over corrections by growth regulating mechanisms in response to low levels of an inhibitory challenge (Smyth 1967, Stebbing 1981a). This interpretation is congruent with the hypothesis that organisms have a definite capacity to regulate their biomass accumulation.

In studies with other organisms a model for growth regulation characterized by negative feedback has proven useful by providing an accurate description of the net result of numerous complex processes (Calow 1973, Ramkrishna 1983). These models assume active regulation of growth at less than maximum rates. Calow & Townsend (1981) distinguished this as an optimum rather than a maximum strategy model for growth. It should be emphasized that evidence for this model is empirical and that the model's purpose is to describe the responses of whole organisms in terms of the functional properties of the underlying biochemical and physiological mechanisms. The model, however, cannot be used to delineate the structural properties of these mechanisms, since a particular response may be elicited from a number of diverse systems (Calow 1976).

2 A CONCEPTUAL MODEL FOR GROWTH REGULATION

The simplest model for growth regulation has several components (Fig.1, Calow 1976, Patten & Odom 1981). There is a 'set point', genetically derived, which determines the 'required' growth rate. Environmental variables which alter metabolic rates may result in a deviation between the actual growth rate and the 'required' rate. Control mechanisms then act; they minimize any differences in these two rates and thereby regulate the actual growth rate.

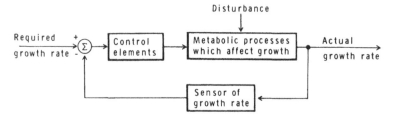

Figure 1. Simplified flow-diagram of model for growth regulation (Stebbing 1981a).

This model, therefore, hypothesizes the existence of internal regulatory mechanisms for growth and is a tool for generating locally consistent hypotheses based on general systems theory (Calow 1976). Two characteristics are inherent in this model. A time lag between detecting changes in actual growth rate and initiating the metabolic machinery to adjust the rate will result in overcorrections or overshoots (Calow 1976). As a consequence the actual growth rate oscillates around the 'required' rate. Also the regulatory mechanism responds to any environmental variable which changes the actual growth rate. Therefore, an organism should respond similarly to a wide range of environmental variables which alter the actual growth rate (Stebbing 1981a).

3 DEMONSTRATION OF GROWTH REGULATION USING TRIBUTYLTIN

In order to examine the suitability of this model, we used the organotin, tributyltin sulfide (TBTS) in a series of experiments to determine if organometals elicit the non-linear growth response observed for petroleum hydrocarbons, copper and a diurnal cyclic temperature regime. TBTS was dissolved in acetone and μl aliquots were added to seawater. Comparisons of effects on growth were made with reference to an acetone control.

Larvae were cultured using techniques described by Costlow & Bookhout (1971), collected daily, rinsed in deionized water, and lyophilized or air dried at 60°C. Samples were subsequently weighed on a Cahn microbalance to the nearest 0.1 μg.

3.1 *Testing the growth regulation model with TBTS*

In these experiments, in order to eliminate any effects of experimental conditions which might alter growth rates, we defined the 'required' growth rate as the actual growth rate of the control larvae. Therefore, our model would predict that the specific growth rate ($\bar{R}$) for larvae exposed to TBTS will oscillate around the $\bar{R}$ for the control group. If the oscillations dampen with time, the effects of stress on metabolism are being effectively counteracted. If the oscillations increase with time, the effects of stress are greater than the organisms's capacity to compensate for them.

Although the TBTS concentrations in this experiment did not result in statistically significant changes in survival or duration of development, changes in growth rate did occur (Fig.2). Control larvae maintained fairly uniform $\bar{R}$ through zoeal development, with daily values ranging between 0.2 and 0.5. The specific growth rate for each treatment, however,

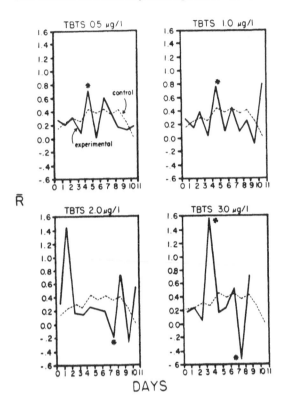

$\bar{R}$

DAYS

Figure 2. Daily specific growth rates ($\bar{R}$ = dw/dt l/w) for *Rhithropanopeus harrisii* zoeae exposed to four concentrations of tributyltin sulfide. Asterisks indicate data points which are significantly different from controls by ANOVA and Student Newman-Keuls multiple range test.

oscillated around $\bar{R}$ for the control as predicted in the regulation model. The number and amplitude of oscillations increased with increasing toxicant concentration and significant overshoots and undershoots were observed. These data suggest that at the higher concentrations of TBTS, growth regulation is less efficient.

3.2 *Hormesis: transient overcorrections from tributyltin*

In another experiment, when the specific growth rates for larvae exposed to 0.5 μg L^{-1} tributyltin oxide, TBTO, were expressed as a percent of $\bar{R}$ for the controls ($\bar{R}\%$), oscillations with decreasing amplitude over time were observed (Fig.3). These oscillations were similar to those found for *R. harrisii* larvae exposed to a cyclic temperature regime (Sanders & Costlow 1981) and to hydroids exposed to several types of toxicants (Stebbing 1979, 1981a,b). Stebbing (1981a) suggested that this ratio ($\bar{R}\%$) reflects the action of the controller (Σ in Fig.1) and used a computer simulation model to demonstrate that $\bar{R}\%$ for hydroids was compatible with the behavior of a control mechanism.

Do these oscillations result in hormesis, the increases in cumulative growth observed in organisms exposed to low levels of toxicants? Quantitatively, the difference between integrations of the area under the oscillation curve (where it is greater than the control) and above the curve (where it is less than the control) should give an estimate of cumulative weight relative to control larvae. If the difference is positive, hormesis is observed, if it is

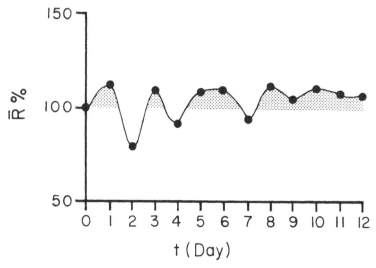

Figure 3. Daily specific growth rates relative to control rates ($\overline{R}\%$), for *R.harrisii* zoeae exposed to tributyltin oxide (0.5 μg L^{-1}). Shaded area represents area under curve which is greater than 100%.

negative, growth has been inhibited and the larvae are smaller than the controls. In Figure 3, the area of the curve which is greater than the control value is shaded. This area is 7% larger than the area of the curve which is less than the control values. The mean megalopa weight for this group was 152.0 + 9.8 μg compared to 141.1 + 6.6 μg for the controls. The cumulative growth for megalopa exposed to TBTO, therefore, was about 7% greater than control growth. The similarity in values suggests that overshoots in $\overline{R}\%$ may result in the increased cumulative growth of zoeae under stress. This integrative approach may be particularly useful for statistical analysis of differences from control values.

4 CONCLUSIONS

The growth response of mud crab larvae to several types of stress is remarkably similar. This response can be described by a model in which growth rate is internally regulated. Growth regulation can explain both linear and non-linear growth responses observed in stressed organisms. This model is a valuable tool for testing hypotheses involving internal regulatory mechanisms; it is not meant, however, to delineate the structural properties of these mechanisms. Many aspects of crustacean growth differ from growth processes in other organisms. However, *R.harrisii* larvae appear to have internal regulatory mechanisms which control biomass accumulation in a manner similar to that in other invertebrates, vertebrates and bacteria.

The significance of these findings is threefold. 1) The capacity to regulate growth should be incorporated into growth models and energy budgets to allow more accurate predictions. This approach has already been used successfully for modeling microbial growth (Remkrishna 1983). 2) A cybernetic perspective can increase our understanding of how larvae cope with and adapt to environmental perturbations. 3) A better understanding of the regulation of growth could provide a sensitive tool for measuring the effects of a specific environmental disturbance on fitness.

REFERENCES

Bookhout, C.G., J.D.Costlow & R.Monroe 1976. Effects of methoxychlor on larval development of mud crab and blue crabs. *Water, Air, Soil Pollut.* 5:349-365.

Calow, P. 1973. On the regulatory nature of individual growth: Some observations from freshwater snails. *Zool. Lond.* 170:415-428.

Calow, P. 1976. *Biological Machines: A Cybernetic Approach to Life.* London: Edward Arnold.

Calow, P. 1978. Ecology, evolution and energetics: A study in metabolic adaptation. In: D.S.McClusky & A.J.Berry (eds.), *Physiology and Behavior of Marine Organisms.* New York: Pergamon Press.

Calow, P. & C.R.Townsend 1981. Resource utilization and growth. In: C.R.Townsend & P.Calow (eds.), *Physiological Ecology: An Evolutionary Approach to Resource Use.* Sunderland, Massachusetts: Sinauer Associates.

Christiansen, M.E. & J.D.Costlow 1975. The effect of salinity and cyclic temperature on larval development of the mud crab, *Rhithropanopeus harrisii* (Brachyura: Xanthridae) reared in the laboratory. *Mar. Biol.* 32:215-221.

Costlow, J.D. & C.G.Bookhout 1971. The effect of cyclic temperatures on larval development in the mud-crab *Rhithropanopeus harrisii.* In: D.J.Crisp (ed.), *Fourth European Marine Biology Symposium,* London: Cambridge University Press.

Costlow, J.D., C.G.Bookhout & R.J.Monroe 1966. Studies on the larval development of the mud crab, *Rhithropanopeus harrisii* (Gould). I. The effect of salinity and temperature on larval development. *Physiol. Zool.* 34:81-100.

Hartnoll, R.G. 1982. Growth. In: L.G.Abele (ed.), *The Biology of Crustacea. Vol.II.* New York: Academic Press.

Hubbel, S.P. 1971. Of sowbugs and systems: The ecological bioenergetics of a terrestrial isopod. In: B.C.Patten (ed.), *Systems Analysis and Simulation Ecology.* New York: Academic Press.

Laughlin, R.B.Jr, J.M.Neff & C.S.Giam 1977. Effects of polychlorinated biphenyls, polychlorinated naphthalenes and phthalate esters on larval development of the mud crab, *Rhithropanopeus harrisii.* In: C.S.Giam (ed.), *Pollutant effects on marine organisms. Deliberations and recommendations of the biological effects program workshop.* New York: D.C.Heath.

Laughlin, R.B.Jr, L.G.L.Young & J.M.Neff 1978. A long-term study of the effects of water-soluble fractions of No.2 fuel oil on the survival, development rate and growth of the mud crab, *Rhithropanopeus harrisii. Mar. Biol.* 47:87-95.

Laughlin, R.B.Jr & J.M.Neff 1979. Interactive effects of salinity, temperature and polycyclic aromatic hydrocarbons on the survival and development rate of larvae of the mud crab, *Rhithropanopeus harrisii. Mar. Biol.* 53:281-291.

Laughlin, R.B.Jr, J.Ng & H.E.Guard 1981. Hormesis: A response to low environmental concentrations of petroleum hydrocarbons. *Science* 211:705-707.

Newell, R.C. 1976. Adaptations to intertidal life. In: R.C.Newell (ed.), *Adaptations to Environment: Essays on the Physiology of Marine Animals.* Boston: Butterworth.

Patten, B.D. & E.P.Odum 1981. The cybernetic nature of ecosystems. *Am. Nat.* 118:886-895.

Ramkrishna, D. 1983. A cybernetic perspective of microbial growth. In: H.W.Blanch, E.T.Pappoutsakis & G.Stephanopoulos (eds.), *Foundations of Biochemical Engineering.* Washington, D.C.: American Chemical Society.

Sanders, B. & J.D.Costlow 1981. Regulation of growth in larvae of the crab *Rhithropanopeus harrisii.* The effect of cyclic temperatures. *J. Therm. Biol.* 6:357-363.

Sanders, B., K.Jenkins, W.Sunda & J.D.Costlow 1983. Free cupric ion activity in seawater: effects on metallothionein and growth in crab larvae. *Science* (in press).

Smyth, H.F.Jr 1967. Sufficient Challenge. *Fd. Cosmet. Toxicol.* 5:51-58.

Stebbing, A.R.D. 1979. An experimental approach to the determination of biological water quality. *Phil. Trans. Roy. Soc. Lond.* B286:465-482.

Stebbing, A.R.D. 1981a. The kinetics of growth control in a colonial hydroid. *J. Mar. Biol. Assoc. UK* 61:35-63.

Stebbing, A.R.D. 1981b. Hormesis – stimulation of colony growth in *Campanularia flexuosa* (Hydrozoa) by copper, cadmium and other toxicants. *Aquatic Toxicol.* 1:227-238.

DARRYL L.FELDER* / JOEL W.MARTIN** / JOSEPH W.GOY*
* Department of Biology, University of Louisiana, Lafayette, USA
** Department of Biological Science, Florida State University, Tallahassee, USA

PATTERNS IN EARLY POSTLARVAL DEVELOPMENT OF DECAPODS

ABSTRACT

Early postlarval stages may differ from larval and adult phases of the life cycle in such characteristics as body size, morphology, molting frequency, growth rate, nutrient requirements, behavior, and habitat. Primarily by way of recent studies, information on these qualities in early postlarvae has begun to accrue, information which has not been previously summarized.

The change in form (metamorphosis) that occurs between larval and postlarval life is pronounced in some decapod groups but subtle in others. However, in almost all the Decapoda, some ontogenetic changes in locomotion, feeding, and habitat coincide with metamorphosis and early postlarval growth. The postmetamorphic (first postlarval) stage, herein termed the decapodid, is often a particularly modified transitional stage; terms such as glaucothöe, puerulus, and megalopa have been applied to it. The postlarval stages that follow the decapodid successively approach more closely the adult form. Morphogenesis of skeletal and other superficial features is particularly apparent at each molt, but histogenesis and organogenesis in early postlarvae is appreciable within intermolt periods. Except for the development of primary and secondary sexual organs, postmetamorphic change in internal anatomy is most pronounced in the first several postlarval instars, with the degree of anatomical reorganization and development decreasing in each of the later juvenile molts. Anatomical change during metamorphosis and the next few postlarval stages usually consists of degeneration of some anatomical features, redirection of some existent structures, and addition of some new structures. Examples of such processes can be seen in the early postlarval development of neurosecretory organs, musculature, muscle innervation, digestive organs, and even pigmentation patterns.

Comparative studies of early postlarvae may be of use in further resolving relationships between decapod taxa. To date, such relationships have been based primarily upon morphological comparisons of larvae or adults.

1 INTRODUCTION

In many instances, literature that compares growth phases of decapod crustaceans makes reference to characteristics of the larval life phase versus characteristics of the adult life phase. By inference, the non-specialist may be led to believe that decapod life histories are indeed partitioned cleanly between these two phases of growth and development, and that structure and function within either of these phases may be subject to relatively little variation. Chapters 1 and 2 of this volume have provided a basis for rejection of such assumptions as they would apply to decapod larvae; those papers demonstrated that some pronounced ontogenetic changes in structure and function occur during the larval phase of growth for a number of decapod taxa. The present chapter documents that the early post-larval stages in the life history of decapod Crustacea may also be unique and varied in many respects; in particular, it documents the degree to which early postlarval stages may differ in structure and function from the adults with which they are frequently grouped.

Lately, a body of information on early postlarval stages has begun to accrue and there has been no previous attempt to review and evaluate specific information on the early post-larval life of decapods. From the late 1960's to the present, a number of papers added appreciably to knowledge of structure and function of these life stages; but, even at this writing, coverage of some subjects is rather limited. The reader may note, for instance, that coverage of certain topics is based upon a rather small set of examples which may not represent broadly all major subtaxa of the Decapoda. Often the coverage is biased toward species of commercial importance or those that are most readily available for laboratory studies. While it may be premature to draw broadly applicable conclusions on the basis of such limited examples, we incorporate them as the existing baseline data to which future findings may be compared. However, even for common commercially important species there remain conspicuous gaps in such fundamental subject areas as early postlarval mor-phology, behavior, and ecology. Authors who have contributed recently to knowledge of early postlarvae in these species have also pointed out subject areas that have not received adequate attention (e.g. Anderson & Linder 1971, Johnson 1975, Seaman & Aska 1974, Serfling & Ford 1975). Fundamental information, such as the number of early postlarval instars and diagnostic characters of these instars, has long remained unavailable for com-mon and well-studied species (Hogarth 1975). For example, morphological characters of postlarvae 10-25 mm in length were just recently described for several well-known *Penaeus* species (Mair 1981).

The paucity of such data perhaps reflects the difficulty of obtaining early postlarvae of known instar and parentage for most decapod species. Whereas larvae can be collected sometimes in large series from the plankton or reared from eggs of identifiable parental females in the laboratory, and whereas the larger adults are often readily collected and easily identified in large series, postlarval series are usually more difficult to obtain. If early postlarvae are collected in the field, they may be difficult to stage and to identify; if reared in the laboratory, they usually must be carried first through an entire larval series and, typically, a metamorphic molt during which there is often high mortality. Thus one can readily appreciate why much of the presently available information on early postlarval stages is biased toward either well-known species for which postlarval series have been thoroughly described or species that are the subject of aquacultural interests and therefore maintained routinely in culture.

In at least some well-studied examples, enough is known of the early postlarval life his-tory to suggest that it be recognized as a delimited phase in the overall life history. For example, a 20-30 day period following metamorphosis to the postlarva in *Penaeus* has been identified as a phase of considerable morphological change (Perez Perez & Ros 1975, Wickins 1976); the 'early period of juvenile development' from the puerulus stage to approximately 1.5 years of age has been termed 'another distinct phase in the life history' of a palinurid lobster (Phillips et al. 1977). However, we will also point out cases in which this early period of development is not clearly demarcated, but rather exists as the earliest component of a continuum toward the development of the mature adult body form. Regardless of definition, the early postlarvae of decapods tend to share a number of biological characteristics. In addition to their smaller body size, they may differ from later stages and ultimately the adults in molting frequency, growth rates, metabolic demands, habitat requirements, behavior, morphology, susceptibility to predation, nutrient requirements, and a host of other characteristics. These unique characteristics may furthermore confer differences in distributions and ecological roles from those of adult and larval stages. Even characters that may clarify phylogenetic relationships can, in early postlarvae, differ from those of both larvae and adults.

In the following treatment of decapod postlarval development, we specifically direct our coverage to early postlarval stages. In general terms, we emphasize characteristics and developmental trends of stages that 1) immediately follow the larval life stages, and 2) differ appreciably in structure and function from mature or nearly mature adults. Thus, we do not, except in making comparisons, address later specializations in secondary sex characters and development to sexual maturity. We also do not analyze the rates of change in overall body size (absolute growth) and allometry (relative growth) of adult structures, because both of these subjects recently have been covered in an excellent review by Hartnoll (1982). Rather, we attempt to compile the presently diffuse and confusing literature so as to gain insight into the diversity and ontogeny of structure and function during early postlarval stages. We intend that this review serve as a current reference on the biology of early decapod postlarvae and hope that it will stimulate further study of this relatively neglected phase in decapod crustacean growth and development.

2 MORPHOLOGICAL TRANSITIONS IN POSTLARVAE

2.1 *Standardization of postlarval nomenclature*

Most crustaceans hatch at an early stage of development when they have relatively few body segments and appendages. They are anamorphic in their postembryonic growth, with the successive addition of new metameres and limbs in their progressive development through a series of larval stages (Snodgrass 1956, Costlow 1968). There is, however, some degree of metamorphosis superimposed on these anamorphic stages of development. Passano (1961) defined crustacean metamorphosis liberally as a change in form at a particular point in the animal's life. If behavioral and physiological changes are included in this definition, metamorphosis may include 1) the gradual and successive changes leading up to the transition, 2) the influence of environmental factors on these changes, 3) the internal reorganization accompanying the more obvious morphological modifications, and 4) the changes in mechanisms controlling and regulating this period of development (Costlow 1968). In the decapod infraorders Anomura, Palinura and Brachyura, where the larval stages are typically pelagic and the adults benthic, change occurs in a well-defined metamorphosis. The larvae in such groups differ more strikingly from adults than in other decapod infraorders, where all stages lead a similar life. However, in almost all the Decapoda, some ontogenetic changes in locomotion, feeding, and habitat coincide with early postlarval growth.

Early carcinologists were reluctant to accept the concept of complete metamorphosis

in the Decapoda (for a historical review see Gurney 1942), and the credit for the definite
proof of crustacean metamorphosis belongs to Thompson (1828, 1829, 1831). In a number of cases developmental stages were given special names of generic significance, for
example *Zoea* (Bosc 1802) and *Megalopa* (Leach 1814), both based on brachyuran forms.
Gurney (1942) proposed a system for classifying decapod developmental phases based in
part upon the method of locomotion; he recognized the nauplius and protozoea with antennal propulsion, zoea with thoracic propulsion, and post-larval form with abdominal
propulsion. Williamson (1957, 1969, 1982a) modified this by including the penaeid protozoea with zoea and changing Gurney's 'post-larva' into megalopa. Gurney's use of the
term post-larva is somewhat ambiguous. Any stage following the larval phase (zoea) can be
termed a postlarva, including juveniles and adults. However, the first postzoeal stage in the
Decapoda is often significantly different from subsequent juvenile stages and is sometimes
considered a larval form (Williamson 1969).

Many names beside megalopa have been used by various workers to describe the postzoeal stages of decapods, including puerulus, nisto, and pseudibacus for Palinuroidea;
glaucothöe for Anomura; eryoneicus for Eryonoidea; mastigopus for Sergestoidea; parva
for Caridea; and grimothea for Galatheidae. Except for Williamson's usage and that of a
few other workers, the term megalopa has almost always signified the first postzoeal phase
of a brachyuran crab. Williamson (1982a) was somewhat ambiguous with his usage of the
term megalopa. While he states that the megalopa stage is confined to the Eumalacostraca,
he uses the same term for Hoplocarida in his Table 1. Kaestner (1970) proposed the term
decapodid for the first postzoeal phase of decapods having the full complement of metameres and appendages and the general characteristics of the order; we follow his usage in
this paper and restrict megalopa to an equivalent name for a brachyuran decapodid.
Throughout this paper, our usage of terms for larval and postlarval decapods will be
defined as follows: nauplius = a larval stage with first three pairs of cephalic appendages
setose and functional, other appendages absent or rudimentary (after Williamson 1969);
zoea = a larval stage with setose natatory exopodites on some or all of the thoracic appendages and with the pleopods absent or rudimentary (after Williamson 1969); postlarva =
any form that occurs after the zoeal stages inclusive of all developmental stages to the
adult; decapodid = the first postlarval stage, the stage that occurs immediately after the
molt from the last larval stage and that has setose natatory pleopods on some or all of
the first five abdominal somites. Terms such as glaucothöe, puerulus, and megalopa will
be considered synonyms for the decapodid stage of Anomura, Palinura, and Brachyura
respectively. Since the decapodid is usually a transitional stage between the zoeal period
of development and juvenile phase of growth, it is subject to considerable variation and
special modification in some decapods, especially in those having abbreviated development
(see Rabalais & Gore, this volume).

2.2 *Early postlarvae of the Dendrobranchiata*

2.2.1 *Penaeoidean decapodid and juvenile*

The Penaeoidea pass through a series of swimming nauplii, protozoeal stages, and a varying number of zoeal stages, before going through a final molt into the decapodid. At this
molt, the mouthparts and maxillipeds assume their basic adult form, the fourth to eighth
thoracic limbs become pereopods with enlarged endopodites and reduced exopodites and
the first five pairs of abdominal appendages become functional pleopods. The swimming

function thus shifts during this molt from the thorax to the abdomen. The postzoeal change in structures is very gradual, and the morphological distinction between the decapodid and subsequent postlarvae is usually not very pronounced (Fig.1,i). Therefore, some workers have recognized more than one decapodid stage in penaeids (Heegaard 1953, Dobkin 1961, Kurata & Pusadee 1974). By our designation, the first postlarva is equivalent to a decapodid and subsequent postlarval stages may be considered early juveniles.

In *Penaeus duorarum,* Dobkin (1961) found that no conspicuous metamorphosis occurs from the last zoea to the first postlarva (decapodid), although the pereopods lose their exopodites and the exopodites of the maxillipeds are either lost or modified. The pleopods do not become biramous until seven or eight molts later. Kurata & Pusadee (1974) succeeded in rearing only the first two postlarval stages in *Metapenaeus burkenroadi;* the gills and epipods of the adult develop in these stages and a temporary degeneration of some of the mouthparts occurs. Later juvenile stages follow these early postlarval stages with further addition of appendages which fit a benthic existence. From laboratory-reared larvae of *Xiphopenaeus kroyeri,* Kurata (1970) described a decapodid that was very similar to the last zoea. That decapodid differs from the last zoea by having hepatic spines, flattened exopodites on the first maxilliped, and long setae on the pleopods. It differs from the subsequent postlarval stages in structure of mouthparts and rigidity of the legs.

2.2.2 *Sergestoidean decapodid and juvenile*

Sergestids pass through a nauplius and several zoeal stages before the molt to the decapodid. At this stage, sometimes termed the mastigopus, the spines of the carapace and abdomen are lost or reduced and the telson approaches the adult form. The fourth and fifth pereopods are lost or reduced to small stumps; these redevelop during subsequent postlarval molts. Relative sizes of adjacent gills also change during postlarval development (Burkenroad 1945, 1981).

Hansen (1922) and Gurney & Lebour (1940) described mastigopus stages for *Sergestes* from planktonic material. Other workers described the complete larval development of sergestids in the laboratory, as in studies by Rao (1968) and Kurata (1970) on *Acetes,* Omori (1971) on *Sergia,* and Knight & Omori (1982) on *Sergestes.* Kurata (1970) described six postlarval stages in *Acetes americanus carolinae,* in which development of the posterior pairs of pleopods and the pleopodal endopodites is gradual; the fourth pereopod redevelops after the sixth postlarval molt (Fig.1,ii). Omori (1971) reported on five postlarval stages in *Sergia lucens;* in the fifth postlarva, the first and second maxillipeds first become chewing and holding appendages. Knight & Omori (1982) described the first two postlarval stages in *Sergestes similis* and observed a progressive reduction of spines on the carapace, abdomen and telson over the course of development from the last zoeal stage to the second postlarva. After the second postlarva, the mouthparts approach the adult form, the pleopods are setose and functional and the endopodites become more setose at each molt.

2.3 *Early postlarvae of the Pleocyemata*

2.3.1 *Caridean decapodid and juvenile*

In the Caridea, the eggs are carried by the females on their pleopods and the larvae hatch as zoeae. Larval development within the Caridea involves a highly variable number of zoeal stages (Knowlton 1974). Carideans are like penaeids and sergestids in that postzoeal development is gradual and morphological distinctions between the decapodid and the

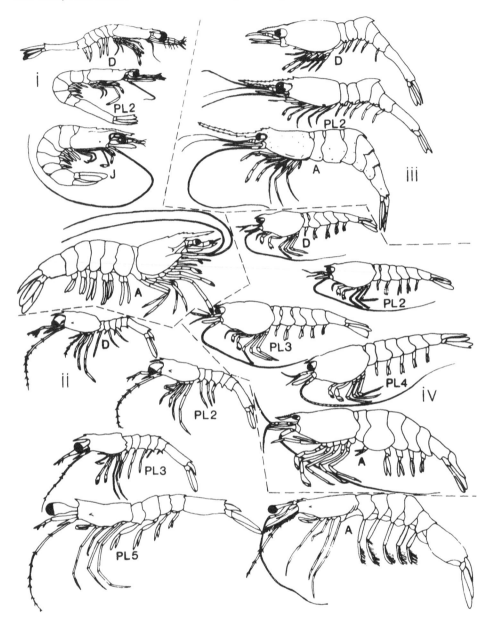

Figure 1. Early postlarval stages of representative decapods. Illustrations adapted from literature, see text. Abbreviations as follows: D – decapodid; PL – postlarva; J – juvenile; A – adult. (i) Family Penaeidae: *Penaeus setiferus*; (ii) Family Sergestidae: *Acetes americanus carolinae*; (iii) Family Oplophoridae: *Systellaspis debilis*; (iv) Family Atyidae: *Caridina denticulata*.

following postlarval stages are not pronounced. At present, there are over 20 recognized families of carideans; in six of these the larval stages are completely unknown. There are many families for which the early zoeal phases are known, but relatively few studies have addressed early postlarval stages. Our short summary of postzoeal caridean development will therefore be restricted to the few comprehensive studies on selected families.

In the Oplophoridae, the decapodid and subsequent postlarval stages have been described for a few species from planktonic material, as in studies by Kemp (1910) on *Systellaspis debilis,* Coutière (1906), Gurney & Lebour (1941) on *Acanthephyra purpurea* and Aizawa (1974) on *A.quadrispinosa.* Kemp (1910) examined a wide size range of *Systellaspis debilis* postlarvae taken from planktonic material collected off the Irish coast and documented changes of the mandibular palp, photophores, rostrum, eyes and branchiae with body size (Fig.1,iii). In *Acanthephyra quadrispinosa,* Aizawa (1974) considered a specimen of 12.9 mm body length to be the decapodid; its short rostrum bears four dorsal and a ventral tooth. In the next few instars eyestalks narrow, the rostrum is elongated, and thoracic exopodites are shortened. A specimen of about 30 mm total length (about half the adult size) resembles an adult, and has a long rostrum with 8-10 dorsal and 5-7 ventral teeth.

Most genera in the family Atyidae live as adults only in freshwater and have a suppressed larval development, but a few species have a complete series of larval stages. The first group is exemplified by *Caridina denticulata denticulata, C.brevirostris, C.denticulata ishigakiensis* and *C.singhalensis,* while examples of atyids with a normal series of larval stages are *Paratya compressa* and *C.pseudogracilirostris* (see Yokoya 1931, and Pillai 1975). In *C.denticulata denticulata* there are few differences between the decapodid, which is the second instar after hatching, and the adult shrimp (Mizue & Iwamoto 1961); the main differences are found in the spination of the rostrum and the shape of the telson and uropods (Fig.1,iv). An interesting aspect of this shrimp's life history is that it does not start to feed until the fourth postlarval stage, subsisting upon yolk reserves up to this point. Shokita (1973a, 1976) found that *C.brevirostris* and *C.denticulata ishigakiensis,* which are both restricted to freshwater, also reach the decapodid after the first post-hatch molt. These shrimps are near the adult form in the decapodid stage but do not start to feed until the next postlarval stage. Secondary sexual structures are developed by the fourth postlarva. *C.singhalensis* hatches in a form that is morphologically postlarval except for its spatulate larval telson lacking uropods and its rudimentary mouthparts (Benzie & de Silva 1983). Development of *Caridina mccullochi* is intermediate between the two basic atyid patterns (Benzie 1982).

The family Pasiphaeidae is generally regarded as primitive because of the exopodites on all pereopods of adults (Fig.2, v). Most genera within the family have large eggs and abbreviated development, although *Leptochela* is an exception. The decapodid in *Leptochela* sp. has pleural armature and carapace spines that are absent in the adult, and exopods of its posterior pereopods not developed (Kurata 1965). Kurata felt that his adult specimens resembled *L.aculeocaudata,* but his figures are closer to *L.sydniensis,* with which the former species has been frequently confused.

In the family Palaemonidae, the molt to the decapodid is often not morphologically conspicuous, but it and subsequent postlarval stages generally take on a more bottom-living habit and progressively attain the characters of the adult shrimp, such as loss of pereopodal exopods. The work of Shokita (1973b, 1977, 1978) on the freshwater species *Macrobrachium shokitai* and *M.asperulum* shows these species exhibit 'abbreviated devel-

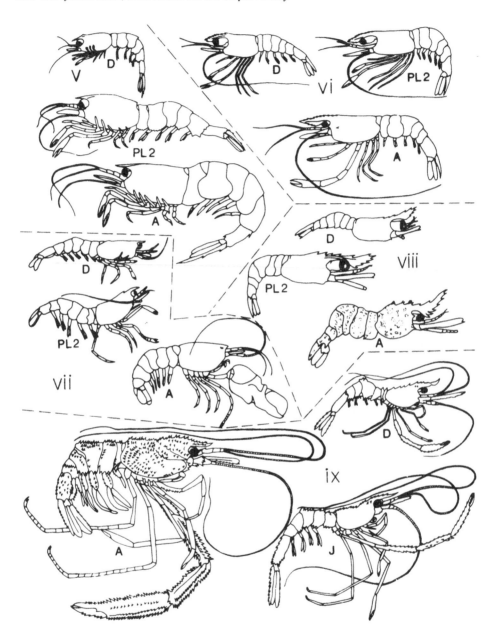

Figure 2. Early postlarval stages of representative decapods. Illustrations adapted from literature, see text. Abbreviations as follows: D – decapodid; PL – postlarva; J – juvenile; A – adult. (v) Family Pasiphaeidae: *Leptochela sydniensis*; (vi) Family Palaemonidae: *Macrobrachium asperulum*; (vii) Family Alpheidae: *Alpheus heterochaelis*; (viii) Family Hippolytidae: *Saron marmoratus*; (ix) Family Stenopodidae: *Stenopus hispidus*.

opment' (Fig.2,vi). By the second postlarval stage, both species are essentially adult in all respects but secondary sexual characters. An even more 'abbreviated' development is seen in *Macrobrachium hendersodayanum* and *Desmocaris trispinosa* that hatch essentially as decapodids (e.g., Jalihal & Sankolli 1975, Powell 1979). At the other extreme, the fresh-water shrimp *Palaemon paucidens* has seven zoeal stages before the molt to the decapodid (Yokoya 1931). A characteristic change of the appendages to resemble the adult form and a considerable reduction in body size occur at the molt to decapodid.

In the estuarine and marine palaemonids, larval development is usually not 'abbreviated'. Gurney (1924) found that the postlarval stages of *Palaemon longirostris* (as *Leander longirostris*) and *Palaemon elegans* (as *L. squilla*) had the adult form in general structure in the decapodid, but with slight differences in the rostrum, antennae, and pereopods. Fincham (1977, 1979a, 1979b) reared these two species and *Palaemonetes varians* and found some variability among individuals, but his results were generally consistent with those that Gurney found using planktonic material. The brackish water shrimp *Leandrites celebensis* passes through seven zoeal stages before becoming a decapodid (Pillai 1974). Another estuarine palaemonid, *Palaemon semmelinkii* goes through 12 larval stages before the adult-like decapodid is reached (Jagardisha & Sankolli 1977).

A large subfamily of palaemonids, the Pontoniinae, is mostly marine, and many of its members are commensal with other invertebrates. Although some larval stages of a few pontoniine shrimp are known, none have to date been reared from eggs to postlarval stages. The majority of species within the subfamily have small eggs, so one would expect a normal sequence of larval stages. A possible exception to this generalization could be *Pontonia minuta* which has very large eggs and hatches as an advanced zoea (Bruce 1972).

The variety of developmental patterns within the snapping shrimp family Alpheidae makes it difficult to propose a generalized developmental scheme that would apply to early postlarvae. Species of *Alpheus* and *Athanas* with a normal sequence of five or more zoeal stages have not yet been reared in the laboratory to the decapodid stage. Therefore, what is known of alpheid postlarvae is based on planktonic material and a few species with abbreviated or direct development. Lebour (1932b) found that the last larval stages of *Alpheus glaber* (as *Alpheus ruber*) and *A. macrocheles* molted to decapodids which were not quite like the adults, although the carapace and chelate legs approached those of the adult. The molt to the second postlarva allows these species to be recognized by the adult characters. Development is direct in *Synalpheus brooksi* (Dobkin 1965), where the hatchling is a decapodid with the carapace covering the eyes as in the adult and with the adult form of appendages. Knowlton (1973) found that the first postlarval form of *Alpheus heterochaelis* assumes the adult form, though some larval characters, such as pereopodal exopods, are retained for another molt (Fig.2,vii). Characteristic asymmetry of the chelipeds occurs much later in postlarval development.

The family Bresiliidae contains the genera *Bresilia, Lucaya,* and *Discias.* Gurney & Lebour (1941) noted that the first postlarva of *Discias atlanticus* was very much like the adult. Bruce (1975) found two small males and a late larva in a sponge off Kenya. Interestingly, the late zoea had adult appendages and the second pleopod had an appendix masculina.

The monogeneric family Rhynchocinetidae exhibits larval characteristics of oplophorids, but certain adult characteristics of palaemonids. Gurney & Lebour (1941) found that *Rhynchocinetes rigens* from plankton material reached the tenth larval stage, with successive molts producing no great change except for the growth of the antennal flagellum. In the decapodid some variation exists in the reduction of the pereopod exopods, and some of the mouthparts are not fully functional.

The larvae of Hippolytidae show such diversity in form that Gurney (1937b), from his studies on the larvae of *Lysmata, Tozeuma, Saron,* and *Latreutes,* suggested the family be divided into several subfamilies. Lebour (1931, 1932a) argued that, based on their larvae, the genera *Caridion* and *Spirontocaris* are closer to the Pandalidae and should be removed from the Hippolytidae. She later (1936, 1940) retracted that opinion after looking at the postlarvae of other *Spirontocaris* spp. and comparing them to species of *Thor.* Lebour (1931) found that after *Caridion gordoni* and *C.steveni* molted to decapodids in the laboratory, the postlarvae retained exopodites on the first four pairs of pereopods. In the second postlarva, both species lost exopodites on the pereopods, the second leg became fully developed, and there was a full adult complement of branchiae. Lebour (1932a, 1936) described the postlarvae of *Spirontocaris cranchii* reared from planktonic larvae. The decapodid was like the adult, with a sharp bend in the abdomen and a long antennal flagellum. The characteristic bifurcate rostrum develops in the fourth postlarva. Lebour (1936) also described the larval and early postlarval stages of *Spirontocaris occulta* and found its decapodid slightly more advanced than that of *S.cranchii.* She noted increases in carpal segments of the second pereopods for both species during early postlarval development. Pike & Williamson (1961) found some postlarval pleopodal characters developing in the last two zoeal stages of *Spirontocaris spinus.* In the decapodid, the pleopods become functional and exopodites are still present on the thoracic appendages. In the second postlarva the exopodites disappear and a supraorbital spine appears on the carapace. In the decapodid of *Thor floridanus* (see Lebour 1940, Broad 1957), there are remnants of exopodites on the first two pereopods and the carpus of the second pereopod is divided into three segments. The second postlarva has the outer antennular flagellum thick at the base, the exopodites have disappeared, and the second pereopod has four carpal segments. Sankolli & Kewalramani (1962) described the larval and early postlarval stages of *Saron marmoratus* reared in the laboratory (Fig.2,viii). Their protozoea is equivalent to the first zoea, while their first postlarva is actually the fifth and last zoeal stage (personal observation of cultured larvae from Hawaii, JWG). Their 'second' postlarva (= decapodid) has functional pleopods and a rudimentary exopodite on the third maxilliped; the exopodites on the pereopods are no longer present. The next postlarval stage has a rostrum characteristic of the adult. Kurata (1968a,b) reared the larval stages and several postlarval stages of *Eualus gracilirostris* and *Heptacarpus futilirostris.* In *Eualus gracilirostris,* the decapodid's rostrum appears as a short spine but it grows longer and develops teeth in successive molts. In *Heptacarpus futilirostris,* the decapodid has remnants of exopodites on the third maxilliped and on the first to fourth pereopods, but lacks epipods. By the second postlarva the exopods are mere rudiments and epipods are present on the third maxilliped and first three pereopods.

The family Pandalidae also shows a great variety of larval types and includes species with abbreviated and extended larval development. In the majority of other decapods, the development of functional pleopods, development of the pereopods, and changes in shape and body proportions provide a clear distinction between zoeae and postlarvae. In the Pandalidae, however, there is not always an abrupt metamorphosis in such characters (Haynes 1976, Rothlisberg 1980).

Kurata (1955) described the larval stages of *Pandalus kessleri* and found it difficult to designate a given stage as postlarval. He considered the fifth stage to be the first postlarva, since it was essentially adult in character except for secondary sexual structures. The first pleopodal appendix interna appears at the sixth postlarva, and the second pleopods acquire appendices masculina at the eighth postlarva. Kurata proposed different appendages attain the postlarval condition at different stages. By his scheme, the postlarval phase begins in Stage IV (= last zoeal stage) in the right carpus of the second pereopod, stage V (= decapodid) in the left carpus of the second pereopod, stage VII (= third postlarva) in the rostrum, sixth somite, telson, antennal scale and third pleopod.

The larvae of *Heterocarpus ensifer* are considered to be among the largest decapod larvae known, with some reaching over 50 mm total length (Gurney & Lebour 1941, Gopala Menon 1972). This is a deep-sea species in which the change from larval to post-larval life involves no great change of habit and none of habitat (Gurney 1942). Therefore, stimulus for metamorphosis might be the onset of developments for sexual maturity. The development of carinae on the larval carapace is restricted to the genus *Heterocarpus,* and can be regarded as an adult character which develops precociously (Gopala Menon 1972). The larvae and early postlarvae of *Pandalus hypsinotus, P.goniurus* and *P.borealis* were described by Haynes (1976, 1978, 1979) from shrimp reared in the laboratory. He considered the molt after zoeal stage VI to be the decapodid for *P.hypsinotus* because at this stage the pleopods were functional and the appendix interna was distinct on all pleopods except the first pair. The second and third postlarvae differ only slightly from the decapodid with the major changes occurring in the gills. In *P.goniurus,* Haynes' stage VI is the decapodid, with a two-segmented mandibular palp, exopodites reduced on the third maxillipeds and pereopods, functional pleopods, and the telson showing the shape and spination of the adult. The second postlarva attains a typical adult rostrum and a three-segmented mandibular palp while it loses the exopodites. *Pandalus borealis* shows a similar developmental pattern to *P.goniurus* but the decapodid shows no trace of the mandibular palp, which appears in the next postlarval stage. Rothlisberg (1980) believed that *Pandalus jordani* did not have a decapodid stage. His first juvenile, though, is equivalent to a decapodid because the mandibular palp appears for the first time, the telson and rostrum have the adult shape, and the exopodites are present as rudiments and are no longer used for locomotion. By the next molt, the mandibular palp is three-segmented and the exopodites are lost. Because of heterochrony in the development of appendages in larval *Pandalus prensor,* Mikulich & Ivanov (1983) felt characters representative of zoeae, decapodids and even juveniles can be seen in a single instar. Natatory setae on the pleopods first appear in the fourth instar, which can be considered a decapodid.

2.3.2 *Stenopodidean decapodid and juvenile*
The Stenopodidea consists of only the Stenopodidae, which has a range of developmental patterns from abbreviated to extended larval development. Stenopodids are unique among the Decapoda in having four pairs of natatory limbs on hatching. Only *Microprosthema* sp. has been reared completely in the laboratory, so descriptions of larvae and postlarvae for other species are based on material from the plankton.

Cano (1892) described and figured the decapodid of *Stenopus spinosus.* He found that the adult characters were to a large extent present in the first postlarva, but the abdomen was still flexed at right angles at the third abdominal somite. Kemp (1910) described the newly hatched young of *Spongicoloides koehleri,* which is a deep water species commensal in sponges. This species shows direct development, and the hatchlings probably represent the decapodid which closely resembles the adult except for its pereopodal exopods and sessile uropods. Gurney & Lebour (1941) described numerous stenopodid larvae and postlarvae of *Stenopus hispidus* (Fig.2,ix). They obtained a decapodid from a larva measuring 21 mm total length, yet they collected larvae of the same species as large as 31 mm in the plankton. The first postlarva, measuring 10 mm in length, had the adult form and coloration. The adult body and appendage spination is gradually acquired with successive molts and remnants of pereopodal exopods are lost. Gurney & Lebour (1941) also described postlarvae of two species they labelled 'Stenopid B' and 'Stenopid C'. From their description of the decapodid of 'Stenopid B' and their notation that it was the most common stenopodid larva encountered from the Bermuda plankton, this larval series probably represents *Microprosthema semilaeve.* Larvae and postlarvae reported as *M.semilaeve,* represent *M.validum* or a new species from the Indian Ocean. Developments beyond the decapodid include increases in rostral and body spination and setal numbers.

2.3.3 *Nephropidean decapodid and juvenile*

The known nephropid lobsters have only three larval stages after emerging from a prezoea of short duration. After molting into the fourth stage (= decapodid), they resemble the adult, yet continue to swim for several days before settling into a benthic habitat. The first pleopod is still absent in the first postlarva.

In the Norway lobster, *Nephrops norvegicus,* the first postlarval stage assumes the general characters of the adult (Farmer 1975) but uropodal exopods remain undivided as in the last larval stage. There are no first pleopods, but throughout the juvenile phase there is a gradual development of these and other external sexual characters. The juvenile stages are bottom dwellers, and construct burrows similar to those of the adults. Uchida & Dotsu (1973) described the larvae and early postlarvae of the tropical lobster *Metanephrops thomsoni.* At decapodid metamorphosis swimming function shifts from the thorax to the abdomen, and by the second postlarva the exopods are completely absent from the pereopods. Compared to *Nephrops norvegicus,* the larval development of *M.thomsoni* is of short duration, nine days as opposed to 2-3 weeks.

The American lobster, *Homarus americanus* (Fig.3,xi) and the European lobster, *H.gammarus,* have been the subject of research since the latter half of the last century. The fourth stage of development for both species is equivalent to the decapodid, because it loses the exopodites on the pereopods and the pleopods are used exclusively for swimming. After molting into the fourth stage, the first postlarva swims for a few days before settling to the bottom. The morphology of the juvenile stages, including the early postlarvae, is similar to that of the adult except for size. Carlsberg et al. (1978) have successfully hybridized these two species. The larvae of the hybrid bear the coloration and morphology of *H.gammarus,* while the early postlarvae are more like that of *H.americanus.* The early postlarvae of *H.americanus* have the claws very similar in morphology, musculature and innervation, but internal changes started in these stages will produce the differentiation into the cutter and crusher claws. The internal reorganization required to produce these specialized claws will be treated later in this chapter.

2.3.4 *Astacoidean decapodid and juvenile*

Freshwater crayfish of the families Astacidae and Cambaridae have a decapodid hatching with all of the appendages of the adult except the first pleopods and uropods. Gurney (1942) felt that, since the uropods do not appear until the second molt, the first three stages correspond to the first three stages of decapods with normal free larvae. The young

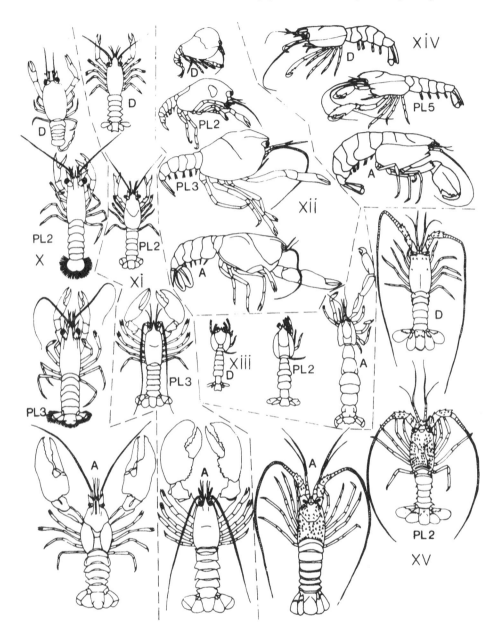

Figure 3. Early postlarval stages of representative decapods. Illustrations adapted from literature, see text. Abbreviations as follows: D – decapodid; PL – postlarva; J – juvenile; A – adult. (x) Family Astacidae: *Pacifastacus leniusculus*; (xi) Family Nephropidae: *Homarus americanus*; (xii) Family Parastacidae: *Engaeus cisternarius*; (xiii) Family Callianassidae: *Callianassa kewalramanii*; (xiv) Family Laomediidae: *Naushonia crangonoides*; (xv) Family Palinuridae: *Panulirus polyphagus*.

when first hatched remain attached for a short time to pleopods of the parent by a filament. This filament is the molted embryonic cuticle which attaches the telson to the empty egg case. When this breaks the hatchlings cling to the pleopods of the adult by means of the specially hooked first chelipeds. At the first molt, the chelipeds loose their hooks and the body becomes more adult-like in shape.

Andrews (1907) conducted a thorough study of the early stages of *Pacifastacus leniusculus* (as *Astacus leniusculus*) and *Orconectes limosus* (as *Cambarus affinis*) (Fig.3,x). The first three stages of these species are similar, but *P.leniusculus* is slightly more advanced in development. In the second stage, *Pacifastacus leniusculus* will take short excursions from the parental female, but *Orconectes limosus* remains firmly clinging onto the adult pleopods. In the third stage, the uropods are free and the gills fully developed. Juveniles spend a short time attached to the parental female and soon take up a free existence. The main feature of the fourth stage is the development of the first pleopods.

2.3.5 *Parastacoidean decapodid and juvenile*

The family Parastacidae is also a freshwater group with an abbreviated developmental sequence in which the hatchlings (= decapodid) remain attached to the adult pleopods for the first three stages. The hatchlings differ from those of astacoid crayfish in that the fourth and fifth legs are modified with recurved hooks for attachment, instead of these hooks being on the first pereopods (Fig.3,xii).

The hatchlings of *Engaeus cisternarius* are soft and inactive, with reduced appendages and sessile eyes (Suter 1977). After 2-3 days, they break the filaments that attached their telsons to the adult's pleopods and grasp the pleopodal setae with small recurved hooks on their fourth and fifth pereopods. However, they remain globular with a large cephalothorax filled with yolk. The second stage, which remains attached to the pleopods by the pereopodal hooks, has a longer rostrum and stalked eyes. The third stage attaches to the adult pleopods by the chelipeds. They start to make short excursions from the parental female and eventually become fully independent, with all appendages completely developed. Suter also compared the similar development of *Engaeus fossor* to that of *E.cisternarius*. The diagnostic characters of *E.fossor* were all well developed in the third stage.

2.3.6 *Thalassinidean decapodid and juvenile*

The thalassinideans are all burrowing forms, and larvae are known for all families. The Thalassinidea do not seem to form a natural group on the basis of developmental stages. The larvae of the Callianassidae and Axiidae are similar and somewhat resemble those of the Nephropsidea. Larvae of the Upogebiidae and Laomediidae resemble more closely those of the Anomura (Gurney 1938, 1942).

Forbes (1973) reported on the unusual abbreviated development of *Callianassa kraussi*, but did not discuss changes in morphology. Sankolli & Shenoy (1975) described two postlarval stages for *Callianassa (Callichirus) kewalramanii* (Fig.3,xiii). The decapodid resembles the adult in general body structure, but has pereopodal exopod rudiments and some larval characters in the telson and maxillipeds. The second postlarval stage approaches the adult in details of appendages, gills, and carapace.

In *Upogebia savignyi,* a sponge commensal, the hatchlings are basically in the adult form (Gurney 1937a). Shenoy (1967) described the first two postlarval stages of *Upogebia kempi.* The decapodid has functional pleopods, reduced fifth pereopods, and exopodites on the first three pereopods. In the second postlarva, the exopodites of the pereopods are lost and a dense fringe of long setae develops on the maxillipeds and first two pairs of pereopods. Goy & Provenzano (1978, 1979) described the larval and early postlarval stages of the laomediid mud shrimp *Naushonia crangonoides* (Fig.3,xiv). The decapodid resembles the adult, but the mouthparts and appendages are not so setose as in the adult. In the first few molts after the decapodid, the morphology does not change conspicuously and there is a gradual development of adult characteristics. In the sixth postlarva, the cervical groove and linea thalassinica are well marked and the full complement of gills is reached.

2.3.7 *Palinuran decapodid and juvenile*

Nothing is known of the larvae or postlarvae in the family Glypheidae or the early larval stages of the family Polychelidae. The superfamily Palinuroidea is comprised of the families Palinuridae, Scyllaridae and Synaxidae and has a characteristic zoeal stage called a phyllosoma (see Phillips & Sastry 1980) and a characteristic decapodid variously called puerulus, nisto or pseudibacus. The last phyllosoma stage, which is usually about 35 mm in length, metamorphoses into the decapodid. Gurney (1942) stated that this metamorphosis is the most profound transformation at a single molt known among Decapoda. There is little pigment and an absence of calcium in the exoskeleton of the early puerulus which causes it to appear almost colorless and transparent. It has a generally smooth, rather than spinose exoskeleton, and relatively large pleopods bearing long setae.

The genus *Eryoneicus* has been established as the postlarval stage of the deep-water lobster family Polychelidae. Gurney (1942) felt that apart from the possession of exopodites on the first two legs these postlarvae have no special larval features. The great dilation of the carapace and thorax and the spination of the abdominal somites probably increase flotation by increasing surface area. The majority of workers feel these stages must settle to the bottom to metamorphose into the young adult. Bernard (1953) thought the adults of *Willemoesia* metamorphosed in the water column, because he found 'soft' small pelagic adult forms and 'hard' adult benthic forms.

Deshmukh (1966) described the puerulus and second postlarva of *Panulirus polyphagus* (Fig.3,xv). The decapodid is transparent, has very long and terminally spatulate antennae, and has the ability to swim upside down by using its large and heavily setose pleopods. After a short time, the decapodid (puerulus) sinks to the bottom, starts to acquire pigmentation, and molts within 3-4 days to a form much more like the adult (Silberbauer 1971).

Compared to the palinurid lobsters, the slipper lobsters or Scyllaridae have not been as thoroughly studied. Dotsu et al. (1966) gave a detailed description of the metamorphosis of *Ibacus ciliatus* and *I.novemdentatus*. They called the decapodid stages 'reptant larvae' and gave a brief description of this stage and young juvenile lobsters that they caught in traps. Lyons (1970) compared decapodids (pueruli) of *Scyllarides nodifer, S.aequinoctialis, Scyllarus depressus, S.americanus* and *S.chacei* (Fig.4,xvi). The transparent decapodids appear to be transitional stages that link planktonic phyllosomas with benthic juveniles and exhibit characters of both. A dramatic reduction in relative size of the pleopods at the molt from the decapodid to the second postlarva was noted in *Scyllarides nodifer, Scyllarus americanus, S.chacei* and *S.depressus*. Pleopods remain reduced in juveniles until after development of genital pores; they rapidly increase in size in successive molts and become fully developed by the time the lobsters reach maturity. More recently, Atkinson & Boustead (1982) described the complete larval development of *Ibacus alticrenatus* from New Zealand. The decapodid is nearly transparent and has vestigial exopodites on the

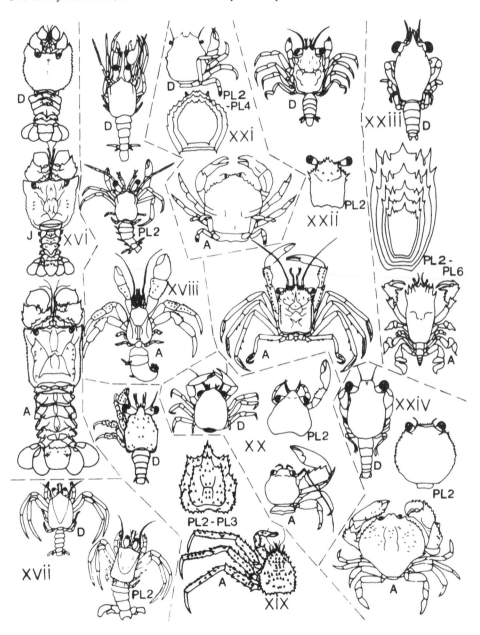

Figure 4. Early postlarval stages of representative decapods. Illustrations adapted from literature, see text. Abbreviations as follows: D – decapodid; PL – postlarva; J – juvenile; A – adult. (xvi) Family Scyllaridae: *Scyllarides nodifer*; (xvii) Family Diogenidae: *Diogenes planimanus*; (xviii) Family Paguridae: *Pagurus beringanus*: (xix) Family Lithodidae: *Paralithodes brevipes*; (xx) Family Porcellanidae; *Petrolisthes lamarckii*; (xxi) Family Dromiidae: *Cryptodromia octodentata*; (xxii) Family Homolidae: *Homola barbata*; (xxiii) Family Raninidae: *Raninoides benedicti*; (xxiv) Family Calappidae: *Cyclöes bairdii.*

pereopods. The pleopods are long, setose and apparently functional for swimming. In the second postlarval stage, the pleopods have become very reduced and the pereopodal exopodites are absent.

2.3.8 *Anomuran decapodid and juvenile*

The diversity of anomuran postlarval stages is even greater than that of some larval series, but very few studies have looked at early postlarval changes past the decapodid. In some families (Coenobitidae, Parapaguridae, Chirostylidae, Galatheidae, Albuneidae, Hippidae) only the early larval stages and decapodids are known, while no larvae have been described to date for the Lomisidae, Pomatochelidae, and Aeglidae.

The diogenid hermit crab, *Paguristes sericeus* was reared through its larval stages in the laboratory, and early postlarval stages were described by Provenzano & Rice (1966). They found that the first crab did not differ much from the decapodid (glaucothoe). Earlier works on the diogenid hermit crabs *Dardanus pectinatus* and *Clibanarius erythropus* by Forest (1954) and Dechancé (1958) showed these species lacked paired sexual appendages in early postlarval stages. In these hermit crabs, the first crab stage and sometimes other postlarvae have non-setose pleopods on the right side. By reduction at postlarval molts, these pleopods on the right side disappear rapidly. In the genus *Paguristes,* however, there are characteristic first pleopods in females and highly modified first and second pleopods in males. Provenzano & Rice (1966) found that the first crabs of *P. sericeus* have paired pleopods. At the second crab stage, the pleopods on the right side disappear along with the left pleopod of the second somite, leaving only single pleopods on the third to fifth somites. In males, this persists until the tenth crab stage when both pairs of sexual appendages appear anew. In the first crab of *Diogenes planimanus,* the uropods and abdomen are relatively symmetrical (Nayak 1981). At the next molt, the abdomen coils, the uropods are asymmetrical, and the pleopods of the right side are reduced to small buds on the second to fifth abdominal somites (Fig.4,xvii).

The metamorphoses of the pagurid hermit crabs *Pagurus annulipes* and *P. longicarpus* were described in the classic work of Thompson (1903), covering both external and internal changes in morphology and anatomy. The decapodid (glaucothöe) has a symmetrical abdomen, paired functional pleopods on the second to fifth abdominal somites, asymmetrical uropods, and thoracic appendages like those of the adult. The telson of the decapodid is rounded, there is an unsegmented mandibular palp, and a small ocular scale is present. The gills are present in the same number and arrangement as the adult. In the second postlarva the abdomen is coiled, the telson notched, the mandibular palp two-segmented, and the ocular scale much like the adult; there are no pleopods on the right side of the abdomen and the second pleopod on the left side is rudimentary. In the next few molts, the pleopod rudiment is lost, except in those crabs that will become females. In the young females, the rudiment of the second pleopod begins to grow to full size after about 30 days of development beyond the decapodid.

The larval development of the hermit crab *Pagurus samuelis* through early postlarval stages was described by Coffin (1960) and Kurata (1968d). There is a full set of functional pleopods on the second to fifth abdominal somites in the glaucothöe stage. The ocular scale is rudimentary, the uropods are slightly asymmetrical, and the mouthparts are not well developed, except for the third maxillipeds. At the molt to the first crab, the abdomen becomes coiled and the ocular scale is well developed. The pleopods are present on only the left side of the second to fifth abdominal somites (Fig.4,xviii).

All adult lithodid crabs are characterized by the lack of uropods and the presence of rudimentary fifth pereopods which are folded under the posterior margin of the carapace. Miller & Coffin (1961) described the developmental stages of *Hapalogaster mertensii.* In the decapodid, the pleopods are functional on the second through fifth abdominal somites, and the fifth pereopods are rudimentary and held beneath the carapace. In the first crab stage, the abdomen is flexed beneath the carapace and the fifth pereopods remain rudimentary. The pleopods are still present but are uniramous and lack setae.

Kurata (1956, 1960, 1964) described the development of *Paralithodes brevipes* (Fig.4, xix), *P.camtschatica, P.platypus* and *Dermaturus mandtii,* all from off Japan. The decapodids (glaucothöes) of these species usually have prominent and pointed rostrums, but no ocular scales. The uropods are symmetrical, and the right cheliped is usually larger than the left one. In the first crab stage, the mouthparts are more like the adult, the abdomen is folded underneath the cephalothorax, the uropods are absent, and the pleopods are rod-like remnants. Kurata (1960) also noticed that in *P.brevipes* and *P.camtschatica* there were occasionally intermediate zoeae in the last larval stage. The decapodids produced by molting of these intermediate zoeae were morphologically intermediate between normal decapodids and first crabs; furthermore, the first crabs produced by molting of the intermediate decapodids were intermediate in form between first and second crabs. In *Cryptolithodes typicus,* the carapace of the decapodid and first crab is triangular with a slightly undulate posterolateral margin, thus differing significantly from the adult (Hart 1965). Uropods do not occur in any stage, larval, postlarval or adult.

There have been numerous larval descriptions of the Porcellanidae (see Gore, Fig.5, this volume), in part because members of this group have only two zoeal stages; however, few descriptions have gone beyond the decapodid. In *Petrolisthes lamarckii,* Shenoy & Sankolli (1975) found that the decapodid resembles the adult in appearance; uropods are first developed at this stage and the statocyst is first implanted in the antennular base. In the second postlarval stage (= first crab), pleopods are reduced in size, the first maxilliped becomes more developed, and setae and spines of the carapace develop (Fig.4,xx).

2.3.9 *Brachyuran decapodid and juvenile*

In the Brachyura or true crabs, zoeal stages progress through a number of molts, at which time appendage setation becomes more complex, segmentation of the abdomen is completed, and abdominal appendages are added. The final zoeal molt gives rise to a decapodid (megalopa) in which the appendages are further modified, the thoracic appendages become functional, and the general morphology and behavior more closely resemble the juvenile crab. Almost all brachyuran megalopae have varying numbers of curved setae, 'brachyuran feelers', on the fifth pereopodal dactyls. The next molt produces the first crab stage. Passano (1961) called the megalopa a transitional form between the planktonic zoea and the bottom living adult, perhaps necessitated by the extreme morphological specialization of the crab form.

The megalopae known from dromiid species are rather homogeneous. They tend to retain the antennal exopod lost in decapodids of most higher crabs; also, the fifth pereopodal coxae are dorsal to the fourth, so the fifth pereopods are carried over the posterior carapace as in adults. Hale (1925) recorded abbreviated development in the two Australian dromiids *Paradromia lateralis* and *Cryptodromia octodentata* (Fig.4,xxi). In both these species, the hatchlings are in the decapodid stage and have well developed swimming pleopods, even though they do not leave the abdomen of the adult female.

There are very few descriptions of homoloid megalopae, but there is a great deal of variation in those described for the families Latreilliidae and Homolidae. Rice (1981b) summarized the main characters for decapodids of these two families. All known homoloid megalopae have well developed uropods with setose endopodites and exopodites. This feature confirms the primitive condition of homoloids shown by their zoeal stages, for setose endopodites of the uropods are found in decapodids of all anomuran groups and in some dromiid crabs. Rice (1964) described the last zoea, megalopa and first crab of *Homola barbata* (Fig.4,xxii). A wider carapace, bidentate rostrum, and gastric spines develop in the transition to first crab megalopae.

Knowledge of larval development of the primitive burrowing crabs within the family Raninidae is limited, so description of postlarval stages is also scanty. Raninid megalopae have the characteristic elongated sub-ovoid carapaces, simple triangular frontal regions, and deflected chelipeds so characteristic of the adults. Knight (1968) described development of *Raninoides benedicti* through the eighth crab stage (Fig.4,xxiii).

On the basis of a number of larval characters, Rice (1980) felt the families of the section Oxystomata are so distinct that members of this section are only united by shared convergent adaptations. There are at present no studies of early postlarval oxystomatous crabs which support this view. Lebour (1944) gave brief descriptions of the megalopae of *Calappa flammea* and *Cycloes bairdii* and provided rather crude illustrations of the first crab stages. The megalopae have somewhat narrowed carapaces that contain 12 or 14 large oil-like globules arranged symmetrically. At the molt to first crab, the carapace becomes more rounded and the chelae take on the shape characteristic of adults (Fig.4,xxiv).

As a result of their low number of larval stages many of the spider crabs have been described through the decapodid (megalopa) and some of the early crab stages. Lebour (1928), in her work on the larval stages of Brachyura around Plymouth, England, gave brief descriptions of the decapodids and early postlarvae for *Maja squinado* (as *Maia squinado*); *Eurynome aspera; Hyas coarctatus; Inachus dorsettensis; I.phalangium* (as *Stenorhynchus phalangium*); *I.leptochirus* (as *I.leptocheirus*); *Macropodia tenuirostris; M.rostrata* and *M.deflexa*. More recent studies of majid larval development have given more morphological details of megalopae and early crab stages. The complete larval development, megalopa and first crab stage of *Epialtus dilatatus* was described by Yang (1968). At molt to first crab, the entire surface of the crab becomes covered with capsules containing 'hooked hair' setae.

Kurata (1969) described larvae of 15 species of spider crabs belonging to 13 genera from Japanese waters. He also described the megalopae in the majority of these species, but he described the first crab stage only in *Achaeus tuberculatus, Pugettia quadridens* and *Acanthophrys longispinosus*. He also characterized the many spines, processes or protuberances that were on the megalopal carapaces. In *Hyas araneus* and *H.coarctatus* it is possible to separate the megalopa of each species by the differences in size of the rostral and dorsal carapace spines (Christiansen 1973).

Yang (1976) described the zoeae, megalopa and first crab stage of the arrow crab *Stenorhynchus seticornis*. This species lacks postlarval fusion of the basal antennal article with the carapace, as typically occurs in majids after metamorphosis. An interesting aspect of this study was that Yang obtained first zoeae from an ovigerous *Stenorhynchus* sp. from deep water. The adult was very similar in morphology to the shallow water *S.seticornis*, but the zoeae of the two *Stenorhynchus* species were readily distinguishable from one another. In contrast, the morphology of the adults of this genus is apparently similar enough to have caused considerable taxonomic confusion.

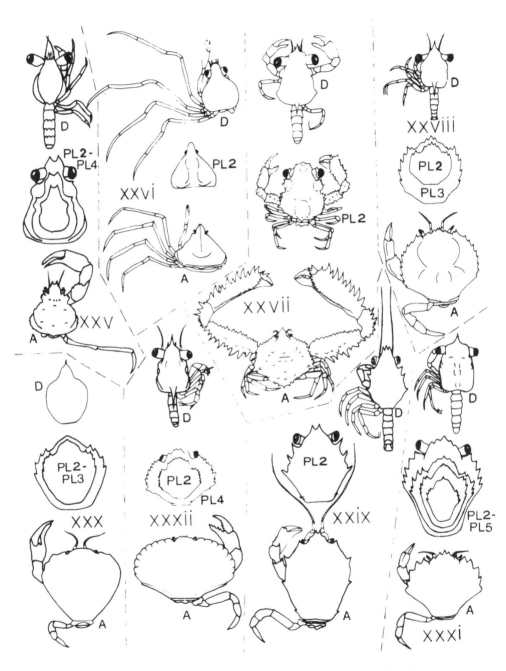

Figure 5. Early postlarval stages of representative decapods. Illustrations adapted from literature, see text. Abbreviations as follows: D – decapodid; PL – postlarva; J – juvenile; A – adult. **(xxv)** Family Majidae: *Inachus dorsettensis*; **(xxvi)** Family Hymenosomatidae: *Trigonoplax unquiformis*; **(xxvii)** Family Parthenopidae: *Parthenope serrata*; **(xxviii)** Family Atelecyclidae: *Atelecyclus septemdentatus*; **(xxix)** Family Corystidae: *Corystes cassivelaunus*; **(xxx)** Family Thiidae: *Thia scutellata*; **(xxxi)** Family Pirimelidae: *Pirimela denticulata*; **(xxxii)** Family Cancridae: *Cancer pagurus*.

The giant West Indian spider crab *Mithrax spinosissimus* has the typical two zoeal stages and megalopa of all majid crabs, but its development is abbreviated since it metamorphoses to the first crab stage in 5-6 days without feeding (Provenzano & Brownell 1977). The mouthparts of both the zoeae and megalopa are nonfunctional and the animal does not seem to take food until the juvenile stages. Ingle (1977) reared the scorpion spider crab, *Inachus dorsettensis* to ninth crab stage in the laboratory (Fig.5,xxv). Ingle concluded that at present the larval and early postlarval stages of *I.dorsettensis* cannot be distinguished from those of *I.phalangium* or *I.leptochirus*, but sexes could be distinguished by the third crab stage. Gill and epipodite development in the megalopal and early crab stages was examined in *Libinia erinacea* by Yang & McLaughlin (1979), who noted that five pairs of laminated gills in the megalopa are supplemented by four additional pairs developed anteriorly in subsequent crab stages. Rathbun (1914) presumed direct development in the majid *Paranaxia serpulifera* with young in the female brood chamber representing two stages by size class. Both had features of the adult, but were not termed megalopae since they did not leave the parent. Two more recent descriptions of majid larvae and postlarvae are those on *Pisa armata* by Ingle & Clark (1980) and *Macropodia rostrata* by Ingle (1982). Others will be discussed in section 4 of this chapter.

The section Oxyrhyncha, besides containing the Majidae, also contains the families Mimilambridae, Hymenosomatidae and Parthenopidae. Nothing is known of mimilambrid development. Numerous larvae of Hymenosomatidae have been described but the only mention of postlarval stages is in the work of Boschi et al. (1969) and Lucas (1971). For *Halicarcinus planatus* Boschi et al. (1969) described a megalopa that has rudimentary mouthparts, but no pleopods or uropods. Lucas (1971) however feels the hymenosomatids do not have a megalopal stage. We propose that his first juvenile crab is equivalent to a decapodid in the form of a benthic megalopa. This is supported by the recent work of Fukuda (1981) on the development of *Trigonoplax unguiformis* (Fig.5,xxvi). He recognized an atypical decapodid (megalopa) in this species, in which the abdomen shows an intermediate form between the last zoea and crab stages and the mouthparts have the characteristic form for a megalopa, even though natatory pleopods are absent. In one of the few developmental studies on parthenopid crabs, Yang (1971) described the megalopa and first crab of *Parthenope serrata* (Fig.5,xxvii). The first crab shows development in the carapace, antennular peduncle, and second maxillipedal podobranch.

The section Cancridea contains the families Atelecyclidae, Cancridae, Corystidae, Pirimelidae and Thiidae. There have been many larvae described for crabs of these families, but not much descriptive work is available on the postlarval stages. Lebour (1928) gave brief descriptions and illustrations of the early postlarval stages of the following cancridean crabs; *Pirimela denticulata, Cancer pagurus, Atelecyclus rotundatus (* as *A.septemdentatus)* (Fig.5,xxviii), *Corystes cassivelaunus* (Fig.5,xxix), and *Thia scutellata (* as *T.polita)* (Fig.5, xxx). In the family Pirimelidae, *Pirimela denticulata* has been reared through the fourth crab stage (Fig.5,xxxi) and *Sirpus zariqueyi* through the first crab stage (Bourdillon-Casanova 1960). In the family Atelecyclidae, the megalopa and the first crab of *Kraussia integra* have been described (Watabe 1971). In the Cancridae, the larval and postlarval development for *Cancer pagurus* has been described to the third crab stage (Ingle 1981) (Fig.5, xxxii). The most conspicuous early postlarval changes in the cancrids include an attenuated and/or subdivided rostrum, broad carapace, and carapace dentition.

The morphology of the larval stages among some representatives of the section Brachyrhyncha is so similar that the distinctions which can be made are rather inconspicuous.

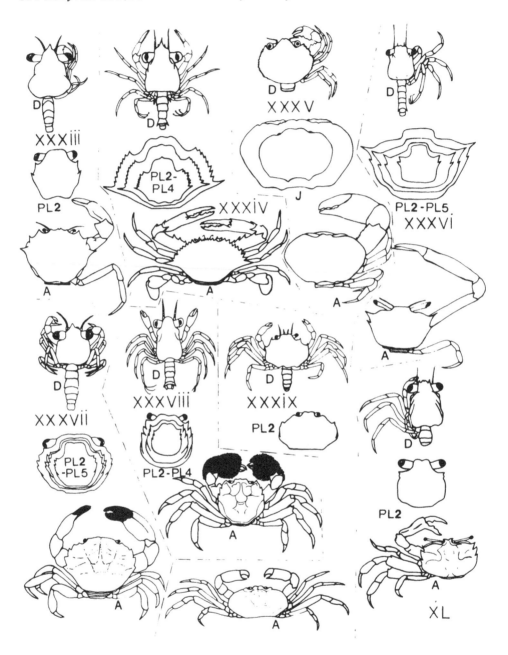

Figure 6. Early postlarval stages of representative decapods. Illustrations adapted from literature, see text. Abbreviations as follows: D – decapodid; PL – postlarva; J – juvenile; A – adult. (xxxiii) Family Geryonidae: *Geryon tridens*; (xxxiv) Family Portunidae: *Portunus pelagicus*; (xxxv) Family Bythograeidae: *Bythograea thermydron*; (xxxvi) Family Goneplacidae: *Goneplax rhomboides*; (xxxvii) Family Xanthidae: *Panopeus turgidus*; (xxxviii) Family Grapsidae: *Eriocheir japonica*; (xxxix) Family Pinnotheridae: *Pinnixa rathbuni*; (xl) Family Ocypodidae: *Macrophthalmus japonicus*.

However, Rice (1980, 1981a) recognized two groups of families based on zoeal characters: 1) a generally primitive group which broadly corresponds to the Cyclometopa or Guinot's (1978) Heterotremata, and 2) a more advanced group which corresponds to the Catometopa or Guinot's Thoracotremata. The decapodids (megalopae) of brachyrhynchan families are also similar, and it is sometimes only in the early crab stages that species begin to take on the characteristics of their respective families.

In the family Geryonidae, the megalopa of *Geryon tridens* has a longer than broad carapace with a ventrally deflected rostrum (Fig.6,xxxiii) (Ingle 1979). There develops in the first crab two strong teeth on each anterolateral margin and a median furrow on the frontal margin. Juveniles of *Geryon* spp. are not readily identifiable, but dactyl flattening (which distinguishes some species) is evident by the third crab stage in *G. tridens.*

In the Portunidae, megalopae and early crab stages have been briefly described for *Portunus pelagicus* (Fig.6,xxxiv) by Prasad & Tampi (1953), Chhapgar (1956), and Yatsuzuka & Sakai (1980), *Charybdis callianassa* by Chhapgar (1956), *Scylla serrata* by Ong (1966), *Charybdis acuta* by Kurata & Omi (1969), *Ovalipes punctatus* and *Thalamita sima* by Muraoka (1969), *Callinectes sapidus* by Kurata (1970), and *Carcinus maenas* and *C. estuarii* (as *C. mediterraneus*) by Rice & Ingle (1975). Aikawa (1937), in his description of the development of *Charybdis bimaculata,* illustrated two different megalopae that molted into two morphologically different first crabs. Since all his material was taken from the plankton, we suggest that he described two closely related portunids. Shen (1935) conducted a very thorough study on postlarval development through the first nine crab stages of *Carcinus maenas.* He examined morphological changes in the pleopods, abdomen and carapace, but also calculated molting frequency and rate of growth. Pleopods of the first pleomere develop only in the male and are present by the second crab stage. By the fifth crab stage sexes can be separated on the basis of abdomen shape.

The new superfamily Bythograeoidea and family Bythograeidae were described by Williams (1980) for a brachyuran crab from the vicinity of submarine thermal vents on the Galapagos Rift (Fig.6,xxxv). *Bythograea thermydron* has some characters of the families Portunidae and Xanthidae, with superficial resemblance to the family Potamidae. Williams also described megalopae and early juveniles of this new species. The megalopae differ from the adult in certain features, especially the well-developed eyes and fully formed normal orbits. Eyestalks degenerate at larger carapace sizes of postlarvae. Williams felt that failure of the interantennular septum to develop may be functionally related to loss of sight, along with heightened dependence on antennular chemoreception in a lightless environment. Both this and large size of the megalopa seem to be neotenic tendencies.

Among the few descriptions of developmental stages for the family Goneplacidae is Lebour's (1928) brief treatment of the megalopa, and several early crab stages of *Goneplax rhomboides* (Fig.6,xxxvi). Paired lateral teeth appear in the first crab stage, and thereafter, the carapace broadens anteriorly to take on the adult state. The adults of the crab *Asthenognathus atlanticus* have been placed with the Pinnotheridae by most workers, but not by Bocquet (1965), who placed them in the Goneplacidae on the basis of larval and postlarval stages since the megalopa has three sensory setae on the fifth pereopod dactyl (lacking in most pinnotherids and some ocypodid megalopae).

The family Xanthidae is one of the largest families in the Brachyura, and there are numerous studies on the larval development of these crabs (see Martin 1984). Most of these descriptions have included the decapodid, but few have gone beyond this stage of development. Postlarval development beyond the decapodid has been briefly described in the following works: Lebour (1928) on *Xantho incisus* and *Pilumnus hirtellus;* Hart (1935) on *Lophopanopeus bellus bellus;* Chamberlain (1961) on *Neopanope sayi;* Kurata (1970) on *Menippe mercenaria, Rhithropanopeus harrisii, Neopanope sayi, Panopeus herbstii, Pilumnus sayi* and *Hexapanopeus angustifrons (as Panopeus occidentalis);* and Kurata et al. (1981) on *Eurytium limosum.* Two studies that provide details of postlarval development were on pilumnid crabs that undergo abbreviated development (Hale 1931, Wear 1967). In *Pilumnus vestitus,* the embryos emerge as soft, flaccid megalopae having a sedentary habit (Hale 1931). The first crab stage has pleopods modified for swimming, but these are lost at the molt to second crab. The adult pleopod buds appear in the third crab stage, and in succeeding stages the abdomen and pleopods progress rapidly toward the adult form. Wear (1967) found that *Pilumnus novaezealandiae* hatches as a megalopa, and *P.lumpinus* hatches as an advanced zoea that molts in 15-30 minutes into a megalopa. The megalopal pleopods in *P.novaezealandiae* begin to atrophy in the first crab and continue through the third crab; thereafter, pleopods of the adult form develop. A comprehensive study of larval and early postlarval stages of *Panopeus bermudensis, P.herbstii, P.turgidus* and *Eurypanopeus depressus* was recently completed (Martin et al. 1984) (Fig.6,xxxvii).

For crabs of the family Grapsidae, larval development and subsequent early postlarval development is known for less than 20% of all species. Early postlarval development of grapsid crabs has been briefly described in works by Hyman (1924b) on *Pachygrapsus marmoratus;* Hart (1935) on *Hemigrapsus nudus* and *H.oregonensis;* Aikawa (1937) on *Plagusia dentipes;* Lebour (1944) on *Planes minutus;* Gamo (1958b) on *Varuna litterata* and *Eriocheir japonica;* Bourdillon-Casanova (1960) on *Brachynotus sexdentatus;* Muraoka (1965, 1967, 1971a,b) on *Plagusia depressa tuberculata, Percon planissimum, Hemigrapsus sanguineus, Gaetice depressus, Pachygrapsus crassipes, P.minutus* and *Grapsus strigosus;* Diaz & Ewald (1968) on *Sesarma ricordi* and *Metasesarma rubripes;* Kurata (1968c) on *Acmaeopleura parvula;* Baba & Miyata (1971) on *Sesarma dehaani;* Baba & Fukuda (1972) on *Chasmagnathus convexus;* Baba & Moriyama (1972) on *Helice tridens wuana;* and Fukuda & Baba (1976) on *Chiromantes bidens.* Both Hyman (1924b) and Aikawa (1937) described two megalopal stages for each of two species, *Pachygrapsus marmoratus* and *Plagusia dentipes.* Their descriptions were based on planktonic material; their second megalopa was apparently *C.bidens,* while their first megalopa probably belonged to a portunid crab. A more thorough study of development was given by Morita (1974) for *Eriocheir japonica,* which was reared through the fifth crab stage (Fig.6,xxxviii). Gill development was complete in the megalopa, and early crab stages were distinguishable by pleopod, maxilla, and carapace shape.

The freshwater grapsid crab *Metopaulias depressus* is confined to water that collects at the base of large tank bromeliads in Jamaica. It has an abbreviated development of two non-feeding zoeal stages and a predominantly benthic megalopa (Hartnoll 1964). The pleopods are rudimentary in the decapodid and completely lost by the first crab stage.

A similar case of abbreviated development occurs in *Geosesarma perracae* (Soh 1969). The non-feeding megalopal decapodid appears to be a juvenile crab but may be considered a modified megalopa since certain larval features remain in the telson, antennules, antennae, and mouthparts.

The Pinnotheridae contains about 26 genera and 220 species, most of which (as adults) are commensals of other invertebrate groups. Larval development is known in less than a quarter of the species, and early postlarval development is known in less than ten species. The Pinnotheridae differ greatly from most other brachyrhynchan larvae in their reduction of the antennae and tendency to lose one or more carapace spines. In these features they resemble the Leucosiidae and Hymenosomatidae. The megalopae tend to lack sensory setae on the fifth pereopod dactyl, though these are present in most other Brachyura. Early postlarval development has been examined briefly in works by Hart (1935) on *Pinnotheres taylori,* Irvine & Coffin (1960) on *Fabia subquadrata,* Roberts (1975) on *Pinnotheres chamae,* Muraoka (1979) on *Pinnixa rathbuni* (Fig.6,xxxix), Yatsuyuka & Iwasaki (1979) on *Pinnotheres* cf. *P.sinensis,* Konishi (1981) on *Sakaina japonica,* Pohle & Telford (1981) on *Dissodactylus crinitichelis,* and Pohle & Telford (1983) on *P.primitivus.* A pinnotherid with abbreviated development is *Pinnotheres moseri*; its megalopal stage is reached in 24-36 hours (Goodbody 1960). According to Faxon (1879) and Hyman (1924a), the last zoea of *Pinnixa sayana* molts directly to the first crab stage. Both authors, however, illustrated this stage with natatory pleopods on the abdomen, so this molt probably represents the decapodid stage.

All genera of the Potamoidea presumably have direct development, with the young hatching in the form of the adult and being carried for a time under the abdomen of the parental female (see Gurney 1942). The mode of development in these crabs indicates a secondary embryonization of the zoeal stages and megalopa. Ancestral larval phases of the Decapoda can still be recognized during this embryonization even with the presence of increased yolk. Pace et al. (1976) described in detail the embryonic development of the freshwater crab *Potamon edulis.* This species exemplifies an extreme form of secondary embryonization in that even the megalopa is imprisoned in the egg and the embryo hatches as a miniature adult.

The family Ocypodidae comprises some 20 genera and 230 species, of which the majority are semiterrestrial. Many of the early postlarval studies are based on megalopae captured in the plankton and the early crabs that were subsequently reared from them in the laboratory. Early postlarval development has been described for ocypodid crabs in works by Kemp (1915) on *Ocypode macrocera,* Rajabai Naidu (1954) on *O.cordimana* and *O. platytarsis,* Gamo (1958a) on *Scopimera globosa* and *Deiratonotus (as Paracleistostoma) cristatum,* Rajabai (1960) on *Dotilla blanfordi,* Kurata (1970) on *Uca pugnax,* Novak & Salmon (1974) on *U.panacea,* and Muraoka (1976) on *U.lactea* and *Macrophthalmus japonicus* (Fig. 6, xl). Some ocypodid megalopae are large in size and usually have special grooves alongside the carapace for folding their appendages tightly against the body. In the first crab, some species begin to burrow into sand and develop tufts of plumose setae between the walking legs. Abbreviated larval development in *Uca subcylindrica* is followed by an atypical *Uca* megalopa and first crab stages in which the carapace is more rounded than typical *Uca* (Rabalais & Cameron 1983).

3 ANATOMY AND FUNCTION IN EARLY POSTLARVAL LIFE

3.1 *Anatomical development*

3.1.1 *General overview*
The morphological changes that take place in transition from larval stages to the decapodid and in the transitions between postlarval stages that follow are often accompanied by considerable changes in behavioral and physiological patterns. These functional changes are related to marked ontogeny in anatomy of organs and tissues during metamorphosis and

early postlarval development (Bellon-Humbert et al. 1978). Unlike external morphogenesis of the exoskeleton, organogenesis and histogenesis need not be confined to changes at a given molt (Steele 1902, Thompson 1903). However, the new level of function can be dependent on endocrine changes associated with molting or exoskeletal modification expressed at the molt.

When a well-demarcated metamorphosis in external features occurs between larval and postlarval developmental phases in decapod crustaceans, it is often accompanied by some pronounced changes in internal anatomy. However, while the decapodid stage may represent a marked change in internal anatomy from the larval state, that internal anatomy often differs from adults. Except for the development of primary and secondary sexual organs, postmetamorphic changes in internal anatomy may be most pronounced in the first several postlarval instars, with the degree of anatomical reorganization and development decreasing in each of the later juvenile molts. Anatomical change during metamorphosis and the next few postlarval molts may in part consist of *degeneration* of some anatomical features, specifically of those that are legacies from larval behavior and physiology. Another facet of anatomical change may involve *redirection* of existent structure; in effect, this would require relatively minor modification of an anatomical feature or its controls in order to transform function from that operative during an earlier stage to that required by later postlarval stages. Finally, there can also be *addition* of anatomical structure; this can involve not only absolute increases or allometric increases in size with overall growth, but also, particularly in the early postlarvae, differentiation of new anatomical features.

The following sections on anatomical development will treat ontogeny of eyestalks and neurosecretory organs, musculature and innervation, the stomach, and a few other organs. Pigmentation will also be included, as it is usually the result of both anatomical structure and products of that structure. Limited reference will be made to gonadal development, a process largely restricted to later stages of development than early postlarvae; however, exceptions will be noted. The most thorough coverage of subjects is divided between references dating from near the turn of the century, which consist of a few meticulously illustrated anatomical monographs (for example, Thompson 1903, Waite 1899), and a contemporary set of papers that in combination provides an excellent coverage of several anatomical features in *Homarus* (reviewed in part by Govind 1982). Apparently, the early initiative in anatomical study of decapod postlarval development was abandoned well before there was representative coverage of the group. Recent efforts, while thorough and exemplary in modern technique, have barely touched upon the diversity of decapod taxa and the spectrum of anatomical structures subject to postlarval change.

3.1.2 *Eyestalks and neurosecretory organs*
The anatomy and function of eyestalks and neurosecretory organs has been relatively well studied in adult decapod crustaceans, but considerably less is known of this subject in early postlarval stages. The studies most applicable to postlarval ontogeny of these anatomical structures are restricted to a few species of the Palaemonidae, the anomuran *Pisidia longicornis,* and a few brachyurans (see LeRoux 1980, 1982 for a compilation of current literature). Observations on decapodid and postlarval histogenesis of these organs is provided for *Palaemon serratus* by Bellon-Humbert et al. (1978) and for several species of North American *Palaemonetes* by Hubschman (1963). Additionally, Hubschman (1971) noted the degeneration of suspected larval glands at metamorphosis to the decapodid in

Palaemonetes. Some assumptions of postlarval histogenesis in the sinus gland and X-organ can be made on the basis of Pyle's (1943) studies of egg anatomy, several larval stages, and adults of *Pinnotheres maculatus* and *Homarus americanus.* However, he did not specifically address histogenesis in early postlarvae. Likewise, our interpretation infers that postlarval stages undergo histogenesis that alters larval conditions to those found in adults (e.g. Orlamunder 1942, Dahl 1957, Matsumoto 1958, Elofsson 1969, Zielhorst & van Herp 1976). Among these, the studies by Matsumoto on *Potamon dehaani,* Pyle on *Homarus americanus,* and Zielhorst & van Herp on *Astacus leptodactylus salinus* are particularly valuable in that they document the appearance of some neurosecretory structures that occur much sooner after hatching in these species with abbreviated development than in those species with a more typical larval life (Bellon-Humbert et al. 1978).

The most pronounced rearrangement and development in anatomy of neurosecretory organs in *Palaemon serratus* and several species of *Palaemonetes* occurs during the transformation from an upside-down swimming larva to a dorsal-side-up postlarva (Bellon-Humbert et al. 1978, Hubschman 1963). Concurrent with this reversal of attitude, some eyestalk features rotate from the larval dorsal position (which would face downward in an upside-down swimming larva) to a ventral position in the dorsal-side-up postlarva; thus a structure that faced downward in the larva, such as the larval sensory pore (LSP; the pore of Hubschman's SPX), would also be directed downward in the postlarvae and adults (Fig.7). In his description of this phenomenon in *Palaemonetes,* Hubschman (1963) did not detail the rotational positioning of internal eyestalk structure during early development. He suggested repositioning occurred by a rotation of the eyestalk on its longitudinal

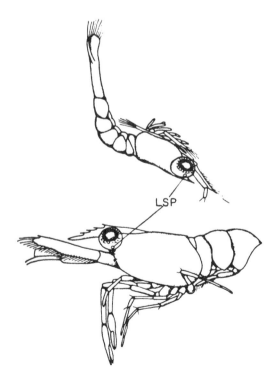

Figure 7. Diagram showing attitude of larval (above) compared to postlarval (below) *Palaemonetes,* with reference to relative position of larval sensory pore (LSP) (slightly modified from Hubschman 1963).

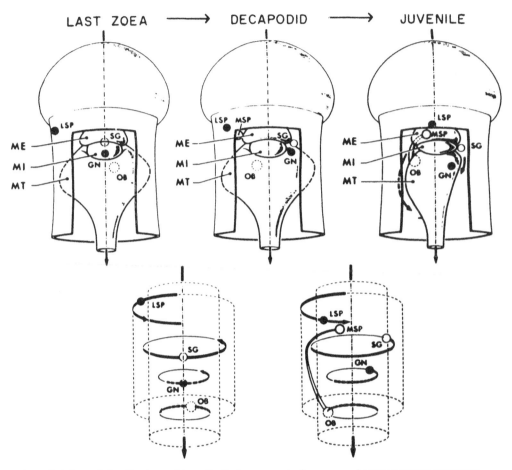

Figure 8. Schematic localization and hypothetical movement of sensory and neurosecretory structures in one right eyestalk (dorsal view) during last zoeal (last mysis), decapodid (postmetamorphic), and juvenile stages of *Palaemon serratus*. GN – giant neuron; LSP – larval sensory pore; ME – medulla externa; MI – medulla interna; MSP – main sensory pore; MT – medulla terminalis; OB – organ of Bellonci; SG – sinus gland (slightly modified from Bellon-Humbert et al. 1978).

axis. Bellon-Humbert et al. (1978) conducted a more thorough analysis of internal movements in eyestalk components during development of *Palaemon*. Their findings suggested that 'rotation of structures occurs around the longitudinal axis of the eyestalk' under the influence of relative growth and torsion in various of the eyestalk medullae (Fig.8). They also documented that the process begins in the larval molt preceding metamorphosis and continues into postlarval developmental stages.

In *Palaemon serratus*, the LSP undergoes its most pronounced movement from a dorsal to ventral position on the surface of the eyestalk (Bellon-Humbert et al. 1978) during the metamorphic molt from the last larva to the decapodid. The first unquestionable appearance of the main sensory pore (MSP) occurs at this stage. Also at this time, some cells begin to take on neurosecretory characteristics at the future locations of the medulla externa (ME) and medulla terminalis (MT) X-organs. The sinus gland (SG) and the giant neuron (GN) have moved to the external dorsolateral side of the eyestalk, but histological features of the SG and GN remain similar to those of the last larval stage. The organ of Bellonci (OB) remains inside the MT during the decapodid stage, and further developments in the OB, SG, X-organs, and medullae are reserved for following postlarval stages.

During these early postdecapodid stages, the eyestalk both grows and acquires the adult arrangement of its components. Medullae also individualize, the eyestalk completes its formation, and the eyestalk straightens on its longitudinal axis and becomes more parallel to the body. These movements, coupled with directional and allometric growth of some structures, bring components into their functional adult configuration (Bellon-Humbert et al. 1978). For example, it is during this phase of growth and development that the OB and the MSP first come into contact, and that the OB first contains stainable granules in vacuoles and extends partially out of the MT. The MSP also becomes more developed, the number of micropores increases, and a secretion product becomes evident at the micropore apertures. Neurosecretory cells become better differentiated into two distinct groups, one in the X-organ at the ventral region of the MT where the OB nerve tract makes its entry (this area termed $MTGX_2$ by Bellon-Humbert et al.) and the other forming the ME X-organ in front of the sinus gland.

Development of the sinus glands to a presumably active condition, with granules in three differently staining regions, becomes evident only after the decapodid stage, even though it is structurally distinguishable in the larval phase of development. Thus, the completion of an affiliation between the sensory pore complex and the distal part of the OB, the connection of the OB with the neurosecretory cells of the medulla terminalis via a nerve tract, the differentiation and development of neurosecretory cells in the MT and the ME X-organs, and the development of granulation in the OB and SG, mark major steps in activation of the sensory and secretory functions of eyestalk organs during early postdecapodid development.

The source of regulatory factors in early postlarvae is unclear. However, Hubschman (1963) demonstrated that in larval *Palaemonetes,* metamorphosis is not under eyestalk control, and he looked elsewhere for the site of regulatory factors that may be analogous to those of insects (Hubschman 1971). While Y-organs resemble prothoracic glands in some respects and are non-nervous secretory organs involved in the control of molting, none are known in larval *Palaemonetes.* Suspected larval glands were found at the base of the first maxilla in all larval stages of five species of *Palaemonetes,* and these deteriorated in the last phase of the terminal larval instar (Hubschman 1971). These glands were completely absent in the decapodid stage, so their activity, if any, could not have extended to postlarval stages. We are therefore left with no evidence of a source for regulatory factors governing early postlarval development in *Palaemonetes* or in *Palaemon.* Hubschman (1963) noted that the sinus gland, in particular that of *Palaemonetes,* does not appear to be functional until after the second or third postlarval molt. Likewise, there is anatomical evidence that the sinus gland in *Homarus americanus* does not function by the decapodid stage (Pyle 1943). When eyestalks were removed from larvae and/or decapodids of an anomuran (LeRoux 1980) and several brachyurans (Costlow 1963, 1966a,b), however, activation of the X-organ and sinus gland complex varied between taxa and sometimes occurred in the decapodid stage. Absence of both eyestalks in decapodids of *Pisidia longicornis, Callinectes sapidus,* and *Sesarma reticulatum* accelerated development, presumably by removing the source of molt inhibiting hormone. On the other hand, removal of eyestalks in larval *Rhithropanopeus harrisii* did not accelerate molting until the third post-larval molt, suggesting that the X-organ and sinus gland complex may not become activated until later in early postlarval development.

3.1.3 *Musculature and innervation*

Relatively little is known on just how embryonic nerves and muscles acquire their adult configurations (Govind 1982). That which is known for the period between embryonic life and the later juvenile stages is largely restricted to the larval stages and primarily represented by studies on the Astacidea. The review of crustacean nerve, muscle, and synapse development by Govind (1982) contains extensive coverage of embryological, larval, and later development of these structures. We confine our treatment solely to the period of early postlarval growth and expand upon coverage of decapod crustaceans.

The specialized asymmetric chelipeds of many species become adult-like during post-larval stages. Such chelipeds may be employed in sexual or agonistic behavior, as in the case of male fiddler crabs (Crane 1975) or crayfish (Stein 1976). Both pre- and postpuber-tal males of decapods may exhibit a positive allometric growth of chelae (Finney & Abele 1981). In other species, both sexes have asymmetric chelipeds and use them in defense or prey capture (Costello & Lang 1979), with mechanical advantage changing to fit feeding strategies that change with maturation (Warner & Jones 1976, Vermeij 1977). Where adult asymmetry exists, it may be fixed or plastic (Przibram 1931). In species having the fixed condition as adults, there may be a given developmental stage when the fate of either cheliped is fixed. For example, in *Homarus americanus,* the fate of chelipeds does not become fixed until the second postlarval stage (Emmel 1908, Lang et al. 1978). In species having a plastic dimorphism, the cheliped of either side can differentiate into either form of the dimorphic pair, regardless of the degree to which development has proceeded beyond early stages and regardless of how differentiated the cheliped pair has become. This was demonstrated in *Alpheus* by Mellon & Stephens (1979). Most commonly, this ability is evident in cases where the major cheliped has been lost and regenerated as a minor cheliped during subsequent molts, while the former minor cheliped has differentiated into the major form during those molts (see Govind & Lang 1978).

In early postlarval stages, decapods may have chelipeds that are indistinguishable from one side of the body to the other. Differentiation in shape, musculature and innervation occurs later. The most thorough studies of this process were conducted on the lobster, *Homarus americanus,* a species with the fixed form of cheliped asymmetry in adults (Costello et al. 1981, Costello & Lang 1979, Govind & Lang 1978, King & Govind 1980, Lang et al. 1977). In the decapodid and following stage, the chelipeds of the two sides are symmetrical both in external appearance and in fiber composition of the closer muscles. Both chelae contain closer muscles that consist of about 30 % short-sarcomere (fast) fibers and have the remainder composed of intermediate- to long-sarcomere (slower) fibers (Fig.9). Late in the second post larval stage, or at the molt to the third postlarval stage, the number of short-sarcomere fibers rapidly increases in one chela and slowly decreases in the other. This trend continues until about the 13th postlarval stage, when the closer muscles of both chelae have attained their adult distributions of fibers. Ultimately (Ogonowski et al. 1980), the 'crusher' chela loses all short-sarcomere fibers while the 'cutter' chela develops a high ratio of short- to long-sarcomere fibers. Concurrent with developmental changes in muscle fiber distributions, a disparity in fatigue resistance develops between the fast axon synapses of the crusher and cutter chelae, and facilitation values for slow axon

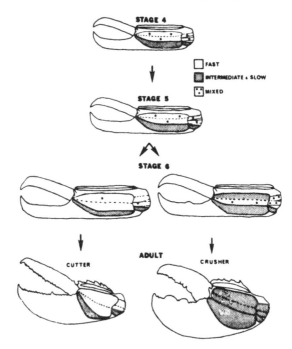

Figure 9. Mesial aspect of lobster (*Homarus americanus*) chelae closer muscles during postlarval development, with differentiation of cutter and crusher chelae (sizes not to scale); from Costello & Long 1979).

synapses in both chelae increases. As a net result, the dorsal bundle of the closer muscle of the crusher chela converts into a tonic motor unit, while the dorsal bundle of the cutter closer muscle retains an enhanced ratio of fast muscle units and becomes a specialized phasic unit (Costello et al. 1981).

Studies of excitatory innervation in chela closer muscles of *Homarus americanus* provided insight into proliferation of axonal tissues and enlargement of synpases in thoracic limbs, concurrent with postlarval growth (King & Govind 1980). Neither the extent of innervation nor the size of excitatory synapses in early postlarvae is as great as in adults, although both are greater in the decapodid stage than in the first larval stage. In the first larval stage, excitatory innervation is restricted to a few discrete muscle fibers. In the decapodid this innervation is more widespread, and in the adult all fibers receive excitatory innervation. Thus, increases in the percentage of innervated fibers coincide with growth of the muscle. The increase in the size of excitatory synapses during this period increases innervation during muscle growth. The size increase in such synapses between the decapodid stage and the adult is in excess of four-fold and is much greater than that which occurs between the first larval stage and the decapodid. King & Govind also noted that a larger proportion of adult synapses had two or more 'presynaptic dense bars' than did either larvae or decapodids. They suggested that adults may therefore also have a 'higher quantal content' for synpatic transmission.

The phenomenon of synaptic growth with development need not be characteristic of decapod species in general. Atwood & Kwan (1976) found that synapses in the limb opener muscle of crayfish remain constant throughout development, while growth of innervation occurs primarily by addition of new synapses. Govind (1982) reviewed the phenomenon of 'sprouting' from nerve terminals. New synapses so produced may be an additional means by which innervation spreads to growing muscle fibers in lobsters.

Concurrent changes in muscular and neural growth occur during early postlarval development of lobsters (Lang et al. 1977). Such anatomical changes underlie allometric growth and the behavioral implications of such growth. In *Homarus americanus* larval and early postlarval stages, a tail-flip (abdominal extension and flexure) escape movement is the usual response to moving visual stimulus, while defensive posturing (display of open chelae) is favored over escape behavior in larger animals. Morphologically, this change in behavioral modes is then facilitated by allometric changes in both chelipeds and abdomen of developing lobsters.

In larval and several early postlarval stages of lobsters, the muscular abdomen comprises a substantial component of the total animal weight (Lang et al. 1977). The abdominal length is also great compared to carapace length (Fig.10). However, the chela weight and length are relatively small compared to the total weight and the carapace length. The ratio of abdominal weight to total animal weight remains nearly constant until an animal is 40-60 mm long. There is, however, strong positive allometric growth in the length and weight of the chela until the animal reaches a total length of about 50 mm and less so thereafter.

Negative allometric growth is characteristic of the abdomen once total length exceeds about 50 mm. Most of the growth in the medial and lateral giant axons has been achieved by the time the animal is 60 mm in carapace length (Lang et al. 1977). Thereafter, because of the relatively small further increase in diameter, and the relationship of velocity to diameter as a power function, the conduction velocity of the giant axons does not substan-

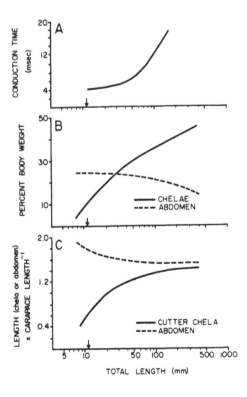

Figure 10. Curves approximating changes during growth of lobster, *Homarus americanus*; arrow indicates approximate size at metamorphosis to decapodid. (A) Conduction time in milliseconds for a spike in medial giant axon to travel from the brain to sixth abdominal ganglion. (B) Relative contribution of chelae and abdomen to total body weight; weight of chelae is combined weight of both claws, severed at autotomy plane; total body weight is weight of intact animal before chelae or abdomen are severed. (C) Cutter chela and abdomen length relative to carapace length; length of cutter chela is propodus length; carapace length measured from eye to posterior margin of thorax (curves are approximations from data points of Lang et al. 1977, originally published under 1977 copyright by the AAAS).

tially increase. In fact, the small increase in axon diameter is less than that needed to compensate for the increase in overall animal length, and there is a striking increase in conduction time. Thus, conduction time from the brain to the telson is 4-5 msec in early postlarval stages and remains relatively constant until the lobster reaches a length of 50 to 60 mm; conduction time then increases and response shifts from escape to defensive display of the chelae (Govind 1982).

There is some evidence that diminishment of tail-flip in adult lobsters has a counterpart in crayfish (Krasne & Wine 1975). However, as pointed out by Lang et al. (1977), crayfish may have relatively smaller chelae than lobsters as adults, and giant axons may differ in relative size between the two animals. Their preliminary experiments with *Procambarus acutus* suggested that at least this species differed from *Homarus* in its behavioral strategy.

The concurrence in changes in behavior and anatomy in astacideans suggests what kinds of anatomical changes should be expected in other decapods that alter behavior in early postlarval life, e.g., the transition from a brachyuran megalopa, with its extended abdomen, to the early crab stage, with a tightly flexed abdomen. It has been documented that relative sizes of abdominal flexor and extensor muscles can change during ontogeny from the megalopa to the adult brachyuran (Trask 1974). Such changes in a brachyuran would almost certainly differ from the ontogeny in shrimp abdomens that retain abdominal swimming as adults, but detailed studies at the neuromuscular level have not been made. Even early postlarval ontogenetic shifts between relative dependence upon the mandibles versus use of the cardiac stomach in mastication, as reported by Factor (1982) and discussed in the next section of this chapter, should be reflected in neural and muscular analogs.

Motor neurons also change during growth. Stephens & Govind (1981) have recently reported on the restriction of innervation fields in the common exciter and common inhibitor motoneurons of deep abdominal extensor muscles during early postlarval development of *Homarus americanus*. Fields that extend to three adjacent abdominal somites during the period from embryonic development through the second postlarva, become restricted to one somite in the third postlarval stage. This is accomplished by selective elimination of functional synapses from an initially widespread distribution. Another mechanism of neuromuscular development that may be operative is the redirection of oscillators, suggest as a possibility in the larval to decapodid stage transition of the lobster *Homarus gammarus* (Laverack et al. 1976). Muscles of thoracic endopodites are small and nonfunctional in the larval stages (Neil et al. 1976). They become well separated from those that cause swimming movements of the exopodites in larval stages and then degenerate in the decapodid (first postlarval) stage. In the decapodid, the exopodites of the pereopods become vestigial, while the endopodites develop as walking legs. Laverack et al. (1976) noted the similarities of patterning and other factors in larval swimming and adult walking of *Homarus*. They suggested that system was an 'oscillator that switches output from one set of muscles to another at the critical molt . . . (from the last larval stage to the decapodid)'. Alternatively, the change in modes of locomotion at metamorphosis could suggest that separate oscillators are present which either start or terminate their activity at the metamorphic molt.

3.1.4 *Stomach, antennal glands, and gonads*

Larval, early postlarval, and adult decapods typically masticate by the stomach and/or the mandibles. Factor (1982) recently compared the relative complexity of the stomach and mandibles in larvae and postlarvae of the xanthid crab *Menippe mercenaria* and reviewed the limited literature that applies to this subject in other decapod species. Considerable change may be expected to occur between planktotrophic larval stages and adults that are dependent upon a wide variety of substantial food items in the benthic environment. However, there is not always a clear trend toward greater complexity with development. For example, the greatest complexity in the foregut of the shrimp *Crangon septemspinosa* occurs in the last larval stage (Regnault 1972), even though a true gastric mill is never developed in larvae or postlarvae of this species.

With the molt from the last larval stage to the decapodid stage in most decapods, the stomach becomes adult-like. However, further changes during early postlarval development may take place in such features as the strength of dentition and ridges, the length and density of setation, the relative sizes of chambers, and the distribution of fine denticles. The most pronounced changes in stomach morphology generally occur with the metamorphic molt. They are particularly pronounced in those species that have a gastric mill as adults, and this is the typical condition in the reptantious brachyurans, anomurans, and macrurans (Factor 1982, Patwardhan 1935). For example, in *Menippe mercenaria*, mastication of food shifts from mandibles to a rudimentary form of the gastric mill at the molt to the decapodid. Simultaneously, the molar process of the mandible becomes more simple (Factor 1982), a simplification largely completed in the early postlarval stages.

A developmental sequence similar to that of *Menippe* was described in *Pagurus annulipes* (as *Eupagurus longicarpus*) by Thompson 1903 (also see Roberts 1970). The full complement of gastric mill components appears at the metamorphic molt. Likewise in the lobster *Homarus americanus*, a full complement of gastric mill teeth first appears in the fourth developmental stage, the decapodid. However, unlike *Menippe* and *Pagurus*, neither the larval nor the postlarval stages of *Homarus* have a molar masticatory process on the mandibles, and it is unclear how food is masticated in the larval stages (Factor 1978, 1981). Thus, in the species that develop a true gastric mill, early postlarval developments are probably limited to further sclerotization and subtle elaboration of an apparatus that is prefigured at the metamorphic molt. From our own observations of stomachs in early postlarval stages in the thalassinidean *Callianassa major*, the processes of sclerotization, definition of setal fields, and chamber modification are completed in early postlarval development.

In the natantious species of the Penaeoidea, Caridea, and Stenopodidea, which typically lack a well-defined gastric mill and depend more fully upon the mandibles for mastication throughout the life history (Patwardhan 1935, Factor 1982), there may also be changes in the stomach during metamorphosis and postlarval development. LeRoux (1971a,b) noted that the pyloric and cardiac chambers of *Palaemonetes varians*, which freely communicate during the larval stages, become more separated after metamorphosis. Although there is no functional difference in these two chambers during the larval stages, after metamorphosis the cardiac chamber becomes specialized. It stores food and mixes that food with enzymes, and the pyloric chamber becomes a straining and (to a limited degree) a triturating apparatus. Heral & Saudray (1979) noted concurrent changes in the mandibles of this species.

As previously noted, metamorphosis of the caridean *Crangon septemspinosa* marks the loss of gastric armature that was progressively developed during the larval phase of growth. However, Regnault (1972) also noted that other structures of the stomach change toward maximal differentiation at the metamorphic molt. While Regnault did not specifically address early postlarval changes in the stomach of *C. septemspinosa*, a number of postlarval elaborations, other than armature, are apparently added to the basic form of the stomach that is prefigured in the decapodid stage.

Much less is known of early postlarval development in the hepatopancreas and other digestive tract features. Thompson (1903) described a complicated metamorphosis in the hepatopancreas of *Pagurus* that takes place within the decapodid (glaucothöe) and the following postlarval stage. During the decapodid there is a reduction of lobes persisting from the zoeae, which gives rise to simple tubular hepatopancreatic masses. Adult-like diverticula develop in the second postlarva, and bilateral symmetry is obscured after the fourth postlarva. Such a developmental sequence may be typical of only those anomuran forms with highly asymmetrical abdomens. Trask (1974) noted, however, that *Cancer anthonyi* also has digestive glands of a simple tubular shape in the decapodid stage. Thompson (1903) observed development of chitinous portions of the gut, achitinous portions of the gut, folds of the gut wall and ceaca or diverticula that develop over the early postlarval stages, and LeRoux (1976a) reported on metamorphic development of the intestinal caecum in *Pisidia longicornis*.

Development of the green glands or antennal glands was treated by Thompson (1903) and Waite (1899) for *Pagurus* and *Homarus* respectively. Both authors pointed out the appearance and development of the nephrosac or vesicle during the decapodid stage, and noted that its growth relative to surrounding structures is a feature of further postlarval development, indicating osmoregulation may be further developed in the postlarval stages. However, the growth of the vesicle need not simply reflect an increased ability to maintain water balance with ontogeny; in fact, the opposite is often observed. For example, Davenport (1972) noted that juveniles of the hermit crab *Pagurus bernhardus* regulate volume in lower salinities than adults, and suggested that it is the size of the aperture of the nephropores in relation to body size that limits this ability in adults.

The reproductive system largely develops after early postlarval stages in decapod crustaceans. Small masses of sex cells may be distinguishable beneath the pericardial septum even in larval stages, and some increase in number apparently occurs during early postlarval development (see Thompson 1903). However, with few exceptions (see LeRoux 1976b), little is known of the earliest developments in internal organs, ducts, and orifices of the reproductive system for most decapods, and attention has usually been directed to noting the earliest morphological developments of externally evident secondary sex characters.

Shape of the abdomen and the form and number of abdominal appendages seem to be among the external characters that are most useful in distinguishing sex at an early postlarval stage (for example, see Forster 1951: Fig.2 and Morita 1974: Figs.20-2, 21-1, 21-2). Later development of the gonads can give rise to organs that are mature in most respects but lack mature sex cells (Knudsen 1960). Further growth of gonads with maturation of the sex cells can, especially in females (Finney & Abele 1981), relate to allometric changes in body form which often mark the attainment of maturity. The degree to which early postlarval stages show development in the gonads and secondary sex characters is probably a function of the number and duration of postlarval stages that occur between the decapodid and the mature adult. Thompson (1903) noted that sexual maturity is probably not reached until the second or third year in the life of *Pagurus*. However, gonads may become mature in crustaceans before secondary sex characters are fully developed (Charniaux-Cotton 1965).

In some small specialized decapod species, sexual maturity may be attained or closely approached shortly after metamorphosis from the larval phase. A particularly striking case of this is represented by the male swarming stages of some commensal pinnotherid crabs. The male and female swarming stages consist of early crab stages. Males at this time may be sexually mature and may precociously inseminate females in which the ovaries are not yet fully developed (Kruczynski 1973, Pearce 1966).

3.1.5 *Pigmentation*

Both coloration and patterns formed of color pigments (chromatophoric and extrachromatophoric) or structural manifestations (scattering, diffraction, and interference of exoskeleton or setation) are usually subject to developmental changes during postlarval life of decapods. Even in the earliest postlarval stages, there is evidence that some color patterns, like many morphological features, are controlled by eyestalk neurosecretory structures. For example, eyestalkless postlarvae of *Pisidia longicornis* do not develop the characteristic color patterns of normal postlarvae (LeRoux 1980). In addition to the endogenous control of color and pattern, exogenous factors (such as decorating materials, epibiotic growths, and precipitates) may also alter apparent coloration, especially in later stages of postlarval development. Also in later stages, certain coloration changes may coincide with maturation, as in carapace whitening of some fiddler crabs (Crane 1975), molt to a 'white' stage in some palinurid lobsters (Sheard 1962), and certain transient patterns in *Ucides* and *Cardisoma.*

Coloration and patterning are unconservative qualities at taxonomic levels above that of species among the decapods. However, at the species level, it is often a useful taxonomic character and may be diagnostic of habitat or behavior. Both habitat and behavior may be altered at metamorphosis, and the degree to which they are altered is often greater over this phase of development than over any other in the life history of a decapod. Thus, striking changes in color and pattern may coincide with this early phase of postlarval development, but literature references are few and are restricted to isolated taxocenes. Occasionally coloration in postlarvae suggests restriction to a unique habitat. For example, Seréne (1961) reported a brachyuran decapodid with camouflage coloration exactly matched to the mantle coloration of a squid (*Loligo*) on which it lived.

Perhaps the most frequently noted case of ontogenetic color change in early postlarvae is that which occurs in the early postlarval development of palinurid and scyllarid lobsters of the genera *Panulirus, Jasus, Ibacus,* and *Scyllarides* (MacGinitie & MacGinitie 1949, Deshmukh 1966, Sorensen 1969, Lyons 1970, Silberbauer 1971, Ritz & Thomas 1973, Serfling & Ford 1975, Berrill 1975, Booth 1979 and Kanciruk 1980). In general, color development in these animals reflects the transition from a planktonic habitat to a benthic one. The decapodid stage, commonly called the puerulus, is the settling stage, transforming from a planktonic swimmer during its early period to a benthic form in its later period. The puerulus is rather transparent. Just prior to ecdysis to the postpuerulus stage, it gains pigment and a more opaque appearance, which may be concurrent with the fading of some isolated larval chromatophores. Both coloration and calcification become more pronounced during the first postpuerulus stage and several stages that follow, and the animal may become cryptically or disruptively patterned and match the benthic substrate and cover in which it has settled. The triggering of pigment deposition may be dependent upon the availability of appropriate substrates. Serfling & Ford (1975) noted that pueruli of *Panulirus interruptus* held without access to vegetation and epifauna typical of the postlarval environment remained transparent for 2-3 weeks and then died instead of molting to the postpuerulus. It also appears that the color and patterns of early postlarvae are not simply developing adult patterns, but rather may serve as camouflage tailored to the juvenile habitat. The adults are often less protectively colored but may instead have an array of behavioral adaptations that are antipredatory (Kanciruk 1980).

Transparency is not the only color of larvae or settling postlarvae. At depth, in poorly lighted crevices, or under the darkness of night, even the apparent red coloration (in light) of some forms may conceal when incident red light is largely absent from the available light spectrum. Also, where seston or substrates may be brightly colored, brightly colored larvae or postlarvae may be effectively concealed.

In terms of ontogenetic color and pattern changes, few references (other than those for lobsters above) document postlarval schemes. Hart (1965) decribed the predominantly scarlet and reddish-orange coloration of zoeae and megalopae in the anomuran *Cryptolithodes typicus.* All postlarval stages seem to be well camouflaged for the rocky subtidal and intertidal habitats in which they live. Knudsen (1960) noted the plasticity of coloration that may occur in early crab stages of several xanthoid crabs. In *Paraxanthias taylori,* the early crab stages resemble the adults in color, while in *Cycloxanthops novemdentatus* and *Lophopanopeus leucomanus leucomanus* the young are of many different bright colors. While young of the latter two species generally occupy the adult habitat, they select the smallest crevices and spaces among coralline coverings and snail tubes atop rocks where bright colors would conceal them. In the case of *Menippe mercenaria,* early crabs and small juveniles may be much darker than adults (Hay & Shore 1918, Wass 1955, Manning 1961, DLF unpublished). These stages often inhabit deeper crevices of rubble and epifaunal interstices than do the lightly colored and less vulnerable adults.

Our own studies of western Atlantic reptantious decapods suggest that it is extremely difficult to generalize about early versus later postlarval colors and color patterns other than to note that early postlarvae may vary, often considerably, from later postlarvae or adults. We have encountered numerous cases in which early postlarval crabs are highly variable in color patterns while adults are somewhat more conservative (for example, in the xanthids *Panopeus, Hexapanopeus, Paractaea,* and *Micropanope*). In other cases, variations in coloration are pronounced in both the adults and the early crab stages (for example, in the xanthid *Xanthodius*). Among the Porcellanidae and the Majidae we have found cases where both adult stages and early postlarvae are not subject to great variation, but where the color of these two phases may be strikingly different (for example, in some *Petrolisthes* and *Mithrax*). These differences seem to be particularly common among shallow water majid crabs. In a few cases, it is uncertain whether color changes reflect an ontogenetic process or just a highly labile coloration response to food and substrates of habitats (for example, in *Acanthonyx* and *Mithrax*). Hines (1982) also noted the transition in color of the majid, *Pugettia producta.* Juveniles recruited onto red algae in the intertidal eelgrass zone were maroon in color, but changed to tan as they grew and migrated to the kelp forest.

Mimetic patterns may be confined to the earliest postlarval stages in some brachyuran species (for example, in *Calappa sulcata*), whereas in others, they may not appear until late during postlarval development. Commonly, crypsis is maintained with pattern change as body size increases and habitat and behavior change. Where adults and early juveniles share the same or a very similar substrate, the growth of the postlarval color markings may be negatively allometric and give adults a pattern of spots or venations that is as fine or finer than that of the early postlarvae. For example, this is characteristic of the portunid *Arenaeus cribrarius.*

3.2 *Behavioral and physiological adaptations to habitats*

Both migration and habitat selection in early postlarvae maximize growth rate, minimize population losses to predation, and facilitate recruitment to adult habitats. Relocation may relate to tolerances and capacities that differ between larvae and adults. Even postlarval metabolism may differ between early and late stages (Sastry & Ellington 1978).

Dietary requirements must be of major importance in inducing or limiting movements and distributions of early postlarvae, for these stages are usually characterized by rapid growth and high metabolic demand. Klein Breteler (1975b) noted the linear relationship between logarithms of oxygen consumption and weight in *Carcinus maenas* over postlarval development. In aquatic invertebrates requirements of food may be 3-4 times those needed for respiration (Jorgensen 1955). Thus, food requirements at early postlarval stages should substantially exceed those of later subadult or adult stages on a per gram organic body weight basis. Regnault (1971) indicated, for postlarval stages of *Crangon septemspinosa,* that the quantity of available food may determine both the molt frequency and final size attained, but optimal growth rate is also dependent upon quality of the diet available to young stages. Recent studies suggest the importance of diverse diets in penaeid postlarvae (Caillouet et al. 1976) and more specifically the benefits of dietary sterols (Teshima & Kanazawa 1971, Teshima 1972, Castell et al. 1975), linoleic acids and phosphatidylcholine (Castell & Covey 1976, Kanazawa et al. 1977, 1979, Conklin et al. 1980, D'Abramo et al. 1981), mineral content and ratios (Gallagher et al. 1978, Boghen & Castell 1981), and protein content or composition (Millikin et al. 1980, Boghen & Castell 1981), for optimal growth of young postlarval stages in a number of decapods. Even dissolved organic matter, such as D-glucose, has been demonstrated to improve nutritional state and growth rates in post-larval penaeid shrimp, albeit most probably by way of associated bacteria that may be used as a source of nutrition (Castille & Lawrence 1979).

Because larvae molt frequently and interrupt feeding to molt, nutritional reserves are more rapidly drained in these stages than in later ones (Bages & Sloane 1981). Specific food requirements appear to be essential during feeding periods, and these needs help determine larval migration routes and habitats. D'Abramo et al. (1981) have suggested, for example, that even small amounts of phosphatidylcholine in the diet may be especially important just prior to molting in juvenile homarid lobsters, before the animals enter pre-molt dependence upon reserves of the hepatopancreas. Even when physical parameters such as temperature, salinity, and substrate grain size vary along with quality and quantity of food, nutrition often appears to be the dominant factor in early postlarval growth (Chittleborough 1975, Klein Breteler 1975a, Caillouet et al. 1976, Labat 1977). Where habitat selection appears to be based upon substrate qualities, food organisms associated with those substrates may, in fact, be the basis for site selection by postlarvae (see Ruello 1973, Caillouet et al. 1976, Regnault 1976, Booth 1979, Aziz & Greenwood 1982). If migrations, as in the case of penaeid shrimp, route early postlarval stages over substrates that are not rich with nutritional organics, growth may be suspended to some degree until nursery areas are reached (Baxter & Renfro 1967). In some cases, metamorphosis itself appears to be stimulated by the presence of appropriate substrates or hosts (Hiro 1937, Serfling & Ford 1975, Castro 1978).

Substrate and cover cannot be completely separated from nutrients as factors in habitat selection. However, cover in nursery areas may diminish predation on penaeid postlarvae (Hackney & Burbanck 1976), and cover per se in part determines habitats selected (Giles

& Zamora 1973, Dall 1981). Both small size and frequency of molt renders postlarvae vulnerable to predation. *Homarus* decapodids construct burrows under obstacles at least two days before molting to second postlarvae, presumably as an antipredatory mechanism (Berrill 1974). Burrows of *Homarus* juveniles appear to serve the same purpose, and may be more complex than adult burrows (Howard & Bennett 1979). Cover is also important for young brachyurans, since their early postlarvae are preyed upon by small fish (Knudsen 1960, Felder & Chaney 1979).

Migration and habitat selection is often either facilitated or constrained by osmoregulatory capacities of specific age groups (Kalber & Costlow 1966, 1968, Kalber 1970, Foskett 1977, Young 1977a,b, Dall 1981). Capacities for postlarval osmoregulation tend to reflect salinity adaptation over the course of range distribution. The largest body of literature on this subject concerns penaeid shrimp. Postlarval penaeids generally migrate into estuaries following larval development in high salinity waters offshore. After a period of growth in inshore waters, mature adults return seaward and release eggs in high salinity water (Perez Farfante 1969, Allen 1972, Young & Carpenter 1977, Young 1978, Dall 1981). Temperature, currents, and local geography can influence salinity preference and periods of peak movement into estuaries (St Amant et al. 1963, Aldrich et al. 1968, Pullen & Trent 1969, Keiser & Aldrich 1976, Staples 1979). Tidal currents and behavioral (swimming) response to salinity appears to facilitate both inshore and offshore movements of *Penaeus duorarum* in Florida (Hughes 1969a,b), but the time of maximum movement may also depend on the coincidence of bright moonlight with nocturnal ebb tides (Beardsley 1970). Chemical factors, other than salinity, in inshore waters can influence movements of penaeid postlarvae (Mair 1980), and endogenous rhythms of feeding and activity may complicate their response to tidal cycles (Hughs 1968, 1969c, Allen 1972). In young penaeids, metabolic rates may not be altered significantly over a broad range of salinity (Gaudy & Sloane 1981) except at extremes (Kutty et al. 1971). The influence of salinity on postlarval growth is not dramatic (Zein-Eldin 1963, Zein-Eldin & Aldrich 1965) and appears to be much less than that of temperature (Zein-Eldin & Griffith 1966). However, movement of penaeid larvae to inshore waters low in salinity reflects a persistent pattern for most coastal species, and postlarvae of many tend to hyperosmoregulate to a greater extent than adults (Dall 1981, Castille & Lawrence 1981a,b).

Ontogenetic changes in osmoregulatory ability are known or suggested by migratory patterns in various representatives of the Caridea. Postlarvae of *Atya bisulcata* move through shallow areas between beach boulders and into the mouths of coastal streams in Hawaii, after metamorphosis from the marine larval phase. The movement coincides with loss of ability to survive in seawater (Ford & Kinzie 1982, Robert A.Kinzie, personal communication). *Macrobrachium rosenbergii,* after larval development and metamorphosis in brackish water, migrates up rivers in the postlarval stage toward adult freshwater habitats (John 1957, Raman 1967, George 1969, Ling 1969). High temperature inhibits this migration, possibly by impeding hyperegulation (Stephensen & Knight 1982). Ontogenetically, the postlarvae show a progressive development of hyperegulatory ability, as evidenced by a simultaneous decrease in the slope of regulation curves and in the isosmotic crossover concentration (Sandifer et al. 1975, Harrison et al. 1981). The hyporegulatory ability of early postlarvae is lost over the course of development to adulthood in *M.rosenbergii,* although it is present in adults of *M.ohione* to some degree (Castille & Lawrence 1981). Migration and osmoregulatory ability appear to be more complex in the genus *Crangon* (Haefner 1969, Labat 1977, Jansen & Kuipers 1980).

Among the Astacidea, adults of both the cambarid *Procambarus clarkii* and the parastacid *Cherax destructor* are more tolerant of salinity than earlier postlarvae (Loyacano 1967, Mills & Geddes 1980). For the thalassinidean *Callianassa islagrande,* preliminary evidence suggests that hyperosmotic and hyperionic regulation may occur in early juveniles while no such ability is evident in adults (Felder 1978). In the hermit crab *Pagurus bernhardus* juveniles regulate body water volume at lower salinities than do adults (Davenport 1972).

Among the brachyuran crabs, some species are known to have postlarvae that differ from adults in osmoregulatory capacities. However, a phenomenon more commonly reported is the difference in osmoregulatory capacities of larvae and adults, as related to their differential distributions (see Kalber & Costlow 1968, Foskett 1977). A number of species migrate, raft, or passively disperse as early postlarvae to nearshore areas or up estuaries (Knudsen 1960, Tagatz 1968, Dudley & Judy 1973, Williams 1971, Klein Breteler 1976, Saigusa 1978), and some stages experience marked salinity changes. Changes in the free amino acid pool may in part facilitate tolerance of salinity changes and estuary penetration (Costlow & Sastry 1966). In *Rhithropanopeus harrisii,* which is highly adapted to estuarine conditions, there is relative absence of salinity effects on molt duration in the postlarval stages (Costlow et al. 1966, Hartnoll 1978). In the semi-terrestrial ocypodid *Uca pugilator,* adults hyper- and hypoosmoregulate, but hypoosmoregulation does not become developed until the first crab (second postlarval) stage (Dietrich 1979). A similar pattern appears in *Uca subcylindrica,* which inhabits semi-arid areas and undergoes a strongly abbreviated larval development. Hypoosmotic regulation is evident at least by the megalopa (decapodid) stage and becomes stronger in the first crab stage (Rabalais 1983).

4 PHYLOGENETIC SIGNIFICANCE OF POSTLARVAL DEVELOPMENT

There is a view that a phylogeny derived from larval characters should coincide with adult phylogeny. This was basically the concept furthered by Gurney (1942), although he did acknowledge (p.30) that 'new characters special to the larva itself may arise and obscure to some extent the primitive form.' In decapods, the adults may be modified for their particular habitat requirements, while the larvae are almost all planktonic and therefore not subject to the same selection pressures as the adults (see Rice 1980, 1981a,b, Williamson 1976, 1982a,b). A second view on larval characters in phylogeny is that larvae and adults are modified independently, and larval groupings need not closely correspond to groupings of adults (Gordon 1955).

We suggest that an understanding of phylogenetic utility of developmental stages lies somewhere between these two viewpoints. There are undoubted examples of convergence among decapod larvae (Williamson 1982b), but there are many cases where larval morphology appears to accurately reflect phylogeny. Early postlarval phases in decapod ontogeny, the decapodid (= first postlarva, megalopa, glaucothöe, puerulus, etc.) and the early juveniles (= early postdecapodids), may sometimes provide more useful indications of phylogenetic origins than could be interpreted from larval or adult stages. Thus, an expanded theory on the phylogenetic significance of developmental stages would necessarily accommodate cases in which recently derived specializations, obscuring of ancestral

Figure 11. An example of convergence among brachyuran larvae. (A-D) family Majidae, subfamily Inachinae; (E) family Homolidae. (A) *Achaeus tuberculatus*; (B) *Macrocheira kaempferi*; (C) *Pleistacantha sanctijohannis* (all after Kurata 1969). (D) *Dorhynchus thomsoni* (after Williamson 1982b); (E) *Homola* cf. *barbata* (after Rice & von Levetzow 1967). Note the closer similarity between *Dorhynchus* and *Homola* than between *Dorhynchus* and other inachine majids.

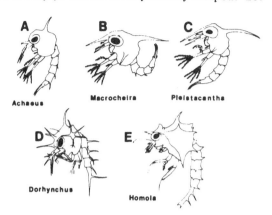

features, and potential for convergence may have been relatively 1) unidirectional, i.e. most pronounced either in the premetamorphic stages (larvae) or the late postlarval stages, 2) bidirectional, i.e. pronounced in both larval and adult stages, 3) non-directional, i.e. characteristic of all stages, or 4) punctuated, i.e. characteristic of isolated developmental stages or sets of stages.

The appeal of decapod larvae in systematics has been partially centered on the hypothesis that convergent evolution among adults may not be manifest extensively in the planktonic zoeae. However, the potential for convergence is probably common among planktonic larvae, as recently derived specializations appear to have extensively modified larval morphology. In the Decapoda many of the extremely spinous larvae such as the 'acanthosoma' stages of sergestoids, the greatly flattened phyllosomas of the Palinuroidea, and the inflated carapaces of the Eryonoidea, are all adaptations for a lengthy pelagic existence. Williamson (1982a) suggested that the flattened bodies of phyllosoma stages may be adaptations enabling them to cling onto scyphozoan hosts rather than as aids for flotation, although it should be noted that these larvae do not attach to the hosts immediately, nor do they remain on the host at all times.

Perhaps the most striking reported example of convergence among decapod larvae is the recent description of the larvae of *Doryhnchus thomsoni* by Williamson (1982b). The unusual larvae of this species (Fig.11) were first collected in 1959, and described by Williamson (1960) as a 'remarkable zoea' probably of the Majidae. The telson and appendages are similar to known majoid zoeas, but the carapace and abdomen bear a large number of spines, unlike any known larvae of the Majoidea. It was suggested, as alternatives, that these unusual larvae might belong to the Tymoloidea (Williamson 1965, Wear & Bathman 1975) or to an undiscovered family outside the Majoidea (Williamson 1976). The multi-spined zoeae are so similar to late zoeae of the Homoloidea that Williamson (1982b) coined the term 'homolomaja' to apply to zoeae which combine characters of the two groups. It is now known that these larvae belong to *D.thomsoni*, a member of the majoid subfamily Inachinae, and a species in which adults 'show no unconventional features' (Williamson 1982b:727). The zoeae and megalopa differ so dramatically from those of other inachine species that assignment even to superfamily level is not immediately obvious. Williamson concludes that the striking differences between the zoea of *D.thomsoni* and zoeae of other Inachinae are the result of recent evolution, but why these recent changes should have produced features resembling those of late homoloid zoeae remains a mystery.

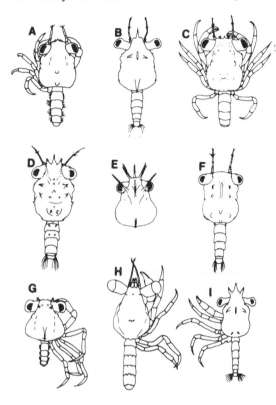

Figure 12. Diversity among megalopae in brachyuran family Majidae. (A) *Dehaanius limbatus* (after Kakati & Sankolli 1975); (B) *Macrocheira kaempferi* (after Kurata 1969); (C) *Libinia dubia* (after Sandifer & Van Engel 1971); (D) *Schizophrys aspera* (after Kurata 1969); (E) *Dorhynchus thomsoni* (after Williamson 1982b); (F) *Pisoides ortmanni* (after Kurata 1969); (G) *Anasimus latus* (after Sandifer & Van Engel 1972); (H) *Inachus dorsettensis* (after Ingle 1977); (I) *Chionoecetes opilio* (after Motoh 1973). Figures slightly modified and not to scale.

Consider now *Dorhynchus thomsoni*. The adults are clearly majids, subfamily Inachinae, but it is equally clear that the zoeae do not conform to the typical majoid zoeal form. Neither does the megalopa stage conform, for it is extremely spinose and unlike other known inachine or even majoid megalopae (Fig.12). The final clue which established the relationship between zoea and adult was provided by neither zoea, megalopa, nor adult, but by the first crab stage which developed from a captive megalopa. Only by the subsequent examination of a series of juvenile crabs was the adult relationship of the first crab stage confirmed.

After consideration of the above and related evidence, we conclude that postlarval stages, either decapodid or early juvenile, deserve closer examination than has traditionally been paid them in studies of the Decapoda. These stages are to various degrees distinct from zoeae and adults in their morphology, physiology, ecology, and behavior. Phylogenetic information may be gleaned from these stages, as well as from larvae and adults.

A well-demarcated decapodid stage is not characteristic of all decapods. In the Penaeidea, Caridea and Stenopodidea there may be no abrupt transformation between zoeal (mysis) and decapodid stages, and the latter may acquire the adult body form by slight and gradual changes (see sections 2.2.1, 2.2.2, 2.3.1 and 2.3.2). Members of the brachyuran family Hymenosomatidae likewise lack a morphologically unique decapodid, as the zoeae metamorphose directly into a decapodid much like a young crab. In a large number of freshwater decapods and highly adapted semi-terrestrial forms with abbreviated development,

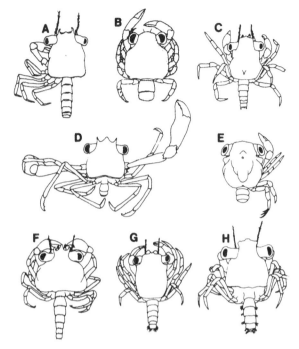

Figure 13. Diversity among megalopae in the brachyuran family Xanthidae. (A) *Hexapanopeus angustifrons* (after Costlow & Bookhout 1966); (B) *Heterozius rotundifrons* (after Wear 1968); (C) *Cycloxanthops truncatus* (after Hong 1977); (D) *Quadrella nitida* (after Garth 1961); (E) *Epixanthus dentatus* (after Saba et al. 1978); (F) *Panopeus herbstii* (after Costlow & Bookhout 1961); (G) *Platyxanthus crenulatus* (after Menu-Marque 1970); (H) *Lophopanopeus leucomanus leucomanus* (after Knudsen 1958). Figures slightly modified and not to scale.

stages are confined to the late developmental period of embryonated eggs, and the young emerge with a body form similar to the adult. However, the young are sometimes erroneously referred to as larvae.

In the remainder of the Decapoda, the decapodid stage is fairly distinct from zoeae and juveniles, although there are exceptions in some additional cases of abbreviated development (see Dobkin 1969, Rabalais & Gore, this volume) and in certain cases among the Caridea where the decapodid differs little from early juveniles. Within some Caridea the decapodid stage differs from most subsequent stages by retaining vestiges of the zoeal epipods on all pereopods, but the carapace appears intermediate between zoeal and juvenile stages. In the remaining decapod groups the decapodid stage (where known) is more clearly distinct from both zoeal and juvenile forms (Figs.1-6). However, among these groups the form of the decapodid varies greatly. Thus, we somewhat disagree with Williamson's (1982a: 84) statement that megalopae 'can usually be identified to an infraorder or superfamily on the basis of adult characters'. While this may hold true at the infraordinal level, it is often impossible to identify (for example) the superfamily of a given brachyuran megalopa, and especially not by strict application of adult characters. The variation of megalopal types in the superfamilies Xanthoidea and Majoidea (Figs.12, 13) illustrates that point.

Although several keys have been published for identification of decapodid stages (e.g. Cook 1965, Dutt & Ravindranath 1975, Mohamed et al. 1968, Williamson 1957), the phylogenetic significance of this stage has only recently been reviewed. Rice (1981b) lamented the incongruence of zoeal and adult classifications and noted (p.1003) that 'since both the adult and the zoeal phases of the Brachyura thus seem to have been affected by convergent evolution, any classificatory evidence from other sources is likely to be of value'; he therefore examined characters of the megalopa stage in representatives of the Podotremata of Guinot (1978). Unfortunately, examination of this stage added little to our knowledge of the evolution of the group. Rice (p.1010) concluded 'this review of podotrematous megalopas has therefore not assisted significantly in assessing either the interrelationships of the constituent groups or their overall relationship with the higher Brachyura'.

Rice's study represents, to our knowledge, the only attempt to employ decapodid characters at levels higher than that of family, although Williams (1980) used megalopal characters in establishing the superfamily Bythograeoidea. Many descriptive works include mention of similarities in this stage at the family level and below. For example, characters of the megalopa of *Pisa* have been used as the basis for hypothesized intergeneric relationships (Ingle & Clark 1980). In another approach, the absence of a unique megalopal morphology in decapodids of the Hymenosomatidae has been pointed out as a diagnostic character of that family (Muraoka 1977). References to identification of crustacean larvae listed by Williamson (1982a:90) provided a good starting point for review of other such papers.

Rice (1981b) noted that the megalopa stage possesses features found in both zoeal and juvenile phases, as well as some unique characters. Phylogenetic analyses of these characters have, in limited attempts to date, proven to be no less difficult than those based upon zoeae. Indeed, as the decapodid stage represents a transition from planktonic to benthic life in almost all decapods, it is likely that convergence has in at least some cases rendered this phase to be of limited use to systematists. However, it is also likely that certain megalopal characters, such as the paired sternal spines in many portunids and the recurved hook on the ischium of the cheliped in many xanthids, will prove to be of some phylogenetic significance. We have not attempted to examine all known decapodid stages in the same manner that Rice (1980) reviewed brachyuran zoeae. Rather, our intent is to encourage the study of this stage by pointing out the potential for useful characters for phylogenetic interpretation. Subsequent studies to Rice's (1981b) review of the pcdotremen megalopae and similar studies of decapodids in other groups should soon provide a better understanding of phylogenetic significance within this stage.

We contend that studies of the decapodid stages do not represent the only, nor necessarily the best, alternative technique to establish relationships on the basis of larval or adult morphologies. The earliest of postdecapodid stages, the early juveniles, may prove to be of equal (or in some cases, greater) phylogenetic significance. Zoeae and adults are variously modified according to their habitat. The postdecapodid juvenile is the first stage where all or almost all larval characters are lost, and the general adult body plan is apparent. However, the morphological attributes of the adult are not yet extensively superimposed on this basic plan.

Several pieces of evidence give credence to a hypothesis of plesiomorphy in early postlarval stages. In general these come from studies that demonstrate either 1) the restriction of presumed 'ancestral' characters to postlarval and sometimes earlier stages, or 2) the absence of presumed 'derived' specializations in early postlarval stages as compared to larvae or adults. Evidence to demonstrate either of these phenomena is limited in scope and has not been specifically used to build support for a theory of plesiomorphy in early postlarval stages.

The retention of ancesteral characters in early postlarval stages is, however, supported by our interpretations of papers concerning such subjects as the ontogeny of the decapod median eye (Elofsson 1963, Paterson 1970). The median eye is located dorsally or anterodorsally to the protocerebrum between a small pair of apical nerves. A mass of sensory cells here usually contains a band or scattered patches of dark pigment, in some respects comparable to the naupliar eye of those crustaceans that exhibit this structure as larvae and/or adults. Paterson (1970) noted that there is a tendency for ontogenetic degeneration of the median eye, especially during postlarval development. Furthermore, the degree of degeneration varies among higher decapod taxa. The median eye of adult palinurids is less degenerate than that of 'other Reptantia so far investigated', but the median eye of adult palinurids is more degenerate than in adult 'Natantia'. However, the median eye in the puerulus (decapodid) stage of palinurids is comparable to that of adult 'Natantia' (Paterson 1970). Thus, there is a strong suggestion that a potentially ancestral feature is, at least by degree, lost over the course of postlarval ontogeny and that the rate and degree of loss may reflect lineages at higher systematic levels.

At lower systematic levels, a number of examples indicate the ontogenetic diminishment of features that characterize probable ancestral stocks. For instance, Williams (1980) has described the gradual diminishment of eyes that takes place during postlarval ontogeny of *Bythograea thermydron.* The gradual degeneration of eyes and orbital features after the megalopa stage, in which eyes are of a typical size for the Brachyura, strongly suggests that this unusual deep-water species most resembles an ancestral form in its early postlarval stages. We have noted similar degeneration of eyes during postlarval development of fossorial forms, especially among the Thalassinidea. Even in behavior patterns, what appear to be ancestral characteristics are sometimes retained in restricted stages of early postlarval development. For example, the tendency for inhabiting gastropod shells in early postlarval stages of the semi-terrestrial coconut crab *Birgus latro* is postulated to be the retention of ancestral behavior from the hermit crab stock that gave rise to this species (Reese 1968).

Other examples of postlarval transitions from ancestral to derived character states may very well be represented among the numerous cases of postlarval character change noted in the earlier portions of this chapter. However, rather than documenting the degeneration or loss of ancestral characters during postlarval ontogeny, the majority of cases document the addition of specialized characters during this period of postlarval development. Gurney (1942:22-27) made a similar observation in his attempt to determine the degree to which ancestral characters 'lost in adults' may be preserved in larval stages. He expressed surprise at the few instances that could be cited and noted that it was much more common to find adult characters 'appearing directly' during postlarval development. It is precisely for this reason — that is, appearance of adult characters late in development — that limited systematic characters are available for use in the identification of early postlarval stages. For example, *Pasiphaea multidentata* and *P. tarda* do not become morphologically distinct until they reach a carapace length of 12 mm, 3 mm less than the size at maturity (Matthews & Pinnoi 1973). In *Paguristes,* even diagnostic characters of the genus may not be present in early postlarval stages (Provenzano & Rice 1966). The same is true of generic and higher level characters in many members of the Cambaridae (Hobbs 1962, 1981).

Similarly, characters traditionally used for justification of phylogenetic placement have been shown to change or develop progressively during postlarval ontogeny of the Brachyura (Morita 1974, Shen 1935, Yang & McLaughlin 1979). In at least one instance, a study of morphogenesis from juveniles to adults corrected a long standing misconception based upon adult features that were erroneously thought to be of major phylogenetic significance. Certain carapacial grooves of adult astacids, that were formerly thought to be remnants of somite boundaries, were found instead to be mechanically induced secondary features

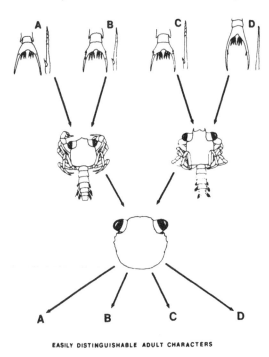

Figure 14. Diagram of ontogenetic transition from zoeal stages to adult stages in (A) *Eurypanopeus depressus*, (B) *Panopeus turgidus*, (C) *Panopeus herbstii* forma *simpsoni*, and (D) *Panopeus bermudensis*. Zoeal stages (above) characterized by armature of telson and antenna; development thereafter transcends one of two megalopal forms and common first crab form before adult characters become evident (after Martin et al. 1984).

EASILY DISTINGUISHABLE ADULT CHARACTERS

which resulted from the attachment of muscles to the carapace. Thus, while these grooves can be used for diagnosis of certain systematic groups (presumably as artificial characters), they are not useful in studies of fundamental tagmatization as it applies to higher phylogeny (Albrecht 1981). The implication is that because the juveniles do not show this derived (albeit mechanically induced) character state to the extent of the adults, early post-larvae are less altered from architypal morphology than are later stages.

Also in favor of this opinion is a recent paper by Martin et al. (1984) in which four species of xanthoid crabs were reared from larvae to adult. Using selected characters of the first, third and fifth crab stages, Martin et al. found that phenotypic divergence increases with ontogeny. In addition, whereas all four species were readily identifiable at zoeal and adult stages, distinction was extremely difficult at the first crab stage (Fig.14). Thus, characters that applied to all four species in the first crab stage were considered plesiomorphic, and the later modifications of older crab stages were treated as derived conditions. If early juvenile stages are, in fact, representative of the more plesiomorphic character stages, then similarities detected in first juvenile morphology could unite natural groupings of the Decapoda and be relatively free of recently derived modifications. Present knowledge of this stage is far from complete. However, Martin et al. (1984) noticed similarities in the first crab stage of known ocypodids, in all known portunids, in known grapsids, and in most majids. Unfortunately, many of the characters employed in that study (e.g. mouthpart morphology) seem to be of limited taxonomic value at the first crab stage, and classificatory schemes based on this stage will have to await detailed descriptions of first juveniles in a larger number of decapod families. In particular, carapacial morphology appears to be a sound character for relating brachyurans to the superfamily level at the first crab stage.

The opposing view is that while larval characters are lost in the first juvenile stage, it is not true that recently derived specializations (and considerable potential for convergence) are absent. In the Pinnotheridae, for example, the widened carapace so typical in the adult is already distinguishable in the decapodid or first crab stage (Muraoka 1979, Kurata 1970). However, the highly specialized adaptations of some stages in commensal species might be expected to obscure plesiomorphy by *punctuated* or *non-directional* specialization which may give rise to such morphological forms as 'invasive stages' early in postlarval ontogeny.

It may also be argued that attainment of a crablike body by convergence in the Brachyura, some Paguroidea, and many Galatheoidea does not gradually develop as ontogeny proceeds but is readily apparent in an early juvenile stage. However, it can at least be noted that the first postlarva in each of these examples is less cancroid in form than are the later stages. A striking example is represented by the ontogeny of *Asthenognathus atlantica*, a goneplacid crab superficially similar to the pinnotherids as an adult. In this species the ratio of carapace width to carapace length changes gradually, from 1.14 in the first crab stage to 1.38 in the more pinnotherid-like (by convergence) fourth crab stage (Bocquet 1965).

Changes in carapace morphology from the first juvenile to later juvenile and adult stages are far more dramatic in other Brachyura. In some species these changes are gradual and subtle over the entire course of postlarval development, while in others the change is more dramatic as adult characters are more abruptly superimposed on an initial brachyurized form (Fig.15). If the morphology of first juvenile stage decapods is going to be useful in systematics, it will probably be in those cases where changes are more dramatic, i.e. where conspicuous modifications greatly mask a more basic form of the first juvenile.

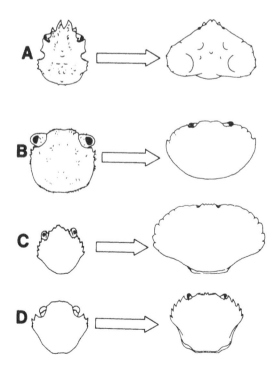

Figure 15. Examples of dramatic change in carapace morphology from juvenile (left) to adult (right) brachyuran crabs. (A) *Eurynolambrus australis* (Majidae) (after Krefft 1952); (B) *Eurytium limosum* (Xanthidae) (after Kurata et al. 1981, Heard 1982); (C) *Cancer pagurus* (Cancridae) (after Lebour 1928, Ingle 1980); *Liocarcinus puber* (Portunidae) (after Lebour 1928, Ingle 1980). Figures slightly modified and not to scale.

Several descriptive studies (e.g. Perez Farfante 1970, Mair 1979, 1981) have dealt with early juvenile stages of economically important species, but morphological comparisons of first stage juveniles have rarely been utilized in systematics. *Dorhynchus thomsoni* provided a good example of one application. For over 20 years both larvae and adults of *D.thomsoni* were known, but it was not known that they constituted life stages of one species until Williamson (1982b) observed a single first crab which died metamorphosing from the megalopa. The link between larvae and adult was realized, and further evidence came from a series of pelagic juveniles in which the characters were intermediate between first crab and adult. Krefft (1952) similarly clarified the systematic position of the majoid *Euryno-lambrus australis* only after observing the characters of the first crab stage. This species had originally been placed in the Pinnotheridae, then transferred to the 'cancerinae' by Miers (1879). Krefft (1952:574) noted that 'these postlarval stages are of the greatest value in the determination of the systematic position of this species'. The 'well-developed, divergent rostra; the incomplete orbit consisting of a supraocular eave' and other characters of the postlarvae positively placed *E.australis* in the Majidae; these and other characters are not readily apparent in the adult (Fig.15). Krefft stated (p.574) that 'the oxyrhynchous characteristics are minimised by this adoption of this cancroid form'.

The probability that decapodid and first juvenile stages are as subject to evolutionary adaptation as zoeae and adults cannot be denied. Indeed, cases may be discovered in which specialization over all life stages has been *non-directional* and pronounced. The extent to which convergence in early postlarval stages has altered basic morphology cannot yet be evaluated for the Decapoda in general. Yet we have shown that, in at least a few cases, examination of these stages has clarified systematic questions unanswerable from studies of larvae and adults. We feel justified, therefore, in pointing out the possible phylogenetic significance of these stages and hope that greater interest in their systematic utility will be forthcoming. Future study of these stages may be undertaken to determine characters for interspecific distinctions, to complete knowledge of life histories, to document ontogenetic changes in discrete organs and functions, to determine rates of growth and development, and to further define ecological roles of postlarvae. Regardless of the impetus for assembly of data, an eventual synthesis of this information could very well lead to a much improved understanding of phylogenetic relationships of the decapod Crustacea.

5 ACKNOWLEDGEMENTS

We sincerely thank R.H.Gore, S.C.Hand, D.L.Lovett, R.B.Manning, N.N.Rabalais, A.J.Provenzano Jr and F.M.Truesdale for their useful comments and criticisms on the manuscript. We also thank The American Association for the Advancement of Science, The Marine Biological Laboratory, Woods Hole, Cambridge University Press, and the Royal Society of London for permission to reproduce illustrations from their respective journals.

REFERENCES

Aikawa, H. 1929. On larval forms of some Brachyura. *Rec. Oceanogr. Wks. Jap.* 2:17-55.
Aikawa, H. 1933. On larval forms of some Brachyura. Paper II: A note on indeterminable zoeas. *Rec. Oceanogr. Wks. Jap.* 5:124-254.

Aikawa, H. 1937. Further notes on brachyuran larvae. *Rec. Oceanogr. Wks. Jap.* 9:87-162.

Aizawa, Y. 1974. Ecological studies of micronektonic shrimps (Crustacea, Decapoda) in the western North Pacific. *Bull. Oceanogr. Res. Inst. Univ. Tokyo* 6:1-84.

Albrecht, H. 1981. Zur Deutung der Carapaxfurchen der Astacidea (Crustaces, Decapoda). *Zool. Scr.* 10:265-271.

Aldrich, D.V., C.E.Wood & K.N.Baxter 1968. An ecological interpretation of low temperature responses in *Penaeus aztecus* and *P.setiferus* postlarvae. *Bull. Mar. Sci.* 18:61-71.

Allen, J.A. 1972. Recent studies on the rhythms of post-larval decapod Crustacea. *Oceanogr. Mar. Biol. Ann. Rev.* 10:415-436.

Anderson, W.W. & M.J.Lindner 1971. Contribution to the biology of the royal red shrimp, *Hymenopenaeus robustus* Smith. *Fish. Bull. US Fish Wildl. Serv.* 69:313-336.

Andrews, E.A. 1907. The young of the crayfishes *Astacus* and *Cambarus*. *Smithson. Contr. Knowl.* 35: 1-79.

Atkinson, J.M. & N.C.Boustead 1982. The complete larval development of the scyllarid lobster *Ibacus alticrenatus* Bate 1888 in New Zealand waters. *Crustaceana* 42:275-287.

Atwood, H.L. & I.Kwan 1976. Synaptic development in the crayfish opener muscle. *J. Neurobiol.* 7: 289-312.

Aziz, K.A. & J.G.Greenwood 1982. Response of juvenile *Metapenaeus bennettae* Racek & Dall 1975 (Decapoda, Penaeidae) to sediments of differing particle size. *Crustaceana* 43:121-126.

Baba, K. & Y.Fukuda 1972. Larval development of *Chasmagnathus convexus* de Haan (Crustacea, Brachyura) reared under laboratory conditions. *Mem. Fac. Educ., Kumamoto Univ.* 21:90-96.

Baba, K. & K.Miyata 1971. Larval development of *Sesarma (Holometopus) dehaani* H.Milne Edwards (Crustacea, Brachyura) reared in the laboratory. *Mem. Fac. Educ., Kumamoto Univ.* 19:54-64.

Baba, K. & M.Moriyama 1972. Larval development of *Helice tridens wuana* Rathbun and *H. tridens tridens* de Haan (Crustacea, Brachyura) reared in the laboratory. *Mem. Fac. Educ., Kumamoto Univ.* 20:49-68.

Bages, M. & L.Sloane 1981. Effects of dietary protein and starch levels on growth and survival of *Penaeus monodon* (Fabricius) postlarvae. *Aquaculture* 25:117-128.

Baxter, K.N. & W.C.Renfro 1967. Seasonal occurrence and size distribution of postlarval brown and white shrimp near Galveston, Texas, with notes on species identification. *Fish. Bull. US Fish Wildl. Serv.* 66:149-158.

Beardsley, G.L. 1970. Distribution of migrating juvenile pink shrimp, *Penaeus duorarum duorarum* Burkenroad, in Buttonwood Canal, Everglades National Park, Florida. *Trans. Am. Fish. Soc.* 99:401-408.

Bellon-Humbert, C., M.J.P.Thyssen & F.van Herp 1978. Development, location and relocation of sensory and neurosecretory sites in the eyestalks during the larval and postlarval life of *Palaemon serratus* (Pennant). *J. Mar. Biol. Ass. UK* 58:851-868.

Benzie, J.A.H. 1982. The complete larval development of *Caridina muccullochi* Roux, 1926 (Decapoda, Atyidae) reared in the laboratory. *J. Crust. Biol.* 2:493-513.

Benzie, J.A.H. & P.K.de Silva 1983. The abbreviated larval development of *Caridina singhalensis* Ortmann 1894 (Decapoda, Atyidae). *J. Crust. Biol.* 3:117-126.

Bernard, F. 1953. Decapoda Eryonidae (*Eryoneicus* et *Willemoesia*). *Dada Rept.* 37:1-93.

Berrill, M. 1974. The burrowing behavior of newly settled *Homarus vulgaris*. *J. Mar. Biol. Ass. UK* 54: 797-801.

Berrill, M. 1975. Gregarious behavior of juveniles of the spiny lobster, *Panulirus argus* (Crustacea: Decapoda). *Bull. Mar. Sci.* 25:515-522.

Bocquet, C. 1965. Stades larvaires et juveniles de *Tritodynamia atlantica* (Th.Monod) (= *Aesthenognathus atlanticus* Th.Monod) et position systematique de ce crabe. *Cah. Biol. mar.* 6:407-418.

Boghen, A.D. & J.D.Castell 1981. Nutritional value of different dietary proteins to juvenile lobsters, *Homarus americanus*. *Aquaculture* 22:343-351.

Booth, J.D. 1979. North Cape – a 'nursery area' for the packhorse rock lobster, *Jasus verrauxi* (Decapoda: Palinuridae). *NZ J. Mar. Freshwat. Res.* 13:521-528.

Bosc, L.A.G. 1802. *Histoire naturelle des Crustaces.* (2 vols.). Paris.

Boschi, E.E., M.A.Scelzo & B.Goldstein 1969. Desarrollo larval del cangrejo *Halicarcinus planatus* (Fabricius) (Crustacea, Decapoda, Hymenosomidae), en el laboratorio, con observaciones sobre la distribucion de la especie. *Bull. Mar. Sci.* 19:225-242.

Bourdillon-Casanova, L. 1960. Le meroplancton du Golfe de Marsielle: les larves de crustacés décapodes. *Recl. Trav. Stat. mar. Endoume.* 30(80):1-286.

Broad, A.C. 1957. Larval development of the crustacean *Thor floridanus* Kingsley. *J. Elisha Mitchell Scient. Soc.* 73:317-328.

Bruce, A.J. 1972. Notes on some Indo-Pacific Pontoniinae XVIII. A redescription of *Pontonia minuta* Baker, 1907, and the occurrence of abbreviated development in the Pontoniinae (Decapoda Natantia, Palaemonidae). *Crustaceana* 23:65-75.

Bruce, A.J. 1975. On the occurrence of *Discias atlanticus* Gurney, 1939 in the western Indian Ocean (Decapoda, Caridea). *Crustaceana* 29:301-305.

Burkenroad, M.D. 1945. A new sergestid shrimp with remarks on its relationships. *Trans. Conn. Acad. Arts Sci.* 36:553-593.

Burkenroad, M.D. 1981. The higher taxonomy and evolution of Decapoda (Crustacea). *Trans. San Diego Soc. Nat. Hist.* 19:251-268.

Caillouet, C.W., J.P.Norris, E.J.Heald & D.B.Tabb 1976. Growth and yield of pink shrimp (*Penaeus duorarum duorarum*) in concrete tanks. *Trans. Am. Fish. Soc.* 105:259-266.

Cano, G. 1892. Sviluppo postembrionale dello *Stenopus spinosus. Boll. Soc. Nat. Napoli* 5:134-137.

Carlsberg, J.M., J.Van Olst & R.F.Ford 1978. A comparison of larval and juvenile stages of the lobster, *Homarus americanus, Homarus gammarus* and their hybrid. *Proc. World Maric. Soc.* 9:109-122.

Castell, J.D. & J.F.Covey 1976. Dietary lipid requirements of adult lobsters, *Homarus americanus* (M.E.). *J. Nutr.* 106:1159-1165.

Castell, J.D., E.G.Mason & J.G.Covey 1975. Cholesterol requirements of juvenile American lobsters, *(Homarus americanus). J. Fish. Res. Bd. Can.* 32:1431-1435.

Castille, F.L.Jr & A.L.Lawrence 1979. The role of bacteria in the uptake of hexoses from seawater by postlarval penaeid shrimp. *Comp. Biochem. Physiol.* 64A:41-48.

Castille, F.L.Jr & A.L.Lawrence 1981a. The effect of salinity on the osmotic, sodium and chloride concentrations in the hemolymph of euryhaline shrimp of the genus *Penaeus. Comp. Biochem. Physiol.* 68A:75-80.

Castille, F.L.Jr & A.L.Lawrence 1981b. A comparison of the capabilities of juvenile and adult *Penaeus setiferus* and *Penaeus stylirostris* to regulate the osmotic, sodium, and chloride concentration in the hemolymph. *Comp. Biochem. Physiol.* 68A:677-680.

Castille, F.L.Jr & A.L.Lawrence 1981c. The effect of salinity on the osmotic, sodium, and chloride concentrations in the hemolymph of the freshwater shrimps, *Macrobrachium ohione* Smith and *Macrobrachium rosenbergii* de Man. *Comp. Biochem. Physiol.* 70A:47-52.

Castro, P. 1978. Settlement and habitat selection in the larvae of *Echinoecus pentagonus* (A.Milne Edwards), a brachyuran crab symbiotic with sea urchins. *J. Exp. Mar. Biol. Ecol.* 34:259-270

Chamberlain, N.A. 1961. Studies on the larval development of *Neopanope texana sayi* (Smith) and other crabs of the family Xanthidae (Brachyura). *Tech. Rep. Chesapeake Bay Inst.* 22:1-37.

Charniaux-Cotton, H. 1965. Hormonal control of sex differentiation in invertebrates. In: R.L.de Haan & H.Ursrung (eds.), *Organogenesis.* New York: Holt, Rinehart & Winston.

Chhapgar, B.F. 1956. On the breeding habits and larval stages of some crabs of Bombay. *Rec. Indian Mus.* 54:33-52.

Chittleborough, R.G. 1975. Environmental factors affecting growth and survival of juvenile western rock lobsters *Panulirus longipes* (Milne Edwards). *Aust. J. Mar. Freshwat. Res.* 26:177-196.

Christiansen, M.E. 1973. The complete larval development of *Hyas araneus* (Linnaeus) and *Hyas coarctatus* Leach (Decapoda, Brachyura, Majidae) reared in the laboratory. *Norw. J. Zool.* 21:63-89.

Coffin, H.G. 1960. The ovulation, embryology and developmental stages of the hermit crab *Pagurus samuelis* (Stimpson). *Walla Walla Coll. Publ. Dept. Biol. Sci. Stat.* 25:1-30.

Conklin, D.E., L.R.D'Abramo, C.E.Bordner & M.A.Baum 1980. A successful purified diet for the culture of juvenile lobsters: The effect of lecithin. *Aquaculture* 21:243-249.

Cook, H.L. 1965. A generic key to the protozoean, mysis, and postlarval stages of the littoral penaeidae of the northwestern Gulf of Mexico. *Fish. Bull. US Fish Wildl. Serv.* 65:437-447.

Costello, W.J., R.Hill & F.Lang 1981. Innervation patterns of fast and slow motor neurones during development of a lobster neuromuscular system. *J. Exp. Biol.* 91:271-284.

Costello, W.J. & F.Lang 1979. Development of the dimorphic claw closer muscles of the lobster *Homarus americans.* IV. Changes in functional morphology during growth. *Biol. Bull.* 156:179-195.

Costlow, J.D. Jr 1963. The effects of eyestalk extirpation on metamorphosis of megalopa of the blue crab, *Callinectes sapidus* Rathbun. *J. Exp. Biol.* 1952:219-228.

Costlow, J.D. Jr 1966a. The effect of eyestalk extirpation on larval development of the crab, *Sesarma reticulatum* Say. In: H.Barnes (ed.), *Some Contemporary Studies in Marine Science.* London: Allen & Unwin.

Costlow, J.D. Jr 1966b. The effects of eyestalk extirpation on larval development of the mud crab, *Rhithropanopeus harrisii* (Gould). *Gen. Comp. Endocr.* 7:255-274.

Costlow, J.D. Jr 1968. Metamorphosis in Crustaceans. In: W.Etkin & L.J.Gilbert (eds.), *Metamorphosis. A Problem in Developmental Biology.* New York: Appleton-Century-Crofts.

Costlow, J.D. Jr & C.G.Bookhout 1961. The larval stages of *Panopeus herbstii* Milne-Edwards reared in the laboratory. *J. Elisha Mitchell Scient. Soc.* 77:33-42.

Costlow, J.D. Jr & C.G.Bookhout 1966. Larval development of the crab, *Hexapanopeus angustifrons. Chesapeake Sci.* 7:148-156.

Costlow, J.D. Jr, C.G.Bookhout & R.J.Monroe 1966. Studies on the larval development of the crab, *Rhithropanopeus harrisii* (Gould). I. The effect of salinity and temperature on larval development. *Physiol. Zoöl.* 39:81-100.

Costlow, J.D. Jr & A.N.Sastry 1966. Free amino acids in developing stages of two crabs, *Callinectes sapidus* Rathbun and *Rhithropanopeus harrisii* (Gould). *Acta Embryol. Morph. Exp.* 9:44-55.

Coutière, H. 1906. Note sur la synonymie et le développement de quelques Hoplophoridae. *Bull. Mus. océanogr. Monaco* 70:1-20.

Crane, J. 1975. *Fiddler crabs of the world.* Princeton: Princeton University Press.

D'Abramo, L.R., C.E.Bordner, D.E.Conklin & N.A.Baum 1981. Essentiality of dietary phosphatidyl-choline for the survival of juvenile lobsters. *J. Nutr.* 111:425-431.

Dahl, E. 1957. Embryology of X-organs in *Crangon allmani. Nature* 179:482.

Dall, W. 1981. Osmoregulatory ability and juvenile habitat preference in some penaeid prawns. *J. Exp. Mar. Biol. Ecol.* 54:55-64.

Davenport, J. 1972. Effects of size upon salinity tolerance and volume regulation in the hermit crab *Pagurus bernhardus. Mar. Biol.* 17:222-227.

Dechancé, M. 1958. Caractérisation de la glaucothöe et des premiers stades Pagure chez *Clibanarius misanthropus* (Risso) (Crust. Décapode Anomoure). *C.R. Hebd. Séanc. Acad. Sci., Paris, Ser.D* 246: 839-842.

Deshmukh, S. 1966. The puerulus of the spiny lobster *Panulirus polyphagus* (Herbst) and its metamorphosis into the postpuerulus. *Crustaceana* 10:137-150.

Diaz, H. & J.J.Ewald 1968. A comparison of the larval development of *Metasesarma rubripes* (Rathbun) and *Sesarma ricordi* H.Milne Edwards (Brachyura, Grapsidae) reared under similar laboratory conditions. *Crustaceana* suppl.2:225-248.

Dietrich, D.K. 1979. The ontogeny of osmoregulation in the fiddler crab, *Uca pugilator. Am. Zool.* 19: 972.

Dobkin, S. 1961. Early developmental stages of the pink shrimp, *Penaeus duorarum* from Florida waters. *Fish. Bull. US Fish Wildl. Serv.* 61:321-349.

Dobkin, S. 1965. The first post-embryonic stage of *Synalpheus brooksi* Coutière. *Bull. Mar. Sci.* 15: 450-462.

Dobkin, S. 1969. Abbreviated larval development in caridean shrimps and its significance in the artificial culture of these animals. *FAO Fish Rep.* 3:935-946.

Dotsu, Y., O.Tanaka, Y.Shojima & K.Sono 1966. Metamorphosis of the phyllosomas of *Ibacus ciliatus* (von Siebold) and *I.novemdentatus* Gibbes to the reptant larvae stage. *Bull. Fac. Fish, Nagasaki Univ.* 21:195-224.

Dudley, D.L. & M.H.Judy 1973. Seasonal abundance and distribution of juvenile blue crabs in Core Sound, NC 1965-68. *Chesapeake Sci.* 14:51-55.

Dutt, S. & K.Ravindranath 1975. Pueruli of *Panulirus polyphagus* (Herbst) (Crustacea, Decapoda, Palinuridae) from east coast of India with a key to known Indo-West Pacific pueruli of *Panulirus* White. *Proc. Indian Acad. Sci.* 82B:100-107.

Elofsson, R. 1963. The nauplius eye and frontal organs in Malacostraca (Crustacea). *Sarsia* 19:1-54.

Elofsson, R. 1969. The development of the compound eye of *Penaeus duorarum* (Crustacea, Decapoda) with remarks on the nervous system. *Z.Zellforsch mikrosk. Anat.* 97:323-350.

Emmel, N.E. 1908. The experimental control of asymmetry at different stages in the development of the lobster. *J. Exp. Zool.* 5:471-484.

Factor, J.R. 1978. Morphology of the mouthparts of larval lobsters, *Homarus americanus* (Decapoda: Nephropidae), with special emphasis on their setae. *Biol. Bull.* 154:383-408.

Factor, J.R. 1981. Development and metamorphosis of the digestive system of larval lobsters, *Homarus americanus* (Decapoda: Nephropidae). *J. Morph.* 169:225-242.

Factor, J.R. 1982. Development and metamorphosis of the feeding apparatus of the stone crab, *Menippe mercenaria* (Brachyura, Xanthidae). *J. Morph.* 172:299-312.

Farmer, A.S.D. 1975. Synopsis of biological data on the Norway lobster, *Nephrops norvegicus* (Linnaeus, 1758). *FAO Fish. Synopsis* 112:1-97.

Faxon, W. 1879. On some young stages in the development of *Hippa, Porcellana* and *Pinnixa. Bull. Mus. Comp. Zool. Harvard* 5:253-268.

Felder, D.L. 1978. Osmotic and ionic regulation in several western Atlantic Callianassidae (Crustacea, Decapoda, Thalassinidea). *Biol. Bull.* 154:409-429.

Felder, D.L. & A.H.Chaney 1979. Decapod crustacean fauna of Seven and One-half Fathom Reef, Texas: Species composition, abundance, and species diversity. *Contr. mar. Sci.* 22:1-29.

Fincham, A.A. 1977. Larval development of British prawns and shrimp (Crustacea: Decapoda: Natantia). 1. Laboratory methods and a review of *Palaemon (Paleander) elegans* Rathke, 1837. *Bull. Br. Mus. Nat. Hist. (Zool.)* 32:1-28.

Finchman, A.A. 1979a. Larval development of British prawns and shrimps (Crustacea: Decapoda: Natantia. 2. *Palaemonetes (Palaemonetes) varians* (Leach, 1814) and morphological variation. *Bull. Br. Mus. Nat. Hist. (Zool.)* 35:163-182.

Fincham, A.A. 1979b. Larval development of British prawns and shrimps (Crustacea: Decapoda: Natantia). 3. *Palaemon (Palaemon) longirostris* H.Milne Edwards, 1837 and the effect of antibiotics on morphogenesis. *Bull. Br. Mus. Nat. Hist. (Zool.).* 37:17-46.

Finney, W.C. & L.G.Abele 1981. Allometric variation and sexual maturity in the obligate coral commensal *Trapezia ferruginea* Latreille (Decapoda, Xanthidae). *Crustaceana* 41:113-130.

Forbes, A.T. 1973. An unusual abbreviated larval life in the estuarine burrowing prawn *Callianassa kraussi* (Crustacea: Decapoda: Thalassinidea). *Mar. Biol.* 22:361-365.

Ford, J.I. & R.A.Kinzie III 1982. Life crawls upstream. *Nat. Hist.* 91:60-67.

Forest, J. 1954. Sur les premiers stades postlarvaires du Pagure *Dardanus pectinatus* (Ortmann). *C.R hebd. Séanc. Acad. Sci. Paris* (D)239:1697-1699.

Forster, G.R. 1951. The biology of the common prawn, *Leander serratus* Pennant. *J. Mar. Biol. Ass. UK* 30:333-360.

Foskett, J.K. 1977. Osmoregulation in the larvae and adults of the grapsid crab *Sesarma reticulatum* Say. *Biol. Bull. Mar. Biol. Lab., Woods Hole* 153:505-526.

Fukuda, Y. 1981. Larval development of *Trigonoplax unguiformis* (Crustacea, Brachura [sic]) reared in the laboratory. *Zool. Mag., Tokyo* 90:164-173.

Fukuda, Y. & K.Baba 1976. Complete larval development of the sesarminid crabs, *Chiromantes bidens, Holometopus haematocheir, Parasesarma plicatum,* and *Sesarma intermedius,* reared in the laboratory. *Mem. Fac. Educ., Kumamoto Univ.* 25:61-75.

Gallagher, M.L., W.D.Brown, D.E Conklin & M.Sifri 1978. Effects of varying calcium/phosphorus ratios in diets fed to juvenile lobsters *(Homarus americanus). Comp. Biochem. Physiol.* 60A:467-471.

Gamo, S. 1958a. On the postlarval stage of *Scopimera globosa* de Haan and *Paracleistostoma cristatum* de Man, Ocypodidae, Brachyuran Crustacea. *Zool. Mag., Tokyo* 67:67-74.

Gamo, S. 1958b. On the postlarval stages of two species of crabs of the subfamily Varuninae, Grapsidae, Brachyuran Crustacea. *Zool. Mag., Tokyo* 67:373-379.

Gaudy, R. & L.Sloane 1981. Effect of salinity on oxygen consumption in postlarvae of the penaeid shrimps *Penaeus monodon* and *P.stylirostris* without and with acclimation. *Mar. Biol.* 65:297-301.

George, M.J. 1969. Genus *Macrobrachium* Bate 1868. *Bull. Cent. Mar. Fish Res. Inst.* 14:179-216.

Giles, J.H. & G.Zamora 1973. Cover as a factor in habitat selection by juvenile brown *(Penaeus aztecus)* and white *(P.setiferus)* shrimp. *Trans. Am. Fish. Soc.* 102:144-145.

Goodbody, I. 1960. Abbreviated development in a pinnotherid crab. *Nature* 185:704-705.

Gopala Menon, P. 1972. Decapod Crustacea from the International Indian Ocean Expedition: the larval development of *Heterocarpus* (Caridea). *J. Zool., Lond.* 167:371-397.

Gordon, I. 1955. Importance of larval characters in classification. *Nature* 176:911-912.

Govind, C.K. 1982. Development of nerve, muscle, and synapse. In: H.L.Atwood & D.C.Sandeman (eds), *The Biology of Crustacea, Vol.3. Neurobiology: Structure and Function.* New York: Academic Press.

Govind, C.K. & F.Lang 1978. Development of the dimorphic claw closer muscles of the lobster *Homarus americanus.* III. Transformation to dimorphic muscles in juveniles. *Biol. Bull.* 154:55-67.

Goy, J.W. & A.J.Provenzano Jr 1978. Larval development of the rare burrowing mud shrimp *Naushonia crangonoides* Kingsley (Decapoda; Tahalssinidea; Laomediidae). *Biol. Bull.* 154:241-261.

Goy, J.W. & A.J.Provenzano Jr 1979. Juvenile morphology of the rare burrowing mud shrimp *Naushonia crangonoides* Kingsley, with a review of the genus *Naushonia* (Decapoda: Thalassinidea: Laomediidae). *Proc. Biol. Soc. Wash.* 92:339-359.

Guinot, D. 1978. Principes d'une classification évolutive des Crustacés Décapodes Brachyoures. *Bull. Biol. Fr. Belg.* 112:211-292.

Gurney, R. 1924. The larval development of some British prawns (Palaemonidae). I: *Palaemonetes varians. Proc. Zool. Soc. Lond.* 1924:297-328.

Gurney, R. 1937a. Notes on some decapod Crustacea from the Red Sea. I: The Genus *Processa. (P. aequimana, Nikoides danae, Hectarthropus).* II: The larvae of *Upogebia savignyi* Strahl. *Proc Zool. Soc. Lond.* (B)1937:85-101.

Gurney, R. 1937b. Larvae of decapod Crustacea. Part IV: Hippolytidae. *'Discovery' Reports* 14:351-404.

Gurney, R. 1938. Larvae of decapod Crustacea. Part V: Nephropsidea and Thalassinidea. *'Discovery' Reports* 17:291-344.

Gurney, R. 1942. *Larvae of Decapod Crustacea.* London: Ray Society.

Gurney, R. & M.V.Lebour 1940. Larvae of decapod Crustacea. Part VI. The genus *Sergestes. 'Discovery' Reports* 20:1-68.

Gurney, R. & M.V.Lebour 1941. On the larvae of certain Crustacea Macrura, mainly from Bermuda. *J. Linn. Soc., Lond., Zool.* 41:89-181.

Hackney, C.T. & W.D.Burbanck 1976. Some observations on the movement and location of juvenile shrimp in coastal waters of Georgia. *Bull. Ga. Acad. Sci.* 34:129-136.

Haefner, P.A. Jr 1969. Osmoregulation of *Crangon septemaspinosa* Say (Crustacea: Caridea). *Biol. Bull. mar. biol. Lab., Woods Hole* 137:438-446.

Hale, H.M. 1925. The development of two Australian sponge-crabs. *Proc. Linn. Soc. NSW* 1:405-413.

Hale, H.M. 1931. The post-embryonic development of an Australian xanthid crab (*Pilumnus vestitus* Haswell). *Rec. S.Aust. Mus.* 4:321-331.

Hansen, H.J. 1922. Crustaces Decapodes (Sergestides) provenant des campagnes des Yachts 'Hirondelle' et 'Princesse Alice' (1885-1915). *Res. Camp. Sci. Monaco* 64:1-232.

Harrison, K.E., P.L.Lutz & L.Farmer 1981. The ontogeny of osmoregulation ability of *Macrobrachium rosenbergii. Am. Zool.* 21:1015.

Hart, J.F.L. 1935. The larval development of British Columbia Brachyura. I: Xanthidae, Pinnotheridae (in part) and Grapsidae. *Can. J. Res.* 12:411-432.

Hart, J.F.L. 1965. Life history and larval development of *Cryptolithodes typicus* Brandt (Decapoda, Anomura) from British Columbia. *Crustaceana* 8:255-276.

Hartnoll, R.G. 1964. The freshwater grapsid crabs of Jamaica. *Proc. Linn. Soc. Lond.* 175:145-169.

Hartnoll, R.G. 1978. The effect of salinity and temperature on the postlarval growth of the crab *Rhithropanopeus harrisii.* In: D.S.McClusky & A.J.Berry (eds.), *Physiology and Behavior of Marine Organisms.* Oxford: Pergamon Press.

Hartnoll, R.G. 1982. Growth. In: L.G.Abele (ed.), *The Biology of Crustacea, Vol.2.* New York: Academic Press.

Hay, W.P. & C.A.Shore 1918. The decapod Crustacea of Beaufort, North Carolina, and the surrounding region. *Bull. US Bur. Fish* 35:369-475.

Haynes, E. 1976. Description of zoeae of coonstripe shrimp, *Pandalus hypsinotus,* reared in the laboratory. *Fish. Bull. US Fish Wildl. Serv.* 74:323-342.

Haynes, E. 1978. Description of larvae of the humpy shrimp, *Pandalus goniurus,* reared in situ in Kachemak Bay, Alaska. *Fish. Bull US Fish Wildl. Serv.* 76:235-248.

Haynes, E. 1979. Description of larvae of the northern shrimp *Pandalus borealis,* reared in situ in Kachemak Bay, Alaska. *Fish. Bull. US Fish Wildl. Serv.* 77:157-176.

Heard, R.W. 1982. Guide to common tidal marsh invertebrates of the Northeastern Gulf of Mexico. *Publications of the Mississippi-Alabama Sea Grant Consortium,* MASGP-79-004:1-82.

Heegaard, P.E. 1953. Observations on spawning and larval history of the shrimp, *Penaeus setiferus* (L.). *Publ. Inst. Mar. Sci. Univ. Tex.* 3:75-105.

Heral, M. & Y.Saudray 1979. Regimes alimentaires et modifications des structures mandibulaires chez les larves, postlarves et adultes de *Palaemonetes varians* (Leach) (Decapoda, Palaemonidae) etudiees a l'aide du microscope eletronique a balayage. *Rev. Trav. Inst. Péches marit.* 43:353-359.

Hines, A.H. 1982. Coexistence in a kelp forest: Size, population dynamics, and resource partitioning in a guild of spider crabs (Brachyura, Majidae). *Ecol. Monogr.* 52:179-198.

Hiro, F. 1937. Studies on animals inhabiting reef corals. I. *Hapalocarcinus* and *Cryptochirus. Palao. Trop. Biol. Stn. Stud.* 1:137-154.

Hobbs, H.H.Jr 1962. Notes on the affinities of the members of the Blandingii section of the crayfish genus *Procambarus* (Decapoda, Astacidae). *Tulane Stud. Zool.* 9:273-293.

Hobbs, H.H.Jr 1981. The crayfishes of Georgia. *Smithson. Contr. Zool.* 318:1-549.

Hogarth, P.J. 1975. Instar number and growth of juvenile *Carcinus maenas* (L.) (Decapoda Brachyura). *Crustaceana* 29:299-300.

Hong, Sung Yun 1977. The larval stages of *Cycloxanthops truncatus* (de Haan) (Decapoda, Brachyura, Xanthidae) reared under the laboratory conditions. *Publs. Inst. Mar. Sci., Nat. Fish. Univ. Busan* 10: 15-24.

Howard, A.E. & D.B.Bennett 1979. The substrate preference and burrowing behavior of juvenile lobsters (*Homarus gammarus* (L.)). *J. Nat. Hist.* 13:433-438.

Hubschman, J.H. 1963. Development and function of neurosecretory sites in the eyestalks of larval *Palaemonetes* (Decapoda, Natantia). *Biol. Bull.* 125:96-113.

Hubschman, J.H. 1971. Transient larval glands in *Palaemonetes.* In: D.J.Crisp (ed.), *Fourth European Marine Biology Symposium.* Cambridge: University Press.

Hughes, D.A. 1968. Factors controlling emergence of pink shrimp (*Penaeus duorarum*) from the substrate. *Biol. Bull.* 134:48-59.

Hughes, D.A. 1969a. Responses to salinity change as a tidal transport mechanism of pink shrimp, *Penaeus duorarum. Biol. Bull.* 136:323-342.

Hughes, D.A. 1969b. On the mechanisms underlying tide-associated movements of *Penaeus duorarum* Burkenroad. *FAO Fish. Rept.* 57:867-874.

Hughes, D.A. 1969c. Evidence for the endogenous control of swimming in pink shrimp, *Penaeus duorarum. Biol. Bull.* 136:398-404.

Hyman, O.W. 1924a. Studies on larvae of crabs of the family Pinnotheridae. *Proc. US Natn. Mus.* 64: 1-7.

Hyman, O.W. 1924b. Studies on larvae of crabs of the family Grapsidae. *Proc. US Natn. Mus.* 65:1-8.

Ingle, R.W. 1977. The larval and post-larval development of the scorpion spider crab, *Inachus dorsettensis* (Pennant) (Family: Majidae) reared in the laboratory. *Bull. Br. Mus. Nat. Hist. (Zool.)* 30:329-348.

Ingle, R.W. 1979. The larval and post-larval development of the brachyuran crab, *Geryon tridens* Krøyer (Family Geryonidae) reared in the laboratory. *Bull. Br. Mus. Nat. Hist. (Zool.)* 36:217-232.

Ingle, R.W. 1980. *British Crabs.* London: British Museum (Natural History) and Oxford University Press.

Ingle, R.W. 1981. The larval and postlarval development of the edible crab, *Cancer pagurus* (Linnaeus) (Decapoda: Brachyura). *Bull. Br. Mus. Nat. Hist. (Zool.)* 40:211-236.

Ingle, R.W. 1982. Larval and post-larval development of the slender-legged spider crab, *Macropodia rostrata* (Linnaeus) (Oxyrhyncha: Majidae: Inachinae) reared in the laboratory. *Bull. Br. Mus. Nat. Hist. (Zool.)* 42:207-225.

Ingle, R.W. & P.F.Clark 1980. The larval and post-larval development of Gibb's Spider Crab, *Pisa armata* (Latreille) (Family Majidae: Subfamily Pisinae), reared in the laboratory. *J. Nat. Hist.* 14:723-735.

Irvine, J. & H.G.Coffin 1960. Laboratory culture and early stages of *Fabia subquadrata* (Dana), (Crustacea, Decapoda). *Walla Walla Coll. Publ. Dept. Biol. Sci. Stat.* 28:1-24.

Jagardisha, K. & K.N.Sankolli 1977. Laboratory culture of the prawn *Palaemon (Paleander) semmelinkii* (de Man) (Crustacea, Decapoda, Palaemonidae). *Proc. Symp. Warm Water Zooplankton Spec. Publs. Natl. Inst. Oceanogr. Goa:*619-633.

Jalihal, D.R. & K.N.Sankolli 1975. On the abbreviated metamorphosis of the freshwater prawn *Macrobrachium hendersodayanum* (Tiwari) in the laboratory. *J. Karnatak Univ.* 20:283-291.

Jansen, G.M. & B.R.Kuipers 1980. On tidal migration in the shrimp *Crangon crangon. Neth. J. Sea Res.* 14:339-348.

John, C.M. 1957. Bionomics and life history of *Macrobrachium rosenbergii* (de Man). *Bull. Cent. Res. Inst. Univ. Kerala, India,* Ser.C, 15:93-102.

Johnson, M.W. 1975. The postlarvae of *Scyllarides astori* and *Evibacus princeps* of the Eastern Tropical Pacific (Decapoda, Scyllaridae). *Crustaceana* 28:139-144.

Jorgensen, C.B. 1955. Quantitative aspects of filter feeding in invertebrates. *Biol. Rev.* 30:391-454.

Kaestner, A. 1970. *Invertebrate Zoology, Vol.3.* New York: Wiley.

Kakati, V.S. & K.N.Sankolli 1975. On the metamorphosis of the spider crab, *Dehaanius limbatus* (A. Milne Edwards) in laboratory (Brachyura Majidae). *Karnatak Univ. J. Sci.* 20:275-282.

Kalber, F.A. 1976. Osmoregulation in decapod larvae as a consideration in culture techniques. *Helgo. wiss. Meeres.* 20:697-706.

Kalber, F.A. & J.D.Costlow Jr 1966. The ontogeny of osmoregulation and its neurosecretory control in the decapod crustacean, *Rhithropanopeus harrisii* (Gould). *Am. Zool.* 6:221-229.

Kalber, F.A. & J.D.Costlow 1968. Osmoregulation in larvae of the land crab, *Cardisoma guanhumi* Latreille. *Am. Zool.* 8:411-416.

Kanazawa, A., A.Tokiwa, S.Kayama & M.Hirata 1977. Essential fatty acids in the diet of the prawn. I. Effects of linoleic and linoleic acids on growth. *Bull. Jap. Soc. Scient. Fish.* 43:1111-1114.

Kanazawa, A., A.Teshima, S.Tokiwa & H.J.Ceccaldi 1979. Effects of dietary linoleic and linoleic acids on growth of prawn. *Oceanol. Acta* 2:41-47.

Kanciruk, P. 1980. Ecology of juvenile and adult Palinuridae (spiny lobsters). In: J.S.Cobb & B.F.Phillips (eds.), *The Biology and Management of Lobsters, Vol.II, Ecology and Management.* New York: Academic Press.

Keiser, R.K. & D.V.Aldrich 1976. Salinity preference of postlarval brown and white shrimp (*Penaeus aztecus* and *P.setiferus*) in gradient tanks. *Tex. A & M Univ. Sea Grant Publ.* 75-208:1-260.

Kemp, S.W. 1910. The Decapoda Natantia of the coast of Ireland. *Scient. Invest. Fish. Brch Ire.* 1908: 1-190.

Kemp, S.W. 1915. Fauna of the Chilka Lake. Crustacea Decapoda. *Rec. Indian Mus.* 5:199-325.

King, J.A. & C.K.Govind 1980. Development of excitatory innervation in the lobster claw closer muscle. *J. Comp. Neurol.* 194:57-70.

Klein Breteler, W.C.M. 1975a. Laboratory experiments on the influence of environmental factors on the frequency of molting and the increase in size at molting of juvenile shore crabs, *Carcinus maenas. Neth. J. Sea Res.* 9:100-120.

Klein Breteler, W.C.M. 1975b. Food consumption, growth and energy metabolism of juvenile shore crabs, *Carcinus maenas. Neth. J. Sea Res.* 9:255-272.

Klein Breteler, W.C.M. 1976. Migration of the shore crab, *Carcinus maenas,* in the Dutch Wadden Sea. *Neth. J. Sea Res.* 10:338-353.

Knight, M.D. 1968. The larval development of *Raninoides benedicti* Rathbun (Brachyura, Raninidae), with notes on the Pacific records of *Raninoides laevis* (Latreille). *Crustaceana* suppl.2:145-169.

Knight, M.D. & M.Omori 1982. The larval development of *Sergestes similis* Hansen (Crustacea Decapoda, Sergestidae) reared in the laboratory. *Fish. Bull. US Fish Wildl. Serv.* 80:217-244.

Knowlton, R.E. 1973. Larval development of the snapping shrimp, *Alpheus heterochaelis* Say reared in the laboratory. *J. Nat. Hist.* 7:273-306.

Knowlton, R.E. 1974. Larval development processes and controlling factors in decapod Crustacea, with emphasis on Caridea. *Thalassia Jugosl.* 10:138-158.

Knudsen, J.W. 1958. Life cycle studies of the Brachyura of Western North America, I. General culture methods and the life cycle of *Lophopanopeus leucomanus leucomanus* (Lockington). *Bull. South. Calif. Acad. Sci.* 57:51-59.

Knudsen, J.W. 1960. Reproduction, life history, and larval ecology of the California Xanthidae, the pebble crabs. *Pac. Sci.* 14:3-17.

Konishi, K. 1981. A description of laboratory-reared larvae of the pinnotherid crab *Sakaina japonica* Serène (Decapoda, Brachyura). *J. Fac. Sci. Hokkaido Univ. Ser.VI. Zool.* 22:165-176.

Krasne, F.B. & J.J.Wine 1975. Extrinsic modulation of crayfish escape behavior. *J. exp. Biol.* 63:433-450.

Krefft, S. 1952. The early post-larval stages and systematic position of *Eurynolambrus australis* M.E. and L. (Brachyura). *Trans. R. Soc. NZ* 79:574-578.

Kruczynski, W.L. 1973. Distribution and abundance of *Pinnotheres maculatus* Say in Bogue Sound, North Carolina. *Biol. Bull.* 145:482-491.

Kurata, H. 1955. The post-embryonic development of the prawn *Pandalus kessleri*. *Bull. Hokkaido Reg. Fish. Res. Lab.* 12:1-15.

Kurata, H. 1956. Larval stages of *Paralithodes brevipes* (Decapoda, Anomura). *Bull. Hokkaido Reg. Fish. Res. Lab.* 14:25-34.

Kurata, H. 1960. Last stage zoea of *Paralithodes* with intermediate form between normal last stage zoea and glaucothoe. *Bull. Hokkaido Reg. Fish. Res. Lab.* 22:49-56.

Kurata, H. 1964. Larvae of the decapod Crustacea of Hokkaido. 6. Lithodidae (Anomura). *Bull. Hokkaido Reg. Fish. Res. Lab.* 29:49-65.

Kurata, H. 1965. Larvae of Decapoda Crustacea of Hokkaido. 11. Pasiphaeidae (Natantia). *Bull. Hokkaido Reg. Fish. Res. Lab.* 30:15-20.

Kurata, H. 1968a. Larvae of Decapoda Macrura of Arasaki, Sagami Bay. I. *Eualus gracilirostris* (Stimpson) (Hippolytidae). *Bull. Tokai Reg. Fish. Res. Lab.* 55:245-251.

Kurata, H. 1968b. Larvae of Decapoda Macrura of Arasaki, Sagami Bay. II. *Heptacarpus futilirostris* (Bate) (Hippolytidae). *Bull. Tokai Reg. Fish. Res. Lab.* 55:253-258.

Kurata, H. 1968c. Larvae of Decapoda Brachyura of Arasaki, Sagami Bay. I. *Acmaeopleura parvula* (Grapsidae). *Bull. Tokai Reg. Fish. Res. Lab.* 55:259-263.

Kurata, H. 1968d. Larvae of Decapoda Anomura of Arasaki, Sagami Bay. I. *Pagurus samuelis* (Paguridae). *Bull. Tokai Reg. Fish. Res. Lab.* 56:160-172.

Kurata, H. 1969. Larvae of Decapoda Macura of Arasaki, Sagami Bay. IV. Majidae. *Bull. Tokai Reg. Fish. Res. Lab.* 57:81-127.

Kurata, H. 1970. Studies on the life histories of decapod Crustacea of Georgia. *Univ. Georgia, Mar. Inst. Publ. Sapelo Island:* 1-274.

Kurata, H. & H.Omi 1969. The larval stages of a swimming crab, *Charybdis acuta. Bull. Tokai Reg. Fish. Res. Lab.* 57:129-136.

Kurata, H. & V.Pusadee 1974. Larvae and early postlarvae of a shrimp, *Metapenaeus burkenroadi*, reared in the laboratory. *Bull. Nasei Reg. Fish. Res. Lab.* 7:69-84.

Kurata, H., R.W.Heard & J.W.Martin 1981. Larval development under laboratory conditions of the xanthid mud crab *Eurytium limosum* (Say, 1818) (Brachyura: Xanthidae) from Georgia. *Gulf Res. Rep.* 7:19-25.

Kutty, M.N., G.Murugapoothy & T.S.Krishnan 1971. Influence of salinity and temperature on the oxygen consumption in young juveniles of the Indian prawn, *Penaeus indicus. Mar. Biol.* 11:125-131.

Labat, J.P. 1977. Ecologie de *Crangon crangon* (L) (Decapoda, Caridea) dans un étang de la côte Languedocienne. *Vie Milieu* 27:273-292.

Lang, F., C.K.Govind & W.J.Costello 1978. Experimental transformation of muscle fibers in lobsters. *Science* 201:1037-1039.

Lang, F., C.K.Govind, W.J.Costello & S.I.Green 1977. Developmental neuroethology: Changes in escape and defensive behavior during growth of the lobster. *Science* 197:682-685.

Laverack, M.S., D.L.Macmillan & D.M.Neil 1976. A comparison of beating parameters in larval and post-larval locomotor systems of the lobster *Homarus gammarus* (L.). *Phil. Trans. R.Soc. Lond.* 274:87-99.

Leach, W.E. 1814. *Malacostraca Podophthalmata Britanniae*. London.

Lebour, M.V. 1928. The larval stages of the Plymouth Brachyura. *Proc. Zool. Soc. Lond.* 1928:473-560.

Lebour, M.V. 1931. The larvae of the Plymouth Caridea. I: The larvae of the Crangonidae. II. The larvae of the Hippolytidae. *Proc. Zool. Soc. Lond.* 1931:1-9.

Lebour, M.V. 1932a. The larval stages of the Plymouth Caridea. III: The larval stages of *Spirontocaris cranchii* (Leach). *Proc. Zool. Soc. Lond.* 1932:131-137.

Lebour, M.V. 1932b. The larval stages of the Plymouth Caridea. IV: The Alpheidae. *Proc. Zool. Soc. Lond.* 1932:463-469.

Lebour, M.V. 1936. Notes on the Plymouth species of *Spirontocaris. Proc. Zool. Soc. Lond.* 1936:89-104.

Lebour, M.V. 1940. The larvae of the British species of *Spirontocaris* and their relation to *Thor* (Crustacea Decapoda). *J. Mar. Biol. Ass. UK* 24:505-514.

Lebour, M.V. 1944. Larval crabs from Bermuda. *Zoologica* 29:113-128.

LeRoux, A.A. 1971a. Étude anatomique et fonctionnelle du proventricule des larves et des juvéniles du premier stade de *Palaemonetes varians* (Leach) (Décapode, Natantia). I. Données anatomiques. *Bull. Soc. Zool. Fr.* 96:127-140.

LeRoux, A.A. 1971b. Étude anatomique et fonctionnelle du proventricule de larves et des juvéniles du premier stade de *Palaemonetes varians* (Leach) (Décapode, Natantia). II. Les phénomènes mécanique de la digestion. *Bull. Soc. Zool. Fr.* 96:141-149.

LeRoux, A.A. 1976a. Altération du caecum intestinal, an cours de la mue de metamorphose, chez *Pisidia longicornis* (Linné) (Crustacé Décapode). *Bull. Soc. Zool. Fr.* 101(5):25-29.

LeRoux, A.A. 1976b. Aspects de la differenciation sexuelle chez *Pisidia longicornis* (Linné) (Crustacea, Decapoda). *C.R. Acad. Sci. Paris* 283:959-962.

LeRoux, A.A. 1980. Effets de l'ablation des pédoncules oculaires et de quelques conditions d'élevage sur le développement de *Pisidia longicornis* (Linné) (Crustacé, Décapode, Anomure). *Arch. Zool. Exp. Gén.* 121:97-214.

LeRoux, A.A. 1982. Les organes endocrines chez les larves des Crustacés Eucarides. Intervention dans la croissance au cours de la vie larvaire et des premiers stades juveniles. *Oceanis* 8:505-531.

Ling, S.W. 1969. The general biology and development of *Macrobrachium rosenbergii* (de Man). *FAO Fish. Rep.* 3(57):589-606.

Loyacano, H. 1967. Some effects of salinity on two populations of red swamp crayfish *Procambarus clarki. Proc. 21st Annu. Conf. Southeast Assoc. Game and Fish Comm:*423-435.

Lucas, J.S. 1971. The larval stages of some Australian species of *Halicarcinus* (Crustacea, Brachyura, Hymenosomatidae). I. Morphology. *Bull. Mar. Sci.* 21:471-490.

Lyons, W.G. 1970. Scyllarid lobsters (Crustacea, Decapoda). *Memoirs of the Hourglass Cruises, Fla. Dept. Nat. Resources Mar. Res. Lab.* 1(4):1-74.

MacGinitie, G.E. & N.MacGinitie 1949. *Natural History of Marine Animals.* New York: McGraw-Hill.

Mair, J.McD. 1979. The identification of postlarvae of four species of *Penaeus* (Crustacea: Decapoda) from the Pacific coast of Mexico. *J. Zool. Lond.* 188:347-351.

Mair, J.McD. 1980. Salinity and water-type preferences of four species of postlarval shrimp (*Penaeus*) from west Mexico. *J. exp. mar. Biol. Ecol.* 45:69-82.

Mair, J.McD. 1981. Identification of small juvenile penaeid shrimp from the Pacific Coast of Mexico. *Bull. Mar. Sci.* 31:174-176.

Manning, R.B. 1961. Some growth changes in the stone crab, *Menippe mercenaria* (Say). *Q. J. Fla Acad. Sci.* 23:273-277.

Martin, J.W. 1984. Notes and bibliography on the larvae of xanthid crabs, with a key to the known xanthid zoeas of the western Atlantic and Gulf of Mexico. *Bull. Mar. Sci.* 34:220-239.

Martin, J.W., D.L.Felder & F.M.Truesdale 1984. A comparative study of morphology and ontogeny in juvenile stages of four western Atlantic xanthoid crabs (Crustacea: Decapoda: Brachyura). *Phil. Trans. R.Soc. Lond.* (B)303:537-604.

Matsumoto, L. 1958. Morphological studies on the neurosecretion in crabs. *Biol. J. Okayama Univ.* 4: 103-176.

Matthews, J.B.L. & S.Pinnoi 1973. Ecological studies on the deep-water pelagic community of Korsfjorden, Western Norway. The species of *Pasiphaea* and *Sergestes* (Crustacea Decapoda) recorded in 1968 and 1969. *Sarsia* 52:123-144.

Mellon, De F. & P.J.Stephens 1979. The motor organization of claw closer muscles in snapping shrimp. *J. Comp. Physiol.* 132A:109-116.

Menu-Marque, S.A. 1970. Desarrollo larval del cangrejo *Platyxanthus crenulatus* (A.Milne Edwards, 1879) en el laboratorio (Decapoda, Brachyura, Xanthidae). *Physis* 29:477-494.

Miers, E.J. 1879. On the classification of the maioid Crustacea or Oxyrhyncha, with a synopsis of the families, subfamilies and genera. *J. Linn Soc. Lond., Zool.* 14:634-673.

Mikulich, L.V. & B.G.Ivanov 1983. The far-eastern shrimp *Pandalus prensor* Stimpson (Decapoda, Pandalidae): Description of laboratory reared larvae. *Crustaceana* 44:61-75.

Miller, P.E. & H.G.Coffin 1961. A laboratory study of the developmental stages of *Hapalogaster mertensii* (Brandt) (Crustacea, Decapoda). *Walla Walla Coll. Publ. Dept. Biol. Sci. Stat.* 30:1-18.

Millikin, M.R., G.N.Biddle, T.C.Siewicki, A.R.Fortner & P.H.Fair 1980. Effects of various levels of dietary protein on survival, molting frequency and growth of juvenile blue crabs (*Callinectes sapidus*). *Aquaculture* 19:149-162.

Mills, B.J. & M.C.Geddes 1980. Salinity tolerance and osmoregulation of the Australian freshwater crayfish *Cherax destructor* Clark (Decapoda: Parastacidae). *Aust. J. Mar. Freshwat. Res.* 30:667-676.

Mizue, K. & Y.Iwamoto 1961. On the development and growth of *Neocardina denticulata* de Haan. *Bull. Fac. Fish. Nagaski Univ.* 10:15-24.

Mohamed, K.H., P.Vedavyasa Rao & M.J.George 1968. Postlarvae of penaeid prawns of southeast coast of India with a key to their identification. *FAO Fish. Rep.* 57:487-504.

Morita, T. 1974. Morphological observation on the development of larva of *Eriocheir japonica* de Haan. *Zool. Mag. (Tokyo)* 83:24-81.

Motoh, H. 1973. Laboratory-reared zoeae and megalopae of zuwai crab from the Sea of Japan. *Bull. Jap. Soc. Scient. Fish.* 39:24-81.

Muraoka, K. 1965. On the post-larval stage of *Plagusia depressa tuberculata* Lamarck (Grapsidae). *Res. on Crust.* 2:83-90.

Muraoka, K. 1967. On the postlarval stages of *Percnon planissimum* (Herbst) Grapsidae. *Res. on Crust.* 3:61-67.

Muraoka, K. 1969. On the post-larval stage of two species of the swimming crab. *Bull. Kanagawa Perfec. Mus.* 1:1-7.

Muraoka, K. 1971a. On the post-larval stage of three species of the shore crab, Grapsidae. *Bull, Kanagawa Perfec. Mus.* 1:8-17.

Muraoka, K. 1971b. On the post-larval characters of the two species of shore crabs. *Res. on Crust.* 4,5: 1-11, 224-235.

Muraoka, K. 1976. The post-larval development of *Uca lactea* (de Haan) and *Macrophthalmus (Mareotis) japonicus* (de Haan) (Crustacea, Brachyura, Ocypodidae). *Zool. Mag.* 85:40-51.

Muraoka, K. 1977. The larval stages of *Halicarcinus orientalis* Sakai and *Rhynchoplax mesor* Stimpson reared in the laboratory (Crustacea, Brachyura, Hymenosomatidae). *Zool. Mag.* 86:94-99.

Muraoka K. 1979. Post-larva of *Pinnixa rathbuni* (Crustacea, Brachyura, Pinnotheridae). *Zool. Mag.* 88:288-294.

Nayak, V.N. 1981. Larval development of the hermit crab *Diogenes planimanus* Henderson (Decapoda, Anomura, Diogenidae) in the laboratory. *Indian J. Mar. Sci.* 10(2):136-141.

Neil, D.M., D.L.Macmillan, R.M.Robertson & M.S.Laverack 1976. The structure and function of thoracic exopodites in the larvae of the lobster, *Homarus gammarus* (L.). *Phil. Trans. R. Soc. London* 274:53-68.

Novak, A. & M.Salmon 1974. *Uca panacea*, a new species of fiddler crab from the Gulf Coast of the United States. *Proc. Biol. Soc. Wash.* 87:313-326.

Ogonowski, M.M., F.Lang & C.M.Govind 1980. Histochemistry of lobster claw-closer muscles during development. *J. Exp. Zool.* 213:359-367.

Omori, M. 1971. Preliminary rearing experiments on larvae of *Sergestes lucens* (Penaeidea, Natantia, Decapoda). *Mar. Biol.* 9:228-234.

Ong, K.S. 1966. Observations on the post-larval history of *Scylla serrata* Forskal reared in the laboratory. *Malay. Agric. J.* 45:429-443.

Orlamunder, J. 1942. Zur Entwicklung und Formbildung des *Birgus latro* L., mit besonderer Berucksichtigung des X-Organs. *Z. wiss. Zool.* 155:280-316.

Pace, F., R.R.Harris & V.Jaccarini 1976. The embryonic development of the Mediterranean freshwater crab, *Potamon edulis (= P.fluviatile)* (Crustacea, Decapoda, Potamonidae). *J. Zool.* 180:93-106.

Passano, L.M. 1961. The regulation of crustacean metamorphosis. *Am. Zool.* 1:89-95.

Paterson, N.B. 1970. The median eye of some South African Palinuridae (Decapoda, Crustacea). *Ann. S.Afr. Mus.* 57:87-102.

Patwardhan, S.S. 1935. On the structure and mechanism of the gastric mill in Decapoda. V. The structure of the gastric mill in Natantous Macrura − Caridea. *Proc. India Acad. Sci.* 1B:693-704.

Pearce, J.B. 1966. The biology of the mussel crab, *Fabia subquadrata*, from the waters of the San Juan Archipelago, Washington. *Pacif. Sci.* 20:3-35.

Perez Farfante, I. 1969. Western Atlantic shrimps of the genus *Penaeus*. *Fish. Bull. US Fish Wildl. Serv.* 67:461-591.

Perez Farfante, I. 1970. Diagnostic characters of juveniles of the shrimps *Penaeus aztecus aztecus, P. duorarum duorarum,* and *P.brasiliensis* (Crustacea, Decapoda, Penaeidae). *Spec. Scient. Rep. US Fish. Wildl. Serv., Fisheries* 599:1-26.

Perez Perez, D. & R.M.Ros 1975. Descripcion y desarrollo de los estadios postlarvales del camaron blanco *Penaeus schmitti* Burkenroad. *Investigaciones Marinas, Univ. de la Habana, Ciencias,* Serie 8(21):1-114.

Phillips, B.F., N.A.Campbell & W.A.Rea 1977. Laboratory growth of early juveniles of the western rock lobster *Panulirus longipes cygnus*. *Mar. Biol.* 39:31-39.

Phillips, B.F. & A.N.Sastry 1980. Larval ecology. In: J.S.Cobb & B.F.Phillips (eds.), *Biology and Management of Lobsters, Vol.2. Ecology and Management*. New York: Academic Press.

Pike, R.B. & D.I.Williamson 1961. The larvae of *Spirontocaris* and related genera (Decapoda, Hippolytidae). *Crustaceana* 2:187-208.

Pillai, N.N. 1974. Larval development of *Leandrites celebensis* (de Man) (Decapoda: Palaemonidae), reared in the laboratory. *J. Mar. Biol. Ass. India* 16:708-720.

Pillai, N.N. 1975. Larval development of *Caridina pseudogracilirostris* reared in the laboratory. *J. mar. Biol. Ass. India* 17:1-17.

Pohle, G. & M.Telford 1981. The larval development of *Dissodactylus crinitichelis* Moreira, 1901 (Brachyura: Pinnotheridae) in the laboratory. *Bull. Mar. Sci.* 31:753-773.

Pohle, G. & M.Telford 1983. The larval development of *Dissodactylus primitivus* Bouvier, 1917 (Brachyura: Pinnotheridae) reared in the laboratory. *Bull. Mar. Sci.* 33:257-273.

Powell, C.G. 1979. Suppression of larval development in the African freshwater shrimp *Desmocaris trispinosa* (Decapoda, Palaemonidae). *Crustaceana* suppl.5:185-194.

Prasad, R.R. & R.R.S.Tampi 1953. A contribution to the biology of the blue swimming crab *Neptunus pelagicus* (L.) with a note on the zoea of *Thalamita crenata* Latreille. *J. Bombay Nat. Hist. Soc.* 51: 674-689.

Przibram, H. 1931. *Connecting Laws in Animal Morphology*. London: University of London Press, Ltd.

Provenzano, A.J.Jr & W.N.Brownell 1977. Larval and early post-larval stages of the West Indian spider crab, *Mithrax spinosissimus* (Lamarck) (Decapoda: Majidae). *Proc. Biol. Soc. Wash.* 90:735-752.

Provenzano, A.J.Jr & A.L.Rice 1966. Juvenile morphology and the development of taxonomic characters in *Paguristes sericeus* A.Milne Edwards (Decapoda, Diogenidae). *Crustaceana* 10:53-69.

Pullen, E.J. & W.L.Trent 1969. White shrimp emigration in relation to size, sex, temperature and salinity. *FAO Fish. Rept.* 57:1001-1014.

Pyle, R.W. 1943. The histogenesis and cyclic phenomena of the sinus gland and X-organ in Crustacea. *Biol. Bull.* 85:87-102.

Rabalais, N.N. 1983. Adaptations of a fiddler crab, *Uca subcylindrica* (Stimpson, 1859) to semi-arid environments. PhD dissertation, Univ. of Texas.

Rabalais, N.N. & J.N.Cameron 1983. Abbreviated development of *Uca subcylindrica* (Stimpson, 1859) (Crustacea, Decapoda, Ocypodidae) reared in the laboratory. *J. Crust. Biol.* 3:519-541.

Rajabai, K.G. 1960. Studies on the larval development of Brachyura. I. The early and postlarval development of *Dotilla blanfordi* Alcock. *Ann. Mag. Nat. Hist.* (13)2:129-135.

Rajabai Naidu, K.G. 1954. The postlarval development of the shore crab, *Ocypoda platytarsis* M.Ed. and *Ocypoda cordimana* Desmarest. *Proc. Indian Acad. Sci.* (B)40:89-101.

Raje, P.C. & M.R.Ranade 1975. Early life history of a stenopodid shrimp *Microprosthema semilaeve* (Decapoda: Macrura). *J. mar. Biol. Ass. India* 17:213-222.

Raman, K. 1967. Observations on the fishery and biology of the giant fresh-water prawn *Macrobrachium rosenbergii* de Man. *Proc. Symp. Crustacea, Mar. Biol. Assoc. India* 2:649-669.

Rao, P.V. 1968. A new species of shrimp, *Acetes cochinensis* (Crustacea, Decapoda, Sergestidae) from southwest coast of India with an account of its larval development. *J. Mar. Biol. Ass. India* 10:298-320.

Rathbun, M.J. 1914. Stalk-eyed crustaceans collected at the Monte Bello Islands. *Proc. Zool. Soc. Lond.* 1914:653-664.

Reese, E.S. 1968. Shell use: An adaptation for emigration from the sea by the coconut crab. *Science* 161:385-386.

Regnault, M. 1971. Croissance an laboratoire de *Crangon septemspinosa* Say (Crustacea Decapoda, Natantia) de la metamorphose a la maturité sexuelle. *Bull. Mus. Natn. Hist. Nat., Paris* 42:1108-1126.

Regnault, M. 1972. Développement de l'estomac les larves de *Crangon septemspinosa* Say (Crustacea, Decapoda, Crangonidae); son influence sur le mode de nutrition. *Bull. Mus. Natn. Hist. Nat., Paris* 67(Zoologie 53):840-856.

Regnault, M. 1976. Influence du substrat sur la mortalité et la croissance de la crevette *Crangon crangon* (L.) en élevage. *Cah. Biol. Mar.* 17:347-357.

Rice, A.L. 1964. The metamorphosis of a species of *Homola* (Crustacea, Decapoda, Dromiacea). *Bull. Mar. Sci.* 14:221-228.

Rice, A.L. 1980. Crab zoeal morphology and its bearing on the classification of the Brachyura. *Trans. Zool. Soc. Lond.* 35:271-424.

Rice, A.L. 1981a. Crab zoeae and brachyuran classification: a re-appraisal. *Bull. Br. Mus. Nat. Hist. Zool.* 40:287-296.

Rice, A.L. 1981b. The megalopa stage in brachyuran crabs. The Podotremata Guinot. *J. Nat. Hist.* 15: 1003-1011.

Rice, A.L. & R.W.Ingle 1975. The larval development of *Carcinus maenas* (L.) and *C.mediterraneus* Czerniavsky (Crustacea, Brachyura, Portunidae) reared in the laboratory. *Bull. Br. Mus. Nat. Hist., Zool.* 28:103-119.

Rice, A.L. & K.G.von Levetzow 1967. Larvae of *Homola* (Crustacea: Dromiacea) from South Africa. *J. Nat. Hist.* 1:435-453.

Ritz, D.A. & L.R.Thomas 1973. The larval and postlarval stages of *Ibacus peronii* Leach (Decapoda, Reptantia, Scyllaridae). *Crustaceana* 24:5-17.

Roberts, M.H. Jr 1970. Larval development of *Pagurus longicarpus* Say reared in the laboratory. I. Description of larval instars. *Biol. Bull.* 139:188-202.

Roberts, M.H. Jr 1975. Larval development of *Pinnotheres chamae* reared in the laboratory. *Chesapeake Sci.* 16:242-252.

Rothlisberg, P.C. 1980. A complete larval description of *Pandalus jordani* Rathbun (Decapoda, Pandalidae) and its relation to other members of the genus *Pandalus*. *Crustaceana* 38:19-48.

Ruello, N.V. 1973. The influence of rainfall on the distribution and abundance of the school prawn *Metapenaeus macleayi* in the Hunter River region (Australia). *Mar. Biol.* 23:221-228.

Saba, M., M.Takeda & Y.Nakasone 1978. Larval development of *Epixanthus dentatus* (White) (Brachyura, Xanthidae). *Bull. Natn. Sci. Mus. Tokyo (Zool.)* 4:151-161.

Saigusa, M. 1978. Ecological distribution of three species of the genus *Sesarma* in winter season. *Zool. Mag.* 87:142-150.

St.Amant, L.S., K.C.Corkum & J.G.Broom 1963. Studies on growth dynamics of the brown shrimp, *Penaeus aztecus*, in Louisiana waters. *Proc. Gulf Caribb. Fish. Inst.* 15th Ann. Sess.:14-26.

Sandifer, P.A., J.S.Hopkins & T.I.J.Smith 1975. Observations on salinity tolerance and osmoregulation in laboratory-reared *Macrobrachium rosenbergii* postlarvae (Crustacea: Caridea). *Aquaculture* 6:103-114.

Sandifer, P.A. & W.A.Van Engel 1971. Larval development of the spider crab, *Libinia dubia* H.Milne Edwards (Brachyura, Majidae, Pisinae) reared in laboratory culture. *Chesapeake Sci.* 12:18-25.

Sandifer, P.A. & W.A.van Engel 1972. Larval stages of the spider crab, *Anasimus latus* Rathbun, 1894 (Brachyura, Majidae, Inachinae) obtained in the laboratory. *Crustaceana* 23:141-151.

Sankolli, K.N. & H.G.Kewalramani 1962. Larval development of *Saron marmoratus* (Olivier) in the laboratory. *J. Mar. Biol. Ass. India* 4:106-120.

Sankolli, K.N. & S.Shenoy 1975. Larval development of mud shrimp *Callianassa (Callichirus) kewalramani* Sankolli, in the laboratory (Crustacea, Decapoda). *Bull. Dept. Mar. Sci. Univ. Cochin* 7:705-720.

Sastry, A.N. & W.R.Ellington 1978. Lactate dehydrogenase during the larval development of *Cancer irroratus:* Effect of constant and cyclic tidal regimes. *Experientia* 34:308-309.

Seaman, W. & D.Y.Aska 1974. Research and information needs of the Florida spiny lobster fishery; Proceedings of a conference held March 12 1974, in Miami. *Fla. State Univ. Syst. Fla., Sea Grant Program Rep.* 74-201:1-64.

Seréne, R. 1961. A megalopa commensal in a squid. *Proc. Pacif. Sci. Congr.* 10:35-36.

Serfling, S.A. & R.F.Ford 1975. Ecological studies of the puerulus larval stages of the California spiny lobster, *Panulirus interruptus*. *Fish. Bull. US Fish Wildl. Serv.* 73:360-377.

Sheard, K. 1962. *The Western Australian Crayfishery*, 1944-1961. Perth: Paterson Brokensha Pty.

Shen, C.J. 1935. An investigation of the post-larval development of the Shore-Crab *Carcinus maenas*, with special reference to the external secondary sexual characters. *Proc. Zool. Soc. Lond.* 1935:1-33.

Shenoy, S. 1967. Studies on larval development in Anomura (Crustacea, Decapoda). II. *Proc Symp. Crust., Mar. Biol. Ass. Mandapan Camp, India* 2:777-804.

Shenoy, S. & K.N.Sankolli 1975. On the life history of a porcellanid crab, *Petrolisthes lamarckii* (Leach), as observed in the laboratory. *J. Mar. Biol. Ass. India* 17:147-159.

Shokita, S. 1973a. Abbreviated larval development of fresh-water atyid shrimp, *Caridina brevirostris* Stimpson from Iriomote Island of the Ryukyus (Decapoda, Atyidae). *Bull. Sci. Engin. Div. Univ. Ryukyus* 16:222-232.

Shokita, S. 1973b. Abbreviated larval development of the fresh-water prawn, *Macrobrachium shokitai* Fujino et Baba (Decapoda, Palaemonidae) from Iriomote Island of the Ryukyus. *Annotnes Zool. Jap.* 46:111-126.

Shokita, S. 1976. Early life-history of the land-locked atyid shrimp *Caridina denticulata ishigakiensis* Fujino & Shokita, from the Ryukyu Islands. *Res. on Crust.* 7:1-10.

Shokita, S. 1977. Abbreviated metamorphosis of land-locked freshwater prawn, *Macrobrachium asperulum* (Von Martens) from Taiwan. *Annotnes Zool. Jap.* 50:110-122.

Shokita, S. 1978. Larval development of interspecific hybrid between *Macrobrachium asperulum* from Taikwan and *Macrobrachium shokitai* from the Ryukyus. *Bull. Jap. Soc. Sci. Fish.* 44:1187-1196.

Silberbauer, B.I. 1971. The biology of the South African rock lobster, *Jasus lalandii* (H.Milne Edwards). I. Development. *S.Afr. Div. Sea Fish., Invest. Rep.* 92:1-70.

Snodgrass, R.E. 1956. Crustacean Metamorphoses. *Smithson. Misc. Collns.* 131(10):1-78.

Soh, Cheng Lam 1969. Abbreviated development of a non-marine crab, *Sesarma (Geosesarma) perracae* (Brachyura, Grapsidae), from Singapore. *J. Zool. Lond.* 158:357-370.

Sorensen, J.H. 1969. New Zealand rock lobster. Distribution, growth, embryology and development. *Fish. Tech. Rep. NZ Mar. Dep.* 29:1-46.

Staples, D.J. 1979. Seasonal migration patterns of postlarval and juvenile banana prawns, *Penaeus merguiensis* de Man, in the major rivers of the Gulf of Carpentia, Australia. *Aust. J. mar. Freshwat. Res.* 30:143-157.

Steele, M. 1902. The crayfish of Missouri. *Publ. Univ. Cincinnati, Bulletin* 10:1-50.

Stein, R.A. 1976. Sexual dimorphism in crayfish chelae: functional significance linked to reproductive activities. *Can. J. Zool.* 54:220-227.

Stephens, P.J. & C.K.Govind 1981. Peripheral innervation fields of single lobster motoneurons defined by synapse elimination during development. *Brain Res.* 212:476-480.

Stephenson, M.J. & A.W.Knight 1982. Temperature and acclimation effects on salinity preferences of post-larvae of the giant Malaysian prawn *Macrobrachium rosenbergii* (de Man). *J. exp. mar. Biol. Ecol.* 60:253-260.

Suter, P.J. 1977. The biology of two species of *Engaeus* (Decapoda: Parastacidae) in Tasmania II. Larval history and larval development, with particular reference to *E.cisternarius*. *Aust. J. Mar. Freshwat. Res.* 28:85-93.

Tagatz, M.E. 1968. Biology of the blue crab, *Callinectes sapidus* Rathbun, in the St Johns River, Florida. *Fish. Bull. US Fish. Wildl. Serv.* 67:17-33.

Teshima, S. 1972. Studies on the sterol metabolism in marine crustaceans. *Mem. Fac. Fish., Kagoshima Univ.* 21:69-147.

Teshima, S. & A.Kanazawa 1971. Biosynthesis of sterols in the lobster, *Panulirus japonica*, the prawn *Penaeus japonicus*, and the crab *Portunus trituberculatus*. *Comp. Biochem. Physiol.* 38B:597-602.

Thompson, J.V. 1828. Zoological researches and illustrations, or natural history of nondescript or imperfectly known animals. I: On the metamorphoses of the Crustacea, and on Zoea, exposing their singular structure and demonstrating that they are not, as has been supposed, a peculiar genus, but the larva of Crustacea. *Cork:* 1-11.

Thompson, J.V. 1829. Addenda to Memoir I: On the metamorphoses of the Crustacea. *Cork:* 63-66.

Thompson, J.V. 1831. Letter to the editor of the 'Zoological Journal' dated Cork, Dec.16, 1830. *Zool. J.* 5:383-384.

Thompson, M.T. 1903. The metamorphosis of the hermit-crab. *Proc. Boston Soc. Nat. Hist.* 31:147-209.

Trask, T. 1974. Laboratory-reared larvae of *Cancer anthonyi* (Decapoda: Brachyura) with a brief description of the internal anatomy of the megalopa. *Mar. Biol.* 27:63-74.

Uchida, T. & Y.Dotsu 1973. Collection of the T.S.Nagasaki Maru of Nagasaki Univ. IV. On larva hatching and larval development of the lobster, *Nephrops thomsoni. Bull. Fac. Fish. Nagasaki Univ.* 36:23-35.

Vermeij, G.J. 1977. Patterns in crab claw size: the geography of crushing. *Syst. Zool.* 26:138-151.

Waite, F.C. 1899. The structure and development of the antennal glands in *Homarus americanus* Milne Edwards. *Bull. Mus. Comp. Zool. Harv.* 35:151-210.

Warner, G.F. & A.R.Jones 1976. Leverage and muscle type in crab chelae (Crustacea: Brachyura). *J. Zool.* 108:57-68.

Wass, M.L. 1955. The decapod crustaceans of Alligator Harbor and adjacent inshore areas of northwestern Florida. *Q. J. Fla. Acad. Sci.* 18:129-176.

Watabe, T. 1971. On the megalopa and first crab stage of *Kraussia integra* (de Haan) Atelecyclidae. *Res. on Crust.* 4, 5:1-11.

Wear, R.G. 1967. Life history studies on New Zealand Brachyura. 1. Embryonic and post-embryonic development of *Pilumnus novaezealandiae* Filhol, 1886 and of *P.lumpinus* Bennett, 1964 (Xanthidae, Pilumninae). *NZ J. Mar. Freshwat. Res.* 1:481-535.

Wear, R.G. 1968. Life history studies on New Zealand Brachyura 2. Family Xanthidae. Larvae of *Heterozius rotundifrons* A.Milne Edwards 1867, *Ozius truncatus* H.Milne Edwards, 1834, and *Heteropanope (Pilumnopeus) serratifrons* (Kinahan, 1856). *NZ J. Mar. Freshw. Res.* 2:293-332.

Wear, R.G. & E.J.Batham 1975. Larvae of the deep sea crab *Cymonomus bathamae* Dell, 1971 (Decapoda, Dorippidae) with observations on the larval affinities of the Tymolinae. *Crustaceana* 28:113-130.

Wickens, J.F. 1976. Prawn biology and culture. *Oceanogr-Mar. Biol. Annu. Rev.* 14:435-507.

Williams, A.B. 1971. A ten-year study of meroplankton in North Carolina estuaries: Annual occurrence of some brachyuran developmental stages. *Chesapeake Sci.* 12:53-61.

Williams, A.B. 1980. A new crab family from the vicinity of submarine thermal vents on the Galapagos Rift (Crustacea: Decapoda: Brachyura). *Proc. Biol. Soc. Wash.* 93:443-472.

Williamson, D.I. 1957. Crustacea, Decapoda: Larvae I. General. *Con. Internat. Pour Explor de la Mer Zooplank. Sheet.* 67:1-6.

Williamson, D.I. 1960. A remarkable zoea, attributed to the Majidae (Decapoda, Brachyura). *Ann. Mag. Nat. Hist.* (13)3:141-144.

Williamson, D.I. 1965. Some larval stages of three Australian crabs belonging to the families Homolidae and Raninidae and observations on the affinities of these families (Crustacea, Decapoda). *Aust. J. Mar. Freshwat. Res.* 16:369-398.

Williamson, D.I. 1969. Names of larvae in the Decapoda and Euphausiacea. *Crustaceana* 16:210-213.

Williamson, D.I. 1976. Larval characters and the origin of crabs. *Thalassia Jugosl.* 10:401-414.

Williamson, D.I. 1982a. Larval morphology and diversity. In: L.G.Abele (ed.), *The Biology of Crustacea, Vol. 2.* New York: Academic Press.

Williamson, D.I. 1982b. The larval characters of *Dorhynchus thomsoni* Thomson (Crustacea, Brachyura, Majoidea) and their evolution. *J. Nat. Hist.* 16:727-744.

Yang, W.T. 1968. The zoeae, megalopa, and first crab of *Epialtus dilatatus* (Brachyura, Majidae) reared in the laboratory. *Crustaceana* suppl.2:181-202.

Yang, W.T. 1971. The larval and postlarval development of *Parthenope serrata* reared in the laboratory and the systematic position of the Parthenopinae (Crustacea, Brachyura). *Biol. Bull.* 140:166-189.

Yang, W.T. 1976. Studies on the Western Atlantic arrow crab genus *Stenorhynchus* (Decapoda Brachyura, Majidae). I. Larval characters of two species and comparison with other larvae of Inachinae. *Crustaceana* 31:157-177.

Yang, W T. & P.A.McLaughlin 1979. Development of the epipodite of the second maxilliped and gills in *Libinia erinacea* (Decapoda, Brachyura, Oxyrhyncha). *Crustaceana* suppl.5:47-54.

Yatsuzuka, K. & N.Iwasaki 1979. On the larval development of *Pinnotheres* aff. *sinensis* Shen. *Rep. Usa Mar. Biol. Inst. Kochi Univ.* 1:76-96.

Yatsuzuka, K. & K.Sakai 1980. The larvae and juvenile crabs of Japanese Portunidae (Crustacea, Brachyura). I. *Portunus (Portunus) pelagicus* (Linn.). *Rep. Usa Mar. Biol. Inst. Kochi Univ.* 2:25-41.

Yokoya, Y. 1931. On the metamorphosis of two Japanese freshwater shrimps, *Paratya compressa* and *Leander paucidens*, with reference to the development of their appendages. *J. Coll. Agric.* 21:75-150.

Young, A.M. 1979a. Osmoregulation in larvae of the striped hermit crab *Clibanarius vittatus* (Bosc) (Decapoda: Anomura: Diogenidae). *Estuar. & coastal mar. Sci.* 9:595-601.

Young, A.M. 1979b. Osmoregulation in three hermit crab species, *Clibanarius vittatus* (Bosc), *Pagurus longicarpus* Say and *P.pollicaris* Say (Crustacea: Decapoda; Anomura). *Comp. Biochem. Physiol.* 63A:377-382.

Young, P.C. 1978. Moreton Bay, Queensland: A nursery area for juvenile penaeid prawns. *Anst. J. Mar. Freshwat. Res.* 29:55-75.

Young, P.C. & S.M.Carpenter 1977. Recruitment of postlarval penaeid prawns to nursery areas in Moreton Bay, Queensland. *Aust. J. Mar. Freshwat. Res.* 28:745-773.

Zein-Eldin, Z.P. 1963. Effect of salinity on growth of postlarval penaeid shrimp. *Biol. Bull.* 125:188-196.

Zein-Eldin, Z.P. & D.V.Aldrich 1965. Growth and survival of postlarval *Penaeus aztecus* under controlled conditions of temperature and salinity. *Biol. Bull.* 129:199-216.

Zein-Eldin, Z.P. & G.W.Griffith 1966. The effect of temperature upon the growth of laboratory-held postlarval *Penaeus aztecus. Biol. Bull.* 131:186-196.

Zielhorst, A.J.A.G. & F.van Herp 1976. Développement du système neurosécréteur du pédoncule oculaire des larves d'*Astacus leptodactylus salinus* Nordmann (Crustacea Decapoda Reptantia). *C.R. Hebd. Séanc. Acad. Sci. Paris* (D)283:1755-1758.

TAXONOMIC INDEX

SUBJECT INDEX

Printed and bound by CPI Group (UK) Ltd, Croydon, CR0 4YY

22/10/2024

01777634-0010